Vehicular Net...

Techniques, Standards, and Applications

OTHER TELECOMMUNICATIONS BOOKS FROM AUERBACH

Broadband Mobile Multimedia:
Techniques and Applications
Yan Zhang, Shiwen Mao, Laurence T. Yang,
and Thomas M Chen, Editors
ISBN: 978-1-4200-5184-1

Carrier Ethernet: Providing the Need for Speed
Gilbert Held
ISBN: 978-1-4200-6039-3

Cognitive Radio Networks
Yang Xiao and Fei Hu, Editors
ISBN: 978-1-4200-6420-9

Converging NGN Wireline and Mobile 3G Networks
with IMS: Converging NGN and 3G Mobile
Rebecca Copeland
ISBN: 978-0-8493-9250-4

Data-driven Block Ciphers for Fast
Telecommunication Systems
Nikolai Moldovyan and Alexander A. Moldovyan
ISBN: 978-1-4200-5411-8

Data Scheduling and Transmission Strategies in
Asymmetric Telecommunication Environments
Abhishek Roy and Navrati Saxena
ISBN: 978-1-4200-4655-7

Encyclopedia of Wireless and Mobile
Communications
Borko Furht
ISBN: 978-1-4200-4326-6

Handbook of Mobile Broadcasting:
DVB-H, DMB, ISDB-T, AND MEDIAFLO
Borko Furht and Syed A. Ahson, Editors
ISBN: 978-1-4200-5386-9

The Internet of Things: From RFID to the
Next-Generation Pervasive Networked Systems
Lu Yan, Yan Zhang, Laurence T. Yang,
and Huansheng Ning, Editors
ISBN: 978-1-4200-5281-7

Introduction to Communications Technologies:
A Guide for Non-Engineers, Second Edition
Stephan Jones, Ron Kovac, and Frank M. Groom
ISBN: 978-1-4200-4684-7

Introduction to Light Emitting Diode Technology
and Applications
Gilbert Held
ISBN: 978-1-4200-7662-2

Ionosphere and Applied Aspects of Radio
Communication and Radar
Nathan Blaunstein and Eugeniu Plohotniuc
ISBN: 978-1-4200-5514-6

MEMS and Nanotechnology-Based Sensors and
Devices for Communications, Medical
and Aerospace Applications
A. R. Jha
ISBN: 978-0-8493-8069-3

Millimeter Wave Technology in Wireless PAN,
LAN, and MAN
Shao-Qiu Xiao, Ming-Tuo Zhou,
and Yan Zhang, Editors
ISBN: 978-0-8493-8227-7

Mobile Telemedicine:
A Computing and Networking Perspective
Yang Xiao and Hui Chen, Editors
ISBN: 978-1-4200-6046-1

Mobile WiMAX:
Toward Broadband Wireless Metropolitan
Area Networks
Yan Zhang and Hsiao-Hwa Chen, Editors
ISBN: 978-0-8493-2624-0

Network Design for IP Convergence
Yezid Donoso
ISBN: 9781420067507

Optical Wireless Communications:
IR for Wireless Connectivity
Roberto Ramirez-Iniguez, Sevia M. Idrus, and Ziran Sun
ISBN: 978-0-8493-7209-4

Packet Forwarding Technologies
Weidong Wu, Editor
ISBN: 978-0-8493-8057-0

Satellite Systems Engineering in an IPv6
Environment
Daniel Minoli
ISBN: 9781420078688

Security in an IPv6 Environment
Daniel Minoli and Jake Kouns
ISBN: 978-1-4200-9229-5

Security in Wireless Mesh Networks
Yan Zhang, Jun Zheng and Honglin Hu, Editors
ISBN: 978-0-8493-8250-5

Unlicensed Mobile Access Technology:
Protocols, Architectures, Security,
Standards and Applications
Yan Zhang, Laurence T. Yang,
and Jianhua Ma, Editors
ISBN: 978-1-4200-5537-5

Value-Added Services for Next Generation
Networks
Thierry Van de Velde
ISBN: 978-0-8493-7318-3

WiMAX/MobileFi:
Advanced Research and Technology
Yang Xiao, Editor
ISBN: 978-1-4200-4351-8

Wireless Quality of Service:
Techniques, Standards, and Applications
Maode Ma, Mieso K. Denko,
and Yan Zhang, Editors
ISBN: 978-1-4200-5130-8

Vehicular Networks

Techniques, Standards, and Applications

Edited by

Hassnaa Moustafa • Yan Zhang

CRC Press
Taylor & Francis Group
Boca Raton London New York

CRC Press is an imprint of the
Taylor & Francis Group, an **informa** business
AN AUERBACH BOOK

CRC Press
Taylor & Francis Group
6000 Broken Sound Parkway NW, Suite 300
Boca Raton, FL 33487-2742

First issued in paperback 2019

ISBN-13: 978-1-4200-8571-6 (hbk)
ISBN-13: 978-0-367-38578-1 (pbk)

Library of Congress Cataloging-in-Publication Data

Vehicular networks : techniques, standards, and applications / editors, Hassnaa
    Moustafa, Yan Zhang.
        p. cm.
    Includes bibliographical references and index.
    ISBN 978-1-4200-8571-6 (hardcover : alk. paper)
      1. Vehicular ad hoc networks (Computer networks) I. Moustafa, Hassnaa. II.
    Zhang, Yan, 1977-

    TE228.37.V43 2009
    388.3'124--dc22   2009000521

Visit the Taylor & Francis Web site at
http://www.taylorandfrancis.com

and the CRC Press Web site at
http://www.crcpress.com

Contents

Preface

Vehicular networks are attracting significant interest in both academia and industry, driven by road safety requirements and intelligent traffic control. Vehicular networks form a novel class of wireless networks and are spontaneously formed between moving vehicles equipped with wireless interfaces of similar or different technologies. Vehicular networks, also known as VANETs (vehicular ad hoc networks), are examples of real applications of ad hoc networks enabling vehicle-to-vehicle and vehicle-to-infrastructure communications. These networks have significantly different characteristics compared to other wireless and mobile networks especially concerning high speed and unpredictable topology and present a very active field of research, development, standardization, and field trials.

VANET technology is entering a critical phase, where academia, industry, and governments worldwide are investing significant time and resources on the large-scale VANET deployment so that its benefits in the road safety and improvement of traffic flow could be leveraged. In this context, many national and international projects in government, industry, and academia are devoted to vehicular networks. These include consortia like Vehicle Safety Consortium (VSC) in the United States, the Car-2-Car Communication Consortium (C2C-CC), the ETSI-ITS in Europe, the Advanced Safety Vehicle Program (ASV) in Japan, and other standardization efforts like IEEE 802.11p (WAVE), and field trials like the large-scale Vehicle Infrastructure Integration Program (VII) in the United States. Besides such efforts, there exists a proliferation of conferences and workshops on the topic of vehicular networks, treating technical, policy, and economic challenges. However, there is a lack of an integrated material discussing the topic in details. This fact creates a strong motivation to write this book.

Vehicular Networks: Techniques, Standards, and Applications is the book to present and discuss the recent advances in the development of vehicular networks and to disseminate the most advanced ideas and solutions in the field. This volume allows for complete cross-referencing on routing, security, medium access, data dissemination, scheduling, mobility, services, market introduction, and the like. In particular, the book is organized into 15 chapters, each covering a unique topic in detail. The book first presents (in Chapters 1 and 2) the fundamentals for vehicular networks together with the ongoing standardization activities and the promising applications. It then covers all important design aspects (in Chapters 3 to 13). Chapter 3 gives a panoramic description for the medium access control (MAC) protocols in vehicular networks and highlights the ongoing IEEE standardization efforts in this context. Chapter 4 discusses the principle of heterogeneous wireless communication for vehicular networks, enabling the communication through different wireless technologies. Chapters 5 and 6 focus on the routing challenge in vehicular networks, where Chapter 5 presents state-of-the-art routing protocols suitable for the vehicular network environment, and Chapter 6 gives a user perspective analysis for routing protocols in vehicular networks. Chapter 7 discusses data dissemination in vehicular networks and its criticality in terms of network load presenting a wide number of algorithms that aim at ensuring reliable and scalable messages dissemination. Chapter 8 focuses on message scheduling within vehicular networks. Regarding the technical challenges related to the mobility feature, Chapter 9 presents IP address autoconfiguration in vehicular networks, discussing the potential approaches and highlighting the ongoing standardization activities. Chapter 10 focuses on network mobility (NEMO) in vehicular networks. Chapter 11 presents a new concept known as mobile ad hoc NEMO (MANEMO) that is promising in future vehicular networks. The MANEMO concept and scenarios are presented together with some candidate solution approaches in vehicular networks. As security presents an extraordinary challenge for the wireless research community, Chapter 12 focuses on security in vehicular networks, giving a panoramic description of the key threats and the security requirements for these networks and presenting a number of solutions aiming for secure communication in vehicular networks. In addition, Chapter 13 focuses on the problem of confidence management in vehicular networks. Chapter 14 presents a basic variant of geocast (use of geographical positions for addressing and routing of data packets) in vehicular networks and explains extensions for various aspects, including reliability and efficiency of data transport, security and privacy, and Internet integration. Finally, Chapter 15 ends the book through a wide discussion on the topic of market introduction for vehicular networks and their deployment strategies.

This book has the following salient features:

■ Provides a comprehensive reference on vehicular networks
■ Presents state-of-the-art techniques for vehicular networks

- Covers basics, techniques, advanced topics, standard specifications, and future directions
- Contains illustrative figures enabling easy reading

We owe our deepest gratitude to all the chapters' authors for their valuable contribution to this book and their great efforts. All of them are extremely professional and cooperative. We wish to express our thanks to Auerbach Publications (Taylor & Francis Group) especially Richard O'Hanley for soliciting the ideas in this book and having accepted to work with us for its publication, and Jennifer Ahringer for her huge efforts in the editorial process. Last but not least, a special thank you to our families and friends for their constant encouragement, patience, and understanding throughout this project.

The book serves as a comprehensive and essential reference on vehicular networks and is intended as a textbook for senior undergraduate and graduate-level courses on vehicular networks. It can also be used as a supplementary textbook for undergraduate courses on wireless or mobile communication and ad hoc networks. The book is a useful resource for the students and researchers to learn vehicular networks. In addition, it will be valuable to professionals in the domain of vehicular networks from both academia and industry and generally has instant appeal to the people who are willing to contribute to vehicular network technology.

We welcome and appreciate your feedback and hope you enjoy reading the book.

Hassnaa Moustafa and Yan Zhang

About the Editors

Hassnaa Moustafa received the Master of Science degree in parallel and distributed systems from the University of Paris XI (Orsay) in France and the PhD degree in computer and networks from Telecom ParisTech in France [formerly the Ecole Nationale Superieure des Telecommunications ("ENST")], and she was involved in teaching activities for graduate and postgraduate students at Telecom ParisTech during her stay. Since January 2005, she has worked at France Telecom R&D (Orange Labs) http://www.francetelecom.com/fr_FR/innovation/. She manages a number of research projects, and her activities are mainly in the area of wireless networks, including ad hoc networks, vehicular networks, hybrid wireless networks, and mesh networks. She worked in the design of routing protocols for ad hoc networks and hybrid wireless networks for a number of years. Her current research activities concern security, authentication, authorization, and accounting (AAA), services access control and IP autoconfiguration. She is also involved in some standardization activities within the IETF.

She has presented papers at a number of international conferences including ICC, Globecom, PIMRC, VTC, Mobicom, and many others, and in a number of international journals. Moustafa has also coauthored a large number of chapters in various books with CRC Press. She served as a TPC member for many international conferences and workshops, such as, MWNS 2009, MUE 2009, MobiSec 2009, PIMRC 2008, WEEDEV 2008, WAMSNET 2008, UIC 2008, NTMS 2008, MWNS 2008, ICCCN 2008, NTMS 2007, ChinaCom 2007, UbiRoads 2007, a TPC cochair for WITS 2008, a publication chair for Co-NEXT 2008, NTMS 2008, and a workshop chair for COMSWARE 2009. Moustafa was publicity chair for 4G WiMAX 2008 and CONET 2008 workshops and a publicity cochair for the UIC-09 conference. She served as a peer reviewer for many international journals, such as, *IEEE Transactions on Wireless Communications, IEEE Transactions on Parallel and Distributed Systems, IEEE Transactions on Mobile Computing, IEEE*

Transactions on Broadcasting, Wiley Wireless Communications and Mobile Computing Journal, Springer Telecommunication Systems Journal, Springer Annals of Telecommunications, Elsevier Computer Networks, and a reviewer for many international conferences, such as, Globecom, VTC, ISCC, GIIS, ASWN, APNOMs, PCAC, ICN, MAAS, Medhoc, and MWCN. Moustafa is a member of IEEE and IEEE ComSoc.

Yan Zhang received a PhD degree in the School of Electrical and Electronics Engineering, Nanyang Technological University, Singapore. Since August 2006, he has worked with Simula Research Laboratory, Norway (http://www.simula.no). He is associate editor of *Security and Communication Networks* (Wiley); on the editorial board of *International Journal of Network Security; International Journal of Ubiquitous Computing, Transactions on Internet and Information Systems (TIIS); International Journal of Autonomous and Adaptive Communications Systems (IJAACS)* and *International Journal of Smart Home (IJSH)*. Zhang currently serves as the book series editor for "Wireless Networks and Mobile Communications" (Auerbach Publications, CRC Press, Taylor & Francis Group). He serves as guest coeditor for *Wiley Security and Communication Networks* special issue on "secure multimedia communication"; guest coeditor for *Springer Wireless Personal Communications* special issue on selected papers from ISWCS 2007; guest coeditor for *Elsevier Computer Communications* special issue on "adaptive multicarrier communications and networks"; guest coeditor for *Inderscience International Journal of Autonomous and Adaptive Communications Systems (IJAACS)* special issue on "cognitive radio systems"; guest coeditor for *The Journal of Universal Computer Science (JUCS)* special issue on "multimedia security in communication"; guest coeditor for *Springer Journal of Cluster Computing* special issue on "algorithm and distributed computing in wireless sensor networks"; guest coeditor for *EURASIP Journal on Wireless Communications and Networking (JWCN)* special issue on "OFDMA architectures, protocols, and applications"; and guest coeditor for *Springer Journal of Wireless Personal Communications* special issue on "security and multimodality in pervasive environments."

Zhang is coeditor for books, including *Resource, Mobility, and Security Management in Wireless Networks and Mobile Communications; Wireless Mesh Networking: Architectures, Protocols, and Standards; Millimeter-Wave Technology in Wireless PAN, LAN, and MAN; Distributed Antenna Systems: Open Architecture for Future Wireless Communications; Security in Wireless Mesh Networks; Mobile WiMAX: Toward Broadband Wireless Metropolitan Area Networks; Wireless Quality-of-Service: Techniques, Standards, and Applications; Broadband Mobile Multimedia: Techniques and Applications; Internet of Things: From RFID to the Next-Generation Pervasive Networked Systems; Unlicensed Mobile Access Technology: Protocols, Architectures, Security, Standards and Applications; Cooperative Wireless Communications;*

WiMAX Network Planning and Optimization; RFID Security: Techniques, Protocols and System-On-Chip Design; Autonomic Computing and Networking; Security in RFID and Sensor Networks; Handbook of Research on Wireless Security; Handbook of Research on Secure Multimedia Distribution; RFID and Sensor Networks; Cognitive Radio Networks; Wireless Technologies for Intelligent Transportation Systems; Vehicular Networks: Techniques, Standards, and Applications; and *Orthogonal Frequency Division Multiple Access (OFDMA).*

Zhang is track cochair for ITNG 2009, publicity cochair for SMPE 2009, publicity cochair for COMSWARE 2009, publicity cochair for ISA 2009, general cochair for WAMSNet 2008, publicity cochair for TrustCom 2008, workshop general cochair for COGCOM 2008, workshop cochair for IEEE APSCC 2008, workshop general cochair for WITS-08, program cochair for PCAC 2008, workshop general cochair for CONET 2008, workshop chair for SecTech 2008, workshop chair for SEA 2008, workshop coorganizer for MUSIC'08, workshop coorganizer for 4G-WiMAX 2008, publicity cochair for SMPE-08, international journals coordinating cochair for FGCN-08, publicity cochair for ICCCAS 2008, workshop chair for ISA 2008, symposium cochair for ChinaCom 2008, industrial cochair for MobiHoc 2008, program cochair for UIC-08, general cochair for CoNET 2007, general cochair for WAMSNet 2007, workshop cochair FGCN 2007, program vice cochair for IEEE ISM 2007, publicity cochair for UIC-07, publication chair for IEEE ISWCS 2007, program cochair for IEEE PCAC'07, special track cochair for "mobility and resource management in wireless/mobile networks" in ITNG 2007, special session coorganizer for "wireless mesh networks" in PDCS 2006, a member of the Technical Program Committee for numerous international conferences, including ICC, PIMRC, VTC, CCNC, AINA, GLOBECOM, ISWCS, and so on. He received the Best Paper Award and Outstanding Service Award in the IEEE 21st International Conference on Advanced Information Networking and Applications (AINA-07). His research interests include resource, mobility, spectrum, data, energy, and security management in wireless networks and mobile computing. Zhang is a member of IEEE and IEEE ComSoc.

List of Contributors

Oreste Andrisano
WiLAB c/o University of Bologna,
   Italy
oreste.andrisano@unibo.it

Roberto Baldessari
NEC, Germany
roberto.baldessari@nw.neclab.eu

Alessandro Bazzi
WiLAB c/o University of Bologna,
   Italy
alessandro.bazzi@unibo.it

Carlos J. Bernardos
University UC3M, Spain
cjbc@it.uc3m.es

Mohamed Salah Bouassida
University of Technology of
   Compiègne (UTC), France
mohamed-salah.bouassida@hds.utc.fr

Gilles Bourdon
Orange Group, France
gilles.bourdon@orange-ftgroup.com

Victor Cabrera
University of Murica, Spain
victorcabrera@um.es

Maria Calderon
University UC3M, Spain
maria@it.uc3m.es

Véronique Cherfaoui
University of Technology of
   Compiègne (UTC), France
veronique.cherfaoui@hds.utc.fr

Zohra-Leïla Cherfi
University of Technology
   of Compiègne (UTC), France
zohra.cherfi@hds.utc.fr

Andrea Conti
WiLAB c/o University of Bologna,
   Italy
a.conti@ieee.org

Thierry Denoeux
University of Technology of
   Compiègne (UTC), France
thierry.denoeux@hds.utc.fr

Bertrand Ducourthial
University of Technology
   of Compiègne (UTC), France
bertrand.ducourthial@hds.utc.fr

Andreas Festag
NEC, Germany
Andreas.Festag@nw.neclab.eu

Moez Jerbi
France Telecom R&D (Orange Labs),
France
moez.jerbi@orange-ftgroup.com

Yacine Khaled
University of Technology
of Compiègne (UTC), France
yacine.khaled@hds.utc.fr

Maria Kihl
Lund University, Sweden
Maria.Kihl@eit.lth.se

Houda Labiod
Telecom ParisTech, France
labiod@enst.fr

Robert Lasowski
Cirquent | Softlab Group, Germany
robert.lasowski@cirquent.de

Long Le
NEC, Germany
Long.Le@nw.neclab.eu

Juan A. Martinez
Univerity of Murica, Spain
juanantonio@dif.um.es

Barbara M. Masini
WiLAB c/o University of Bologna,
Italy
barbara.masini@unibo.it

Nader Moayeri
NIST, United States
nader.moayeri@nist.gov

Hassnaa Moustafa
France Telecom R&D (Orange Labs)
France
hassnaa.moustafa@orange-
ftgroup.com

Yi Qian
NIST, United States
yqian@nist.gov

Francisco J. Ros
University of Murica, Spain
fjrm@dif.um.es

Pedro M. Ruiz
University of Murica, Spain
pedrom@dif.um.es

Juan A. Sanchez
University of Murica, Spain
jlaguna@dif.um.es

Christoph Schroth
University of St. Gallen, Switzerland
schroth@unisg.ch

Sidi Mohammed Senouci
France Telecom R&D (OrangeLabs),
France
sidimohammed.senouci@orange-
ftgroup.com

Mohamed Shawky
University of Technology of
Compiègne (UTC), France
Mohamed.shawky@hds.utc.fr

Ignacio Soto
University UC3M, Spain
isoto@it.uc3m.es

Markus Strassberger
BMW Group, Germany
Markus.Strassberger@bmw.de

Christian Tchepnda
France Telecom R&D (Orange Labs),
    France
christian.tchepnda@orange-
    ftgroup.com

Ryuji Wakikawa
Toyota ITC, Japan
ryuji@jp.toyota-itc.com

Wenhui Zhang
NEC, Germany
Wenhui.Zhang@nw.neclab.eu

Chapter 1

Introduction to Vehicular Networks

Hassnaa Moustafa, Sidi Mohammed Senouci,
and Moez Jerbi

Contents

Intervehicle communication (IVC) is attracting considerable attention from the research community and the automotive industry, where it is beneficial in providing intelligent transportation system (ITS) as well as drivers and passengers' assistant services. In this context, vehicular ad hoc networks (VANETs) are emerging as a new class of wireless network, spontaneously formed between moving vehicles equipped with wireless interfaces that could have similar or different radio interface technologies, employing short-range to medium-range communication systems. A VANET is a form of mobile ad hoc network, providing communications among nearby vehicles and between vehicles and nearby fixed equipment on the roadside.

This chapter gives an overview of vehicular networks (also known as VANETs), showing their potential architectures and possible deployment scenarios. Vehicular network, benefits and real-life applications are presented from a network operator's view, giving potential service examples. A number of technical challenges in vehicular network deployment are discussed; most of these challenges are detailed in the following chapters. Moreover, the role of the involved actors (networks operators, car manufacturers, service providers, and governmental authorities) is shown, as well as the related standardization activities. Finally, some related projects are highlighted.

1.1 Vehicular Network Definition, Architectures, and Deployment Scenarios

1.1.1 What Are Vehicular Networks?

Vehicular networks are a novel class of wireless networks that have emerged thanks to advances in wireless technologies and the automotive industry. Vehicular networks are spontaneously formed between moving vehicles equipped with wireless interfaces that could be of homogeneous or heterogeneous technologies. These networks, also known as VANETs, are considered as one of the ad hoc network real-life applications, enabling communications among nearby vehicles as well as between vehicles and nearby fixed equipment, usually described as roadside equipment. Vehicles can be either private, belonging to individuals or private companies, or public transportation means (e.g., buses and public service vehicles such as police cars). Fixed equipment can belong to the government or private network operators or service providers.

Indeed, vehicular networks are promising in allowing diverse communication services to drivers and passengers. These networks are attracting considerable attention from the research community as well as the automotive industry. High interest for these networks is also shown from governmental authorities and standardization organizations. In this context, a dedicated short-range communications (DSRC) system has emerged in North America, where 75 MHz of spectrum was approved by the U.S. FCC (Federal Communication Commission) in 2003 for such type of communication that mainly targets vehicular networks. On the other hand, the Car-to-Car Communication Consortium (C2C-CC) has been initiated in Europe by car manufacturers and automotive OEMs (original equipment manufacturers), with the main objective of increasing road traffic safety and efficiency by means of intervehicle communication. IEEE is also advancing within the IEEE 1609 family of standards for wireless access in vehicular environments (WAVE).

1.1.2 Vehicular Network Architectures

Vehicular networks can be deployed by network operators and service providers or through integration between operators, providers, and a governmental authority. Recent advances in wireless technologies and the current and advancing trends in ad hoc network scenarios allow a number of deployment architectures for vehicular networks, in highway, rural, and city environments. Such architectures should allow communication among nearby vehicles and between vehicles and nearby fixed roadside equipment. Three alternatives include (i) a pure wireless vehicle-to-vehicle ad hoc network (V2V) allowing standalone vehicular communication with no infrastructure support, (ii) a wired backbone with wireless last hops that can be seen as a WLAN-like vehicular network, (iii) and a hybrid vehicle-to-road (V2R) architecture that does not rely on a fixed infrastructure in a constant manner, but can exploit it for improved performance and service access when it is available. In this latter case, vehicles can communicate with the infrastructure either in a single hop or multihop fashion according to the vehicles' positions with respect to the point of attachment with the infrastructure. Actually the V2R architecture implicitly includes V2V communication.

A reference architecture for vehicular networks is proposed within the C2C-CC, distinguishing between three domains: in-vehicle, ad hoc, and infrastructure domain [1]. Figure 1.1 illustrates this reference architecture. The in-vehicle domain refers to a local network inside each vehicle logically composed of two types of units: (i) an on-board unit (OBU) and (ii) one or more application unit(s) (AUs). An OBU is a device in the vehicle having

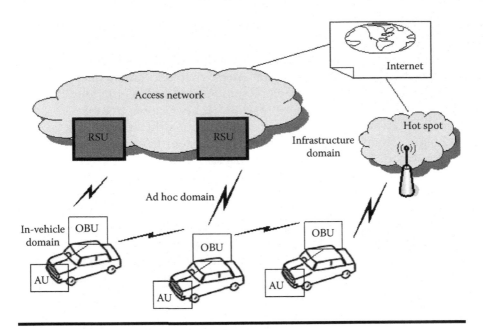

Figure 1.1   C2C-CC reference architecture.

communication capabilities (wireless and/or wired), while an AU is a device executing a single or a set of applications while making use of the OBU's communication capabilities. Indeed, an AU can be an integrated part of a vehicle and be permanently connected to an OBU. It can also be a portable device such as a laptop or PDA that can dynamically attach to (and detach from) an OBU. The AU and OBU are usually connected with a wired connection, while wireless connection is also possible (using, e.g., Bluetooth, WUSB, or UWB). This distinction between AU and OBU is logical, and they can also reside in a single physical unit.

The ad hoc domain is a network composed of vehicles equipped with OBUs and roadside units (RSUs) that are stationary along the road. OBUs of different vehicles form a mobile ad hoc network (MANET), where an OBU is equipped with communication devices, including at least a short-range wireless communication device dedicated for road safety. OBUs and RSUs can be seen as nodes of an ad hoc network, respectively, mobile and static nodes. An RSU can be attached to an infrastructure network, which in turn can be connected to the Internet. RSUs can also communicate to each other directly or via multihop, and their primary role is the improvement of road safety, by executing special applications and by sending, receiving, or forwarding data in the ad hoc domain.

Two types of infrastructure domain access exist: RSU and hot spot. RSUs may allow OBUs to access the infrastructure and, consequently, to be connected to the Internet. OBUs may also communicate with Internet via public, commercial, or private hot spots (Wi-Fi hot spots). In the absence of RSUs and hot spots, OBUs can utilize communication capabilities of cellular radio networks (GSM, GPRS, UMTS, WiMax, and 4G) if they are integrated in the OBU.

1.1.3 Possible Deployment Scenarios for Vehicular Networks

Regarding the C2C CC reference architecture together with the advances in heterogeneous communication technologies, vehicular networks potentially have two main types of communication scenarios: car-to-car (C2C) communication scenario and car-to-infrastructure (C2I) communication scenario.

These types of communication scenarios allow a number of deployment options for vehicular networks. Vehicular network deployment can be integrated into wireless hot spots along the road. Such hot spots can be operated individually at home or at office or by wireless Internet service providers or an integrated operator. On the other hand, vehicular network deployment can be integrated into the existing cellular systems. Vehicles can even communicate with other vehicles directly without a communication infrastructure, where vehicles can cooperate and forward information on behalf of each other. We notice that a combination of these deployment cases is also possible.

Moreover, future architecture for intelligent transportation systems (ITS) considers vehicles as active nodes that are responsible for collecting and forwarding critical information. Consequently, vehicular network coexistence with sensor network would potentially take place, where vehicles would be able to collect and process information by means of intelligent sensors and to exchange information with other nodes (fixed or mobile) in a global communication system.

1.2 Special Characteristics of Vehicular Networks

Vehicular networks have special behavior and characteristics, distinguishing them from other types of mobile networks. In comparison to other communication networks, vehicular networks come with unique attractive features, as follows [2]:

- Unlimited transmission power: Mobile device power issues are usually not a significant constraint in vehicular networks as in the case of classical ad hoc or sensor networks, since the node (vehicle) itself can provide continuous power to computing and communication devices.
- Higher computational capability: Indeed, operating vehicles can afford significant computing, communication, and sensing capabilities.

■ Predictable mobility: Unlike classic mobile ad hoc networks, where it is hard to predict the nodes' mobility, vehicles tend to have very predictable movements that are (usually) limited to roadways. Roadway information is often available from positioning systems and map-based technologies such as GPS. Given the average speed, current speed, and road trajectory, the future position of a vehicle can be predicted.

However, to bring its potency to fruition, vehicular networks have to cope with some challenging characteristics [3], which include

■ Potentially large scale: Unlike most ad hoc networks studied in the literature that usually assume a limited network size, vehicular networks can in principle extend over the entire road network and so include many participants.
■ High mobility: The environment in which vehicular networks operate is extremely dynamic and includes extreme configurations: on highways, relative speeds of up to 300 km/h may occur, while density of nodes may be 1–2 vehicles 1 km on low busy roads. On the other hand, in the city, relative speeds can reach up to 60 km/h and nodes' density can be very high, especially during rush hour.
■ Partitioned network: Vehicular networks will be frequently partitioned. The dynamic nature of traffic may result in large intervehicle gaps in sparsely populated scenarios and hence in several isolated clusters of nodes.
■ Network topology and connectivity: Vehicular network scenarios are very different from classic ad hoc networks. Since vehicles are moving and changing their position constantly, scenarios are very dynamic. Therefore the network topology changes frequently as the links between nodes connect and disconnect very often. Indeed, the degree to which the network is connected is highly dependent on two factors: the range of wireless links and the fraction of participant vehicles, where only a fraction of vehicles on the road could be equipped with wireless interfaces.

1.3 Vehicular Network Potential Applications and Services

Vehicular network applications range from road safety applications oriented to the vehicle or to the driver, to entertainment and commercial applications for passengers, making use of a plethora of cooperating technologies.

The primary vision of vehicular networks includes real-time and safety applications for drivers and passengers, providing safety for the latter and giving essential tools to decide the best path along the way. These applications thus aim to minimize accidents and improve traffic conditions by providing drivers and passengers with useful information including collision warnings, road sign alarms, and in-place traffic views.

Nowadays, vehicular networks are promising in a number of useful driver- and passenger-oriented services, which include Internet connections facility exploiting an available infrastructure in an "on-demand" fashion, electronic tolling system, and a variety of multimedia services. As well as, a variety of communication networks, such as 2-3G, WLANs IEEE 802.11a/b/g/p, and WiMAX, can be exploited to enable new services designed for passengers apart from the safety applications, such as infomobility and entertainment applications, which can rely on the vehicular network itself.

Regarding the discussed applications' potential, vehicular networks open new business opportunities for car manufacturers, automotive OEMs, network operators, service providers, and integrated operators in terms of infrastructure deployment as well as service provision and commercialization. For safety-related applications, the network operator can assure the authentication of each participant through playing the role of a trusted third party that authenticates the participating nodes, or even having the role of a certification authority issuing a certificate to each participant in order to prove the authenticity of them later during the communication. On the other hand, in nonsafety-related applications, network operators and/or service providers, besides network access and services' provision, can have the role of authorizing services' access and billing users for the consumed services. However, one should notice that ad hoc systems still require a certain level of penetration and necessitate high vehicle density for more reliable communication. Also, the investment cost for new communication infrastructure for vehicular networks is high, where as on the other hand cellular communication systems offer a high coverage along roads and have a reliable authentication and security mechanism. Consequently a number of technical challenges needs to be resolved in order to help the evolution of vehicular networks for wide-scale deployment. The following section discusses some of these challenges.

1.4   Technical Challenges

Vehicular networks' special behavior and characteristics create some challenges for vehicular communication, which can greatly impact the future deployment of these networks. A number of technical challenges need to be resolved in order to deploy vehicular networks and to provide useful services for drivers and passengers in such networks. Generally speaking, scalability and interoperability are two important issues that should be satisfied, and the employed protocols and mechanisms should be scalable to numerous vehicles and interoperable with different wireless technologies. The following subsections discuss a number of these challenges; more details are given in the following chapters.

1.4.1 Reliable Communication and MAC Protocols

Similar to ad hoc networks, vehicular networks experience multihop communication, which can potentially extend the network operator fixed infrastructure and thus provide virtual infrastructure among the moving vehicles. Indeed, multihop wireless communication represents a major challenge on the reliability of communication. Consequently, efficient MAC (medium access control) protocols need to be in place, while adapting to the highly dynamic environment of vehicular networks, and considering message priority of some applications (e.g., accident warnings). In spite of the dynamic topology and the high mobility, fast association and low communication latency should be satisfied between communicating vehicles in order to guarantee (i) service reliability for safety-related applications while taking into consideration the time sensitivity during message transfer, and (ii) the quality and continuity of service for nonsafety applications. Moreover, MAC protocols should take into consideration the heterogeneous communication that is liable to take place between different wireless technologies (e.g., Wi-Fi and GSM) in vehicular networks.

1.4.2 Routing and Dissemination

Vehicular networks differ from conventional ad hoc wireless networks by not only experiencing rapid changes in wireless link connections, but also having to deal with different types of network densities [4]. For example, vehicular networks on freeways or urban areas are more likely to form highly dense networks during rush hour traffic, while vehicular networks are expected to experience frequent network fragmentation in sparsely populated rural freeways or during late night hours. Moreover, vehicular networks are expected to handle a wide range of applications ranging from safety to leisure. Consequently, routing and dissemination algorithms should be efficient and should adapt to vehicular network characteristics and applications, permitting different transmission priorities according to the application type (safety-related or not). Until now, most of vehicular network research has focused on analyzing routing algorithms to handle the broadcast storm problem in a highly dense network topology [5,6], under the oversimplified assumption that a typical vehicular network is a well-connected network in nature. So far, the penetration of vehicular network technology is somewhat weak, and hence these networks should rely on an existing infrastructure support for wide-scale deployment. However, in the future, these networks are expected to observe high penetration with lesser infrastructure support, and hence it is important in this case to consider the disconnected network problem, which is a crucial research challenge for developing a reliable and efficient routing protocol that can support highly diverse network topologies.

As for message dissemination, the dissemination algorithms should depend on the network density as well as the application type. For example, message dissemination in safety-related applications should be mostly broadcast-like, in a way to assure the message propagation to the required cluster of vehicles without causing a broadcast storm. In nonsafety-related applications, message transfer through unicast or multicast transmission is more suitable.

1.4.3 Security

Vehicular communication security is a major challenge, having a great impact on the future deployment and application of vehicular networks. Indeed, security and privacy are major concerns in the development and acceptance of services and should not be compromised by ease-of-use of service discovery protocols. As the demand for service discovery is growing, passengers may use services in foreign networks and create immense security problems for themselves and for other network users. Consequently, it is important to propose innovative solutions for secure communication between participants as well as authorized and secure service access. To enhance the vehicular network access ubiquity, these solutions should take advantage of (i) the ad hoc multihop authentication and communication concepts, which on one hand allow secure communication and on the other hand extend the infrastructure coverage with the minimum deployment cost for the network operator, and (ii) the distributed-based authentication. Appropriate security architectures should be in place providing communication between vehicles and allowing different service access. A set of security mechanisms suitable for any vehicular network environment should be developed, providing trust, authentication, access control, and authorized and secure service access. In this context, authentication optimization is important to be studied for both infrastructure-based and infrastructure-less communications, aiming to facilitate the reauthentication process that may need to take place during the vehicle mobility.

Moreover, node behavior is an important issue that can threaten the security of communication and service delivery in vehicular networks and hence is worth consideration. Due to the open and dynamic environment of vehicular networks, nodes cooperation is an important aspect that should be satisfied for allowing successful communication between vehicles. We notice that nodes may behave selfishly by not forwarding messages for others in order to save power and bandwidth or just because of security and privacy concerns. Consequently, appropriate mechanisms should be developed to detect selfishness and enforce node cooperation in vehicular network environment.

1.4.4   IP Configuration and Mobility Management

The potential vehicle-to-infrastructure architecture is promising in allowing vehicular Internet access as well as provision of Internet-related services to drivers and passengers. However, two technical challenges exist under this issue: IP address configuration and mobility management. These challenges can threaten the service quality and the service continuity. Regarding the vehicular network characteristics, IP address configuration should be carried out in an automatic and distributed manner. So far, there is no standard for IP autoconfiguration in ad hoc networks, and hence the problem becomes complex for vehicular networks. We notice a considerable work in progress by a number of standardization bodies aiming to resolve this problem. Besides the Internet Engineering Task Force (IETF) efforts through the Autoconf WG for developing IPv6 solutions for ad hoc networks including vehicular network scenarios, all of the international committees defining architectures for vehicular communication have included a native IPv6 stack in their protocol stacks, namely, IEEE 1609, ISO TC 204 (CALM), C2C-CC, and the newly formed ETSI TC ITS.

As for mobility management, this is a crucial problem for nonsafety applications, where messages dissemination is not broadcast-based. Indeed, the absence of mobility management mechanism threatens service commercialization in vehicular networks and loses the benefit of the vehicle-to-infrastructure architecture since all Internet-related services would guarantee neither service quality nor their continuity.

1.4.5   Application Distribution

From a general view, we can notice that building distributed applications involving passengers in different vehicles requires new distributed algorithms. As a consequence, a distributed algorithmic layer is required for managing the group of participants and ensuring data sharing among distributed programs. Such algorithms could assimilate the neighborhood instability to a kind of fault. However, the lack of communication reliability necessitates employing fault-tolerant techniques. An important requirement, in this context, is allowing mobile participants in vehicular networks to have service access with an acceptable quality level while facilitating the message exchange between vehicles.

1.4.6 Business Models

Business models represent an important challenge for service commercialization in vehicular networks. As a matter of opening a new business opportunity, business models should be rentable for telecom operators and service provides aiming to promoting services and attracting clients. It is also important that business models be affordable and attractive to clients, taking into account the cooperation between mobile clients in vehicular networks, where nodes can be compensated (rewarded) according to their participation. Special payment strategies could be proposed, in this context, for encouraging the cooperation between mobile nodes, where a sort of remuneration can be done for each participant according to his contribution. Consequently, special accounting mechanisms and tailored billing systems are needed, which also assure interdomain accounting. However, processing delay constraints should be considered as well as the need for authentication and integrity, where the operator could assure the authentication, authorization, and secure communication between clients in a way that protects the clients' data and allows for billing the used services.

1.5 Vehicular Network Evolution and Progress

1.5.1 Main Actors

Vehicular networks present a highly active field of research, development, standardization, and field trials. Throughout the world, there are many national and international projects in governments, industry, and academia devoted to such networks. These include the consortia like Vehicle Safety Consortium—VSC (United States) [7], Collision Avoidance Metrics Partnership. CAMP (United States) [8], Car 2 Car Communication Consortium—C2C-CC (Europe) [9], Advanced Safety Vehicle—ASV Program (Japan) [10], a lot of standardization efforts as we will see in the following section, and field trials like the large-scale Vehicle Infrastructure Integration Consortium (VIIC) United States [11].

The Vehicle Infrastructure Integration initiative was first launched by the U.S. Department of Transportation (USDOT) during the ITS World Congress in 2003. Then the Vehicle Infrastructure Integration Consortium was formed in early 2005 by a group of light-duty vehicle manufacturers to actively engage in the design, testing, and evaluation of a deployable VII system for the United States. USDOT's VII program is divided into three phases: (i) Phase I—operational testing and demonstration, (ii) Phase II—research in the areas of enabling technology, institutional issues, and applications to support deployment, and (iii) Phase III—technology scanning to determine potential new technology horizons for VII. The first experimental results presented in the symposium IEEE Wivec2007 in Detroit showed the strong viability of VII.

In Japan, we notice large initiatives and advanced ITS solutions such as VICS (Vehicle Information and Communication System), AHS (Advanced Cruise-Assist Highway System, since 1996), DSSS (Driving Safety Support Systems, since 2002), and ASV (Advanced Safety Vehicle, since 1991). In the milestones of ITS-Safety 2010 project, a large-scale verification testing on public roads is scheduled in 2008 and a nation-wide deployment in 2010.

We notice that the target applications in the United States include safety, traffic efficiency, electronic toll collect (ETC), and customer relationship management (CRM). On the other hand, in Europe, less roadside infrastructure is expected than in the United States and consequently the target applications are safety and traffic efficiency. In this context, the Car 2 Car Communication Consortium [9] is a nonprofit organization initiated by six European car manufacturers (Audi, BMW, DaimlerChrysler, Fiat, Renault, and Volkswagen) with the aim to develop an open industrial standard for intervehicle communication to ensure pan-European interoperability, using wireless LAN technology (WLAN IEEE 802.11 standards). More details are given in the following subsection.

Moreover, the telcos with their large existing infrastructures also give special attention to the development of vehicular networks. Orange Labs, Telcom Italia, AT&T labs, or Deutsche Telekom all take part in the development of the technology via partnerships with industries, universities, and their own R&D teams. In fact, they see such networks as a natural evolution or extension of the current wireless systems, while representing a low-cost solution that improves the performance of telco networks by overcoming the limitations of using multihop technology and giving a potential of new business (develop customer loyalty, catch new customers).

Many industries, and companies, involved in the consortiums cited above, are investing enormous sums for the development of new ITS solutions. Some of them (such as, Dash, Google, and TomTom) are interested in particular in real-time infotainment and guidance of the travelers. They do more than only charting the roads; they also allow the drivers or passengers to receive real-time information about the traffic, to connect to the Internet, and to have other useful information (such as the nearest gas station, restaurants and cinemas) on the way. Another example is Microsoft, which proposed to carmakers and their suppliers a new version of its operating system capable of managing all the embedded systems within the car [12]. Being a technology always under development, vehicular networks belong to the main areas of research topics. Large number of conferences and workshops dealing with this is proof. A set of universities and research institutes (UCLA, Karlsruhe University, Stanford University, INRETS, etc.) take part in the optimization of several challenges encountered in these networks, among which are routing and data dissemination, Phy/Mac, security, self-organization, and the like.

1.5.2 Main Standardization Activities

In 1999, the U.S. Federal Communication Commission allocated 75 MHz of dedicated short-range communication (DSRC) spectrum at 5.9 GHz (5.850–5.925 GHz) to be used exclusively for vehicle-to-vehicle and infrastructure-to-vehicle communications in North America. The primary purpose was to enable public safety applications that save lives and improve traffic flow. Private services are also permitted in order to lower cost and to encourage DSRC development and adoption. The DSRC spectrum is divided into seven 10-MHz wide channels. Channel 5885–5895-MHz is the control channel, which is generally restricted to safety communications only. The two channels at the edges of the spectrum are reserved for future advanced accident-avoidance applications and high-powered public safety usages. The rest are service channels and are available for both safety and non-safety usage.

On the other hand, in Japan, the allocated frequency bands, namely for DSRC, range from 5.770 to 5.850 GHz.

As for Europe, one obstacle to introduce VANETs for road safety was the lack of a dedicated frequency spectrum. Compared to North America and Japan, the process for frequency allocation is considerably complex and time consuming since all European countries and their national authorities are involved. Major steps taken after a few years of work for frequency regulation and redeployment are analysis of spectrum requirements, request for the proposed spectrum, study of compatibility aspects, and recommendation of policies for harmonized spectrum usage. A decision by the European Commission to designate the spectrum has been carried out and the spectrum has been allocated in August 2008 [13], [14] and is on its way for implementation by the EU countries (at the time this book was written). Eventually, the frequency bands 5875–5905 MHz for road safety, an additional 20 MHz above this band as future extension, and 5855–5875 MHz for nonsafety will be available. The allocated frequency of 50 MHz and optional the additional 20 MHz are similar to the 75 MHz ITS band in North America.

1.5.2.1 IEEE

The DSRC radio technology is essentially IEEE 802.11a adjusted for low overhead operations in the DSRC spectrum, and it is being standardized as IEEE 802.11p (at the time this book was written). The overall DSRC communication stack between the link layer and applications is being standardized by the IEEE 1609 working group. Hence, IEEE 1609 is a higher-layer standard on which IEEE 802.11p is based. Indeed, the IEEE 1609 family of standards for wireless access in vehicular environments consists of four standards: (i) IEEE P1609.1—WAVE Resource Manager defines the basic application platform and includes application data read/write protocol between RSU and OBU, (ii) IEEE P1609.2—WAVE Security Services defines the 5.9-GHz DSRC Security, anonymity, authenticity, and confidentiality, (iii) IEEE P1609.3—WAVE Networking Services defines network and transport layer services, including addressing and routing, in support of secure WAVE data exchange, and (iv) IEEE P1609.4—WAVE Multichannel Operations provides DSRC frequency band coordination and management, where it manages lower-layer usage of the seven DSRC channels, and integrates tightly with IEEE 802.11p.

1.5.2.2   C2C-CC

A major driving force for vehicular communication based on WLAN technology in Europe is the C2C-CC [9], a consortium of car manufacturers, suppliers, and research institutes. The C2C-CC assimilates developments from various European R&D projects, creates system and protocol specifications, and provides a framework for system prototyping. In 2007, the C2C-CC took a substantial step forward and published its "manifesto" describing the main concepts of the system, covering system and protocol architecture, use cases, and communication protocols. A core concept of C2C-CC's networking approach is based on wireless ad hoc and multihop communication utilizing geographical addressing and routing. The consortium is looking forward to allowing interoperability among cars from different car manufacturers and suppliers of on-board and roadside units. In this context, the C2C-CC is concerned with real-life demonstrations of safety applications for tangible ad hoc networks.

1.5.2.3   ETSI

The European Telecommunications Standards Institute (ETSI) has recently created a new technical committee TC ITS [15] in order to develop standards and specifications for ITS services [16]. The TC ITS is organized in five working groups: WG1—User and Application Requirements, WG2—Architecture and Cross-Layer Issues, WG3—Transport and Network, WG4—Media and Related Issues, and WG5—Security. The working groups have already agreed on a number of work items for various aspects of vehicular communication including media, networking, and security and safety applications. In WG3, the current focus is on specification of ad hoc networking based on geographical addressing and routing. In order to allow for use of different media, the specification distinguishes between media-independent and media-dependent network functions. The specifications are backed by other work groups, which specifically address media and security issues, such as a European profile standard of IEEE 802.11 for ITS. The technical committee is developing a road map for standardization developments for the coming years in order to achieve a complete set of standards ranging from communication architecture to protocol specifications together with formal test procedures. ITS-related work within ETSI is led by ETSI ERM TG37 (Electromagnetic Compatibility and Radio Spectrum Matters), which works in close cooperation with other ETSI committees and with other standards development organizations (SDOs) notably ISO TC204. ERM TG37 contributes to the development process standards of being led by ISO TC204 and will develop complementary ETSI standards as appropriate.

1.5.2.4 ISO

The worldwide ISO TC204/WG16 has produced a series of draft standards known as CALM (Continuous Air-Interface, Long and Medium Range [17]). The goal of CALM is to develop a standardized networking terminal that is capable of connecting vehicles and roadside systems continuously and seamlessly. This would be accomplished through the use of a wide range of communication media, such as the mobile, cellular, and wireless local area networks, and the short-range microwave (DSRC) or infrared (IR). CALM provides universal access through a number of complimentary media and links them with modern Internet protocols, adaptation layers, and management entities. The CALM architecture separates service provision from medium provision via an IPv6 networking layer, with media handover, and will support services using 2G, 3G, 5 GHz, 60 GHz, MWB (802.16e, 802.20, and HC-SDMA). It will be able to include other technologies as they evolve by use of common service access protocols and the IPv6 networking.

The CALM [17] concept, that ETSI is also helping to develop, is now at the core of several major EU sixth framework research and development projects such as SAFESPOT [18] and CVIS [19], which will test CALM solutions. In the United States the VII initiative will be operating using IEEE 802.11p/1609 standards at 5.9 GHz, which are expected to be aligned with CALM 5.9-GHz standards, although the IEEE standards do not have media handover.

1.5.3 Related Projects

The earliest research in intervehicular communications was conducted by JSK (Association of Electronic Technology for Automobile Traffic and Driving) of Japan in the early 1980s [20]. This work treated intervehicular communications primarily as traffic and driver information systems incorporated in ATMs (asynchronous transfer mode).

From the 1990s through 2000, American PATH [21] and European "Chaffeur" [22] projects investigated and deployed automated platooning systems through the transmission of data among vehicles.

Recently, the promises of wireless communications to support vehicular safety applications have led to several national and international projects around the world. Since 2000, many European projects (CarTALK2000, FleetNet, etc.), supported by automobile manufacturers, private companies, and research institutes, have been proposed with the common goal to create a communication platform for intervehicle communication.

■ The IST European project CarTALK2000 [23] was focused on new driver assistance systems, which are based upon intervehicle communication. The main objectives were the development of cooperative driver assistance systems and the development of a self-organizing ad hoc network as a communication basis, with the aim of preparing a future standard.

■ The FleetNet project in Germany [24], supported by six manufacturers and three universities from 2000 through 2003, produced important results in several research areas, including the experimental characterization of vehicular networks, the proposal of novel network protocols (MAC, routing), and the exploration of different wireless technologies.

1.5.3.1 Recent Projects

At the time this book was written, many activities in research and development of vehicular networks were ongoing. In Europe, major R&D projects were being initiated to constitute the basis of a Europe-wide intelligent transportation system, for example, NoW [25], CVIS [19], SAFESPOT [18], COOPERS [26], GeoNet [27], and GST [28].

■ NoW [25].
Network on Wheels (NoW) is a German project, successor of the project FleetNet-Internet on the Road [24], which mainly works on communication aspects for vehicle-to-vehicle and vehicle-to-roadside communication based on WLAN technology. The specific objective of the NoW project is the development of a communication system that integrates both safety [such as extended electronic break light (EEBL)] and nonsafety applications (such as car-to-home applications). Started in 2004, the final project presentation in May 2008 demonstrated a consolidated technical basis, which serves as reference for planned field. One of the main outcomes of the project is a prototype software platform for car-to-car and car-to-infrastructure communication (http://c2x-sdk.neclab.eu). This platform provides the protocol stack and an open API and offers a toolkit for application design, implementation, and testing.

- CVIS [19].
  Cooperative Vehicle Infrastructure Systems (CVIS) project aims at developing a communication system that is capable of using a wide range of wireless technologies, including cellular networks (GPRS, UMTS), wireless local area networks (WLAN), short-range microwave beacons (DSRC) and infrared. Access to these wireless technologies is based on the new international "CALM" standard [17], which allows future vehicular networking implementation to be integrated with the CVIS platform via standardized CALM service access points. A framework for open application management (FOAM) is defined that connects the in-vehicle systems, roadside infrastructure, and back-end infrastructure, which is necessary for cooperative transport management.
- SAFESPOT [18].
  SAFESPOT provides cooperative systems for road safety, referred to as smart vehicles on smart roads, to prevent road accidents by developing a safety margin assistant that detects potentially dangerous situations in advance and extends the drivers' awareness of the surrounding environment in space and time. This assistant represents an intelligent cooperative system utilizing vehicle-to-vehicle and vehicle-to infrastructure communication based on WLAN technology (IEEE 802.11p).
- COOPERS [26].
  Cooperative Systems for Intelligent Road Safety (COOPERS) project focuses on the development of innovative telematics applications on the road infrastructure with the long-term goal of a cooperative traffic management between vehicle and infrastructure. COOPERS attempts to improve road sensor infrastructure and traffic control applications, develops a communication concept and applications able to cope with the requirements for infrastructure-to-vehicle communication, and demonstrates results at major European motorways with high-density traffic.
- GeoNet [27].
  The EU project GeoNet (http://www.geonet-project.eu) started in February 2008 and implements a reference system for vehicular ad hoc networking using concepts for geographical addressing and routing. Particular focus lies on integration of geonetworking with IPv6 and solutions for IP mobility support. In GeoNet, a vehicle is regarded as a mobile network, where the NEMO protocol handles Internet connectivity of the nodes in the mobile network with intermittent access to roadside units. For wide deployment of the project results, it is planned to provide the GeoNet implementations to other R&D projects.

- GST [28].
  Global System for Telematics (GST) project creates an open and standardized end-to-end architecture for automotive telematics services. The project targets infrastructure-oriented services typically provided by a network operator, such as emergency call services, enhanced floating car data services, safety warnings, and information services.

1.6 Conclusions

Intervehicular communication (IVC) is becoming a reality, driven by navigation safety requirements and by the investments of car manufacturers and public transport authorities. Its opportunities and areas of applications are growing rapidly and include many kinds of services with different goals and requirements. However, it does pose numerous unique and novel challenges from network evolution to event detection and dissemination, making research in this area very attractive. Consequently, IVC is attracting a considerable attention from the research community and the automotive industry, where it is beneficial in providing intelligent transportation system as well as driver and passenger assistant services. In this context, vehicular ad hoc networks are emerging as a new class of wireless networks, spontaneously formed between moving vehicles, and allowing for a number of useful services for drivers and passengers, ranging from road safety applications to entertainment applications. These networks are promising for network operators, service providers, and for a number of industrial and telecom companies in terms of opening new business opportunities.

However, the penetration of vehicular network technology is still weak, and hence there is a need for infrastructure support to help its penetration. At the same time, deploying new infrastructure for these networks necessitate a lot of investment and high cost. It is more economical to rely on the existing infrastructure (owned by network operators, for instance) for accelerating the penetration of such technology with the least cost.

Although many standard organizations are involved in the study and standardization of IVC, vehicular ad hoc networks are considered as a technology under development that merits a lot of research and field trials. Besides the ongoing standardization activities, a number of technical challenges, as discussed in the following chapters, need to be resolved aiming for wide-scale deployment of these networks in the near future. Still, many topics in this field are currently under discussion, such as allocation of a protected frequency band for road safety in Europe, potential usage of the IEEE 802.11p/WAVE standard, integration of multiple wireless technologies, data security, congestion control, data transport, and others.

In addition to technical breakthroughs, the phase of market introduction is critical for the success of this new technology. Also, car manufactures like BMW, Mercedes, Fiat, Ford, Toyota, and Nissan, are currently prototyping vehicles equipped with Wi-Fi (802.11a/b/g) and DSRC technologies, which are expected to be on the road within the next 3–5 years.

References

[1] Car 2 Car Communication Consortium Manifesto, work in progress, May 2007.

[2] M. Nekovee, "Sensor networks on the road: the promises and challenges of vehicular ad hoc networks and vehicular grids," In *Proc. of the Workshop on Ubiquitous Computing and e-Research*, Edinburgh, UK, May 2005.

[3] J. Blum, A. Eskandarian, and L. Hoffmman, "Challenges of intervehicle ad hoc networks," *IEEE Trans. Intelligent Transportation Systems* 5(4) (December 2004):347–351.

[4] O. K. Tonguz and G. Ferrari, *Ad Hoc Wireless Networks: A Communication-Theoretic Perspective*, New York: John Wiley & Sons, 2006.

[5] N. Wisitpongphan, O. Tonguz, J. Parikh, F. Bai, P. Mudalige, and V. Sadekar, "On the Broadcast Storm Problem in Ad Hoc Wireless Network," *IEEE Wireless Communications*, December 2007.

[6] M. Torrent-Moreno, D. Jiang, and H. Hartenstein, "Broadcast reception rates and effects of priority access in 802.11-based vehicular ad-hoc networks," In *Proc. of ACM International Workshop on Vehicular Ad Hoc Networks (VANET 2004)*, Philadelphia, PA, October 2004.

[7] Vehicle Safety Communications (VSC) Projects, http://www.car-to-car.org/fileadmin/downloads/security_2006/sec_06_04_laberteaux_CAMP.pdf (accessed November 2008).

[8] Collision Avoidance Metrics Partnership—CAMP (U.S.), http://www.its.dot.gov/cicas (accessed November 2008).

[9] Car 2 Car Communication Consortium, http://www.car-to-car.org (accessed November 2008).

[10] Advanced Safety Vehicle (ASV) Program, http://www.its.go.jp/ITS/indexHBook.html (accessed November 2008).

[11] The Vehicle Infrastructure Integration (VII), http://www.its.dot.gov/vii (accessed November 2008).

[12] Microsoft—Helping Carmakers and Suppliers Drive Innovation, http://www.microsoft.com/windowsautomotive/default.mspx (accessed November 2008).

[13] Andreas Festag, "Vehicular Ad Hoc Networks Research: Europe," *IEEE Comsoc AHSNTC Newsletter* 1(1) (2008), http://ieee-comsoc.jot.com/WikiHome/AdHocTC/AHSNTC (accessed November 2008).

[14] Commission decision on the harmonized use of radio spectrum in the 5 875-5 905 MHz frequency band for safety-related applications of Intelligent Transport Systems (ITS) on the harmonized use of radio spectrum in the 5 875-5 905 MHz frequency band for safety-related applications of intelligent

transport systems (ITS). Official Journal of the European Union, August 2008. http://www.erodocdb.dk/Docs/doc98/official/pdf/2008671EC.PDF (accessed November 2008).

[15] The European Telecommunications Standards Institute (ETSI), http://www.etsi.org (accessed November 2008).

[16] ETSI TR 102 492-1/2: "Electromagnetic compatibility and radio spectrum matters (ERM); Intelligent Transport Systems (ITS); Part 1: Technical characteristics for pan-European harmonized communications equipment operating in the 5 GHz frequency range and intended for critical road-safety Applications," June 2005. Part 2: "Technical Characteristics for Pan European harmonized communications equipment operating in the 5 GHz frequency range intended for road safety and traffic management, and for non-safety related ITS applications," March 2006.

[17] CALM (Communications Air-Interface, Long and Medium Range), http://www.calm.hu (accessed November 2008).

[18] SAFESPOT (Cooperative vehicles and road infrastructure for road safety), http://www.safespot-eu.org (accessed November 2008).

[19] CVIS (Cooperative Vehicle-Infrastructure Systems), http://www.cvisproject.org (accessed November 2008).

[20] S. Tsugawa, "Issues and recent trends in vehicle safety communication systems," *IATSS Research* 29(1), 2005.

[21] J. Hedrick, M. Tomizuka, and P. Varaiya, "Control issues in automated highway systems," *IEEE Control Systems Magazine* 14(6), 1994.

[22] O. Gehring and H. Fritz, "Practical results of a longitudinal control concept for truck platooning with vehicle to vehicle communication," In *Proc. of the IEEE Conference on Intelligent Transportation Systems*, Boston, 1997.

[23] Safe and comfortable driving based upon inter-vehicle communication, http://www.cvisproject.org (accessed November 2008).

[24] FleetNet project—Internet on the road, http://www.cvisproject.org/en/links/cartalk_2000.htm (accessed November 2008).

[25] NoW (Network-on-Wheels), http://www.network-on-wheels.de.

[26] COOPERS (Cooperative Systems for Intelligent Road Safety), http://www.coopers-ip.eu.

[27] GeoNet (Geographic Network), http://www.geonet-project.eu.

[28] GST (Global Systems for Telematics), http://www.gstforum.org.

Chapter 2

Vehicular Network Applications and Services

Maria Kihl

Contents

Vehicular networks will not be commercially deployed unless there are applications that benefit from them. Since the first research projects on vehicular networks, see, for example, the PROMETHEUS project [1,2], the main focus has been on increasing traffic safety. With the help of vehicular communication, the number of accidents will decrease and human lives will be saved.

Therefore, the two application categories that have attracted the most attention during the years are related to *public safety*, for example, collision avoidance applications and *vehicular traffic coordination*, where vehicles coordinate their movements with each other. Some considerable work has also been done on applications related to *road traffic management*, which can help alleviate congested roads and thereby decrease the number of accidents, as well as decreasing the travel time.

Numerous other applications have also been proposed, which relate to other objectives than traffic safety. Many of these applications can be categorized as *comfort applications*. The objective of these applications is to improve the travel comfort for both the drivers (e.g., with information from roadside restaurants) and the passengers (e.g., with Internet access and video-on-demand systems).

The main objective of the papers proposing various vehicular applications is usually to propose a data dissemination technique for this particular application. Very few papers, if any, have been focused on *how* the applications should be implemented in reality. In this chapter, we will, therefore, describe some proposed solutions for the envisioned applications and, thereby, also discuss suitable communication architectures and protocols for these applications.

2.1　Basic Concepts

In this section, we will describe some basic concepts that are important for understanding vehicular network applications.

2.1.1　On-Board Equipment

It is always assumed that the vehicles that are part of a vehicular network have some form of on-board equipment (OBU). Following is a list of hardware that most papers assume available in each, so-called, *equipped* vehicle:

- A central processing unit (CPU) that implements the applications and communication protocols.
- A wireless transceiver that transmits and receives data to and from the neighboring vehicles and roadside.

- A GPS receiver that provides relatively accurate positioning and time synchronization information.
- Appropriate sensors to measure the various parameters that have to be measured and eventually transmitted.
- An input/output interface that allows human interaction with the system.

Realistically, only a few vehicles will be initially equipped. However, in this chapter we will assume that all vehicles on the road have the relevant equipment. It is out of the scope of this chapter to discuss the application performance when only a part of the vehicles are equipped.

2.1.2 Addressing

For most applications, some kind of addressing scheme is necessary. Most vehicular networks can be classified as ad hoc networks, meaning that the nodes (vehicles and roadside stations) organize themselves in a network. Therefore, the same addressing schemes as in ad hoc networks can be used [3]:

- **Fixed** addressing means that each node has a fixed address assigned by some mechanism at the moment it joins the network; the node uses this address while part of the network. This is the most common addressing scheme in the Internet (with mobile IP being the exception). Most ad hoc networking applications and protocols assume a fixed addressing scheme.
- **Geographical** addressing means that each node is characterized by its geographical position. As the node moves, its address changes. Additional attributes may be used to further select a subset of target vehicles. Examples of such attributes are
  - The direction of movement of the vehicle
  - The road identifier (e.g., number, name, etc.)
  - The type of vehicle (e.g., trucks, 18 wheelers, etc.)
  - Some physical characteristics (e.g., taller than, weighing more than, or with a speed higher than)
  - Some characteristic of the driver (e.g., beginner, professional, etc.)

2.1.3 Data Dissemination

The objective of all applications described in this chapter is to disseminate data between vehicles, potentially including roadside stations. Some important concepts related to data dissemination are described below.

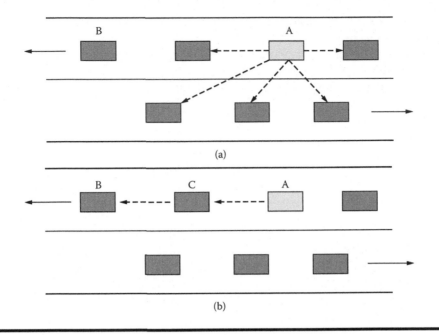

Figure 2.1   Single-hop (a) and multihop (b) data dissemination.

First, data dissemination can be *single-hop* or *multihop* (see Figure 2.1). Single-hop dissemination is usually implemented with broadcast on the MAC layer, as seen in Figure 2.1(a) where vehicle A can send a message only to the cars, which are in its transmission range (i.e., vehicle B never receives the message). Single-hop dissemination is also used when there are roadside base stations that control the communication. Multihop dissemination is closely related to vehicular ad hoc networks (VANETs). Data should be transmitted in several hops, where intermediate vehicles act as relays, as seen in Figure 2.1(b) where vehicle C can relay the message such that vehicles not in the transmission range of vehicle A (e.g., vehicle B) can also receive the message. Therefore, a multihop system requires a network layer capable of multihop routing. Hybrid variants have also been proposed, where, for example, data is disseminated in multihop to the closest base station, which then transmits the data to relevant vehicles using single-hop dissemination.

Second, data can be disseminated in *unicast, multicast,* or *broadcast.* In unicast, there is one sender and one receiver of the data. Note that this is irrespective of addressing scheme, and also geographical addresses could be used for unicast. In multicast, there is one sender and one or several receivers. The sender usually does not know exactly how many receivers there are, only that it sends the data to a specific destination *group.* In vehicular networks, several applications related to public safety require data to be disseminated to, for example, all vehicles in a specific area that are driving in a specific direction, which would correspond to a multicast group. In broadcast, the data should be disseminated to *all* vehicles. However, because a vehicular network can span whole continents, broadcast is usually only performed in a certain area, called the zone-of-relevance (ZOR) [2].

2.1.4 Network Access Technologies

Today, there are several communication standards that may be used as access networks for vehicular network applications. All these standards have advantages and disadvantages depending on the type of application and considered scenario. In vehicular communication systems, the fast and slow fading effects can be expected to cause radio channel problems as both the transmitter and the receiver are moving, potentially toward each other.

IEEE 802.11 is currently the most widely used wireless local area network standard in the world. Therefore, it should come as no surprise that 802.11 is the network access technology most commonly assumed in vehicular networks. However, it has been shown, see for example [5], that, with realistic propagation models, the bit-error rate (BER) of 802.11 can be very high, imposing significant challenges to higher layers. The problem can be traced to the design of 802.11a that was optimized for local area networks with no or low mobility. However, in vehicular networks, the mobility can be very high and the equalization techniques designed for low mobility can barely handle the resulting fast fading conditions. Therefore, a new standard, IEEE 802.11p is currently under development, which is intended for vehicular communication.

The cellular networks cover large areas and may be a good solution for vehicular networks when vehicles are outside major cities and highways. However, the cellular systems have not been designed and provisioned for simultaneous utilization by a large number of users for long periods of time at high traffic volumes. These inherently single-hop networks rely heavily on centralized infrastructure (base stations) to coordinate the transmissions of the mobile nodes. Research projects in Germany and Italy have evaluated GSM/GPRS and 3G systems for IVC systems [6,13,14]. The main argument for using a mobile telephony standard for vehicular communication systems is that the infrastructure is already there and in the near future most vehicles in Europe will have access to these networks. Furthermore, 3G systems support long-range communications, offer quality of service guarantees, and are designed for high-speed mobility. Finally, the freedom to assign different spreading factors to different stations, allows for trade-offs between the robustness of the transmission and the capacity allocated to each station.

Other potential network access technologies are Bluetooth, mainly intended for short-range communication, and several new protocols, specifically designed for vehicular networks. See "A survey and qualitative analysis of MAC protocols for vehicular ad hoc networks" [7] for a good overview. In this chapter, we will only separate between access technologies suitable for multihop ad hoc networks, as IEEE 802.11, and access technologies requiring infrastructure, as cellular networks.

2.1.5 Communication Architectures

The objective of communication architecture is to provide communication facilities for a wide range of applications. Communication architectures are usually based on the layered OSI-model, where each layer provides certain functions. For vehicular networks, the main challenge is that different applications will have very different quality of service (QoS) demands. Therefore, they will probably require different network access technologies, addressing schemes, and protocols.

"(Auto) Mobile communication in a heterogenous and converged world" [11] was the first paper to propose a general communication architecture for vehicular communication systems, incorporating several network access technologies. An illustration of the concept is shown in Figure 2.2. The vehicle will have several network access technologies installed, examples like IEEE 802.11, UMTS/CDMA2000, and Bluetooth are shown in the figure. IEEE 802.11 could be used either in ad hoc mode or with roadside base stations. The *communication* network manager will handle the

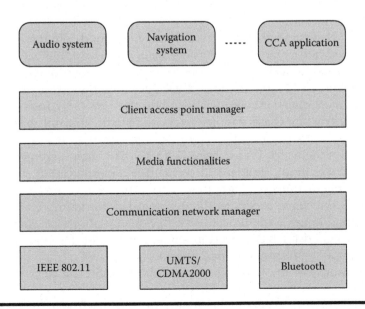

Figure 2.2   A communication architecture.

communication with different networks. The *media* functionalities can include entities like email or application servers. The *client* access point manager represents the interface to the client applications. In the figure, some examples of applications are shown.

Currently, the communication architecture *wireless access in vehicular environment (WAVE)*, based on IEEE 802.11p and IEEE P1609, is under standardization [25]. The architectural ideas come from the internationally agreed standard *dedicated short-range communication (DSRC)* [26]. The main focus of DSRC is on the lower layers (MAC/PHY); however, WAVE also includes higher-layer protocols based on IEEE P1609. Figure 2.3 shows a schematic view of the architecture. WAVE supports both IP and non-IP based applications. The non-IP based applications, for example, emergency warning messages, are supported by the WAVE short message protocol (WSMP). IEEE 802.11p has support for application priorities, which means that high priority data, for example, related to collision avoidance, will be transmitted with minimum latency.

Also, a new communication architecture, *continuous air interface for long and medium distances (CALM)* is under development by ISO. CALM will incorporate WAVE, and it will also provide seamless communication in a heterogenous network environment, with several network access technologies. Further, it will support both geographical and fixed addressing.

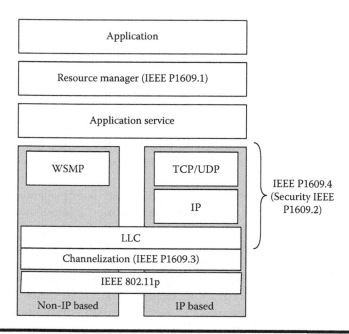

Figure 2.3   Schematic view of the WAVE architecture. WSMP = WAVE Short Message Protocol.

2.1.6 Geocasting

Location-based multicast routing, also called *geocasting* [15], is suitable for several applications. In geocasting, all or some vehicles in a certain area should receive the data. The decision about which vehicles should receive the data is not based on their addresses, instead other metrics as location, direction, or speed are used. Geocasting based on selective broadcast may be a good solution in order to achieve low latencies in, for example, public safety applications. Selective broadcast uses flooding on the MAC or network layer. However, because a pure flooding scheme easily can cause broadcast storm problems [4] in a dense network, for example, on a highway during rush hour, an intelligent forwarding mechanism is used. Each vehicle makes a local decision whether to forward a message or not. The objective is to minimize the number of unneccesary retransmissions of the data. Vehicles close to the source could, for example, cancel their forwarding if they hear another vehicle further away forward the data. The selective broadcast schemes have gained much attention, and several variants of these schemes have been proposed, see, for example, [8–10].

2.2 Public Safety Applications

Public safety applications are geared primarily toward avoiding accidents and loss of life of the occupants of the vehicles. The major characteristic for these applications is that data should be disseminated fast and reliably to a large number of vehicles within a certain ZOR. Therefore, geocasting is the commonly proposed data dissemination method for public safety applications. There are two main classes of public safety–related applications, *cooperative collision avoidance* and *emergency warning message*.

In a cooperative collision avoidance (CCA) application, the objective is to avoid collisions, either chain collisions on highways or head-on collisions on smaller roads. Vehicles can automatically stop if they receive a collision message or sense a collision by rapidly decreasing speed of neighboring vehicles. Obviously, all cooperative collision avoidance applications have very strict real-time demands, both on communication reliability and latency. S. Biswas et al. [9] argue that a collision avoidance application requires delivery latencies of less than 100 ms if chain collisions are to be avoided.

In an emergency warning message (EWM) application, vehicles send out warnings about accidents or hazardous road conditions to other vehicles driving in or approaching the area. These applications may require that the warning message "stays" in the relevant area for a long time. Papers proposing EWM applications sometimes also propose a mechanism for keeping the warning message in the area.

2.2.1 Cooperative Collision Avoidance

In a cooperative collision avoidance (CCA) application, vehicles communicate in order to identify a potential collision situation and then either inform the driver or act automatically. A CCA application may also be invoked when there *is* a collision, so as to avoid other vehicles from driving into the crashed vehicles. The scenario for CCA applications is usually one of the two shown in Figure 2.4. Multilane highways [Figure 2.4(a)] will always have high risks of chain collisions and accidents due to lane changes. On a multilane higway during rush hour, thousands of vehicles will need to be informed in a very short time in case of a potential collision situation. On smaller single-lane roads [Figure 2.4(b)] the main danger is instead head-on collisions. Here, only a few vehicles are involved. However, the relative speed of the vehicles is usually very high, due to the fact that they travel in opposite directions.

Basically all papers proposing data dissemination methods for CCA applications assume a multihop vehicular ad hoc network based on IEEE 802.11. Infrastructure-based solutions, with either cellular networks or roadside stations, will not have low-enough transmission latencies. For scenarios with multilane highways, the proposed data dissemination method is usually some kind of selective broadcast [9]. The dissemination area may be rather limited, because only the closest vehicles will need the information in order to automatically avoid a collision.

Avoidance of collisions is an important part of the efforts to increase road traffic safety, and a CCA application may save human lives. However, the latency and reliability demands are very high. On dense highways, with high speeds, the radio conditions will impose great challenges for this category of applications. An alternative to a pure CCA application is, therefore, an EWM application that disseminates warning messages either *after* an accident has occured or when hazardous road conditions are detected. The EWM applications are described in the following section.

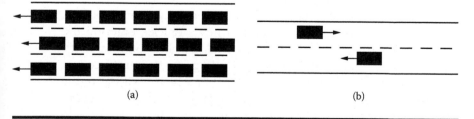

(a)   (b)

Figure 2.4 Two scenarios for collision avoidance applications: (a) Avoidance of chain collisions on multilane highways; (b) Avoidance of head-on collisions on smaller roads.

2.2.2 Emergency Warning Message

The objective of an emergency warning message application is to disseminate warning information to vehicles in the relevant area. EWM applications have received much attention, because they can be rather easily deployed with current network access technologies, both infrastructure based, as cellular networks, and ad hoc network based, as IEEE 802.11. There are basically two categories of EWM applications: *instant EWM* and *abiding EWM*.

2.2.2.1 Instant EWM

In an instant EWM application, see for example [8,10,12], a warning message should be disseminated to all vehicles in the ZOR, which in this case is the surrounding area. One example is an application that sends out a warning message when it senses that a vehicle is involved in an accident, usually by rapidly decreasing speed. Note, that the sending vehicle itself may not be colliding, because drivers close to an accident probably hit the brakes as well.

When the message has been disseminated to all vehicles in the ZOR, which may be several kilometers in diameter, the message will "disappear." Therefore, both IEEE 802.11 in an ad hoc network or a cellular network can be used as network access technology. In case an ad hoc network is used, a selective broadcast algorithm is usually proposed for data dissemination [8,10]. In case a cellular network is used [13], the vehicle sends the warning message to its base station, and then the cellular network broadcasts the message to all vehicles in the ZOR. The main advantage of the cellular network is that the warning message can also be disseminated when there is a low density of vehicles, which means that the distance between vehicles is larger than the transmission range of IEEE 802.11. However, the main disadvantage with this solution is that the cellular network, at the moment, is not designed to handle this type of message dissemination and that it may be congested during rush hours on multilane highways.

2.2.2.2 Abiding EWM

The objective of an abiding EWM application is to warn other drivers about an accident or hazardous road condition during a longer time period. The message needs to "stay" in the ZOR and new vehicles entering the area should receive the message. An abiding EWM application may use a dissemination method called abiding (or stored) geocast, introduced in "Performance evaluation of stored geocast" [17], in which a virtual traffic sign application was proposed. The basic principle is shown in Figure 2.5. All vehicles entering the destination region should receive the warning message. From the beginning, the warning message has been disseminated to all vehicles in the region using an ordinary geocasting algorithm. After this the informed vehicles need to detect a new vehicle entering the

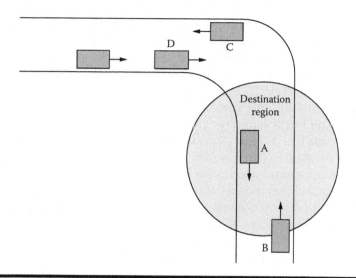

Figure 2.5 **An illustration of abiding geocast.**

region. This can be implemented with some kind of neighbor information exchange algorithm, typically using HELLO messages. When a new
vehicle is detected, it will, therefore, receive the warning message by the
other vehicles. In Figure 2.5, vehicle A will inform vehicle B that is entering the region. One simple improvement of the algorithm is to let vehicles
that have left the region but that still are in the surrounding area, inform
other vehicles approaching the region. In this case, vehicle C could inform
vehicle D, because it will soon enter the region.

One challenge for abiding EWM applications is to ensure a reliable dissemination of warning messages on roads with low traffic, for example,
during night times. In "Time-stable geocast for ad hoc networks and its
application with virtual warning signs," [16] a solution is proposed that
assumes some infrastructrure. Here it is assumed that the message first is
transmitted to a central server. The server is then responsible for disseminating the message in the region. This solution probably requires a cellular
network–based access technology, where the vehicle transmits the warning
message to its base station, which obviously is in or close to the region. The
base station can then easily transmit the message to all other vehicles in or
approaching the region because they will connect with the base station.

2.3 Vehicular Traffic Coordination

Vehicular traffic coordination has been in focus for almost 20 years, since
the idea of vehicular communication was introduced. The objective of the
applications in this category is to assist the drivers in different ways. In

particular on multilane highways with heavy traffic, actions like passing and lane change may be the cause of serious accidents. Some assistance from the vehicles may reduce or eliminate risks during these maneuvers. The ultimate vision is of course self-driving vehicles that coordinate themselves with vehicular communication.

The PATH project in California has for a long time worked on the so-called platooning, where vehicles form tight columns by closely following each other on highways [18]. Vehicle platooning may increase the throughput on the highways as well as decrease the risk of accidents. However, the technical challenges are enormous. For example, high-speed closed-loop control is of paramount importance for this application. Vehicles will need to adjust their speed and direction continuously depending on the location and speed of surrounding vehicles. Therefore, the demands on the latency and communication reliability are very strict.

Most work on traffic coordination applications has been focused on the control algorithms [19, 20]. Distributed control is a major technical challenge for this group of applications. There exist well-known linear models for car-following systems [21]; however, they are out of the scope of this book. The challenge is to coordinate the control actions for all the vehicles in the system. Vision-based methods can also be incorporated, so that the vehicles avoid obstacles on the road [22]. In this case, the application uses cameras and image-processing, and, thereby, it can identify obstacles as pedestrians or parked cars.

One major challenge for these applications, apart from the distributed control, is the communication system requirements. If vehicular communication is discussed, most papers assume single-hop short-range communication systems, like radar, infrared, or ultrawideband (UWB) [21]. However, this would imply that the traffic coordination applications cannot be a part of general communication architecture for vehicular communication. Very few papers (see, for example, [23, 24]) have used IEEE 802.11 in their systems. One problem is that IEEE 802.11 may not be able to provide the communication reliability and latency required; however, more research is needed on this subject.

However, another major challenge for this group of applications will be human behavior. Do drivers want their vehicles to act automatically during lane changes or overtaking? And what about letting go of the steering wheel completely, as will be the case in vehicle platooning? It would be very dangerous if drivers started to interfere with the coordination application. Today, some cruise control systems adjust the speed depending on the speed of the vehicle in front, in order to keep a safe distance from other vehicles. However, it will probably take many years before drivers would like to leave the control of the vehicle to a computer system.

2.4 Road Traffic Management

Road traffic management applications are focused on improving traffic flow, thus reducing both congestion problems and travel times. The main difference between road traffic management applications and public safety applications is the real-time requirements. As discussed before, public safety applications have strict demands on latency and communication reliability because if a message is delayed or lost, a serious collision may occur. The real-time demands on road traffic management applications are less strict. The objective is to provide the drivers with information regarding the traffic, either in the surrounding area, or in a specific location, for example, an intersection. Therefore, the latencies can be longer and some data losses are acceptable.

2.4.1 Traffic Monitoring

One application in this category is *traffic monitoring* that can provide high-resolution, localized, timely traffic information for several miles around the current location of the vehicle [27–29]. The basic function of a traffic monitoring application is as follows. All vehicles are assumed to be equipped with a navigation system or at least a digital map with standardized identifiers for each road segment (see an illustration in Figure 2.6). Also, it is assumed that the vehicles have sensors that can measure the relevant metrics, for example, speed or temperature. Each vehicle collects data, for example, the speed on its current road segment, and then, at regular time intervals, transmits this data to all vehicles in a suitable ZOR, which can be several kilometers in diameter. A vehicle that receives traffic data about a certain road segment, stores the data in a table, potentially after some

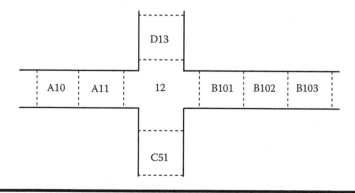

Figure 2.6   A digital map with road segment identifiers.

data treatment, for example, aggregation with other data about the same road segment. The stored information can then be used to either simply inform the driver or to improve the performance of the vehicle's navigation system [30].

Traffic monitoring can be implemented with multihop dissemination or with base stations that relay data coming from the vehicles.

Infrastructure-based traffic monitoring systems are already in use. Traffic information, regarding accidents and road work, is broadcasted in several places in the world, and used by some advanced navigation systems. Therefore, these systems could be improved to also incorporate information from the vehicles [30].

If multihop data dissemination is used, a geocasting protocol is probably a suitable solution [31]. The geocasting protocol can disseminate the data to all vehicles in the ZOR.

One issue here is the risk of packet congestion when there is high density of vehicles in the area. Therefore, the geocasting algorithm should include mechanisms for minimizing the transmission overheads. Also, the routing protocol should include prioritization of data, so that public safety–related data is not delayed due to traffic monitoring data.

Instead of a geocasting algorithm, a so-called diffusion mechanism can be used [27–29]. Diffusion works at the application layer, thus not being a "true" multihop mechanism. In a diffusion mechanism, each vehicle has a table with aggregated data about each road segment. At regular intervals, the vehicle broadcasts (single-hop) its table to all neighbors. A vehicle that receives a data message, aggregates the data with its own table, and so on. The result is that the data is "diffused" in the network, each vehicle having more accurate information on the state of the nearby traffic (and relatively outdated information from distant regions). The main advantage of the scheme is its simplicity and good fit for the needs of traffic-monitoring applications. In "Specification and Performance Evaluation of Two Zone Dissemination Protocols for Vehicular Ad-Hoc Networks," [29] a comparison between a geocasting protocol and a diffusion mechanism in a highway scenario was presented. The assumption in this paper was that the relevance of information about a particular phenomenon decreases with the distance to that phenomenon, which should be true for a traffic-monitoring application. The authors showed that a diffusion mechanism can achieve good results at a much lower network load than when using a geocasting algorithm.

2.4.2 Intersection Assistance

Several papers have proposed various applications related to intersections, because many accidents occur there, both involving other vehicles and pedestrians. The objective of the applications is usually to assist the drivers at the intersections [35], in order to avoid accidents with other vehicles [32,33] or pedestrians [34]. In "Communication-based intersection assistance," Bluetooth is used as network access technology due to the short communication distances at intersections [32]. However, IEEE 802.11 is probably a better choice, due to its applicability to a broader range of applications. Also, the intersection assistance application could be communicating with the traffic lights, in order to optimize the traffic light scheduling [35]. One envisioned scenario is that emergency vehicles, for example, ambulances, could communicate with the traffic lights, thereby getting green lights on all intersections.

2.5 Comfort Applications

The main focus of comfort applications is to make travel more pleasant. This class of applications may be motivated by the desire of the passengers to communicate either with other vehicles or with ground-based destinations, for example, Internet hosts or the public service telephone network (PSTN). Multimedia files, for example, DVDs, music, news, audio-books, prerecorded shows, can be uploaded to the car's entertainment system. Also, various traveler information applications belong to this category. For example, the driver could recieve local information regarding restaurants, hotels, and the like, when the vehicle approaches a town. Another example is advertisements of gas stations or restaurants along the road. Most work in this category has been focused on Internet access because Internet is the key technology to most comfort applications.

Today, some vehicle manufacturers provide Internet access in the vehicles via cellular networks. In-vehicle communication with IEEE 802.11 allows all passengers in the vehicle to access the Internet. Also, in-vehicle TV is supported, via the ordinary broadcast TV providers. Therefore, a more comprehensive solution that also includes travel information and other comfort applications is feasible via the same access technology because all the necessary components are already there.

Of course, several papers have proposed IEEE 802.11-based solutions as well, see, for example, [36]. The main problem with IEEE 802.11 is the limited radio range. It would be very costly to place enough base stations along the roads in order to provide full coverage for IEEE 802.11. Therefore, a hybrid solution may be feasible, where multihop communication can be provided to the closest base station, also called *gateway* (see Figure 2.7 [37,38]). In this case, the vehicles form an ad hoc network in order to help each other with the data transfer to and from the gateways. Figure 2.7 shows that unicast communication between two vehicles is also

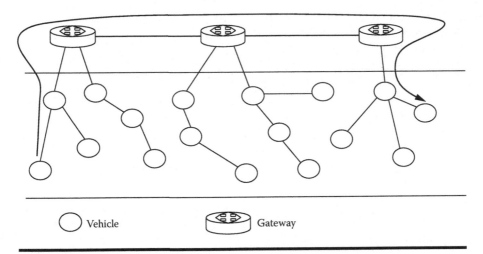

Figure 2.7 A hybrid solution for Internet access.

feasible, via the gateways. However, because this solution requires new infrastructure to be installed as well as new protocols and standards for the ad hoc communication, it is at the moment not a competitor to the cellular networks.

2.6 Conclusions

In this chapter, we have described and categorized the commonly proposed applications for vehicular networks.

It is clear that most research has been focused on traffic safety applications with the objective to decrease the number of traffic deaths and injuries. However, the technical challenge in collision avoidance and traffic coordination is a major issue. These applications require very low latency and very high communication reliability. Neither cellular networks nor IEEE 802.11 can at the moment provide the required QoS for high vehicle speeds.

On the other hand, comfort applications, such as Internet access, are already available in the vehicles via cellular networks. Here, the driving force has been the industry that can see business advantages when providing these applications to the market. However, these solutions have been implemented without any standards or international agreements, making it difficult to exploit them for a wider range of applications.

One issue that has not been solved is the security of intervehicle communication systems. If deployed, a vehicular ad hoc network will potentially involve thousands of vehicles, each with a unique owner. The issues of personal integrity and data confidentiality will be enourmous. Some work has been performed in this area (see, for example, [39–41]) however, more work is definitely required before a complete security architecture can be standardized.

References

[1] M. Williams, "PROMETHEUS-The European research programme for optimising the road transport system in Europe," In *Proc. of the IEE Colloquium on Driver Information* (Digest No. 127), 1988.

[2] W. Kremer, "Realistic simulation of a broadcast protocol for an inter vehicle communication system (IVCS)," In *Proc. of the 41st IEEE Vehicular Technology Conference*, St. Louis, MO, 1991.

[3] I. Chlamtac, M. Conti, and J. J.-N. Liu, "Mobile ad hoc networking: Imperatives and Challenges," *Ad Hoc Networks* 1(1), 2003.

[4] S. Ni, Y. Tseng, Y. Chen, and J. Sheu, "The broadcast storm problem in a mobile ad hoc network," In *Proc. of the ACM International Conference on Mobile Computing and Networking*, Seattle, WA, 1999.

[5] M. Torrent-Moreno, D. Jiang, and H. Hartenstein, "Broadcast reception rates and effects of priority access in 802.11-based vehicular ad-hoc networks," In *Proc. of the First ACM Workshop on Vehicular Ad Hoc Networks*, Philadelphia, PA, 2004.

[6] M. Lott, R. Halfmann, E. Schulz, and M. Radimirsch, "Medium access and radio resource management for ad hoc networks based on UTRA TDD," In *Proc. of the 2nd ACM International Symposium on Mobile Ad Hoc Networking & Computing*, Long Beach, CA, 2001.

[7] H. Menouar, F. Filali, and M. Lenardi, "A survey and qualitative analysis of MAC protocols for vehicular ad hoc networks," *IEEE Wireless Communications* 13(5), 2006.

[8] A. Benslimane, "Optimized dissemination of alarm messages in vehicular ad-hoc networks (VANET)," In *Proc. of the 7th IEEE International Conference on High Speed Networks and Multimedia Communications*, Toulouse, France, 2004.

[9] S. Biswas, R. Tatchikou, and F. Dion, "Vehicle-to-vehicle wireless communication protocols for enhancing highway traffic safety," *IEEE Communication Magazine.* 44(1), 2006.

[10] L. Briesemeister, L. Schäfers, and G. Hommel, "Disseminating messages among highly mobile hosts based on inter-vehicle communication," In *Proc. of the IEEE Intelligent Vehicle Symposium*, Dearborn, MI, 2000.

[11] W. Kellerer, C. Bettstetter, C. Schwingenschlögl, P. Sties, and H. Vögel, "(Auto) Mobile Communication in a Heterogeneous and Converged World," *IEEE Personal Communications* 8(6), 2001.

[12] X. Yang, J. Liu, F. Zhao, and N. H. Vaidya, "A vehicle-to-vehicle communication protocol for cooperative collision warning," In *Proc. of the First Annual International Conference on Mobile and Ubiquitous Systems: Networking and Services*, Boston, MA, 2004.

[13] O. Andrisano, R. Verdone, and M. Nakagawa, "Intelligent transportation systems: The role of third-generation mobile radio networks," *IEEE Communications Magazine* 38(9), 2000.

[14] B. M. Masini, C. Fontana, and R. Verdone, "Provision of an emergency warning system through GPRS: Performance evaluation," In *Proc. of the 7th IEEE International Conference of Intelligent Transportation Systems*, Washington, DC, 2004.

[15] C. Maihöfer, "A survey of geocast routing protocols," *IEEE Communications Surveys & Tutorials* 6(2), 2004.

[16] C. Maihöfer and R. Eberhardt, "Time-stable geocast for ad hoc networks and its application with virtual warning signs," *Computer Communications* 27, 2004.

[17] C. Maihöfer, C. Cseh, W. Franz, and R. Eberhardt, "Performance evaluation of stored geocast," In *Proc. of the IEEE 58th Vehicular Technology Conference*, Orlando, FL, 2003.

[18] P. Varayia, "Smart cars on smart roads: Problems of control," *IEEE Transactions on Automatic Control* 38(2), 1993.

[19] J. Yan and R. B. Bitmead, "Coordinated control and information architecture," In *Proc. of the 42nd IEEE Conference on Decision and Control*, Maui, HI, 2003.

[20] J. A. Fax and R. M. Murray, "Information flow and cooperative control of vehicle formations," *IEEE Transactions on Automatic Control* 49(9), 2004.

[21] P. Caravani, E. De Santis, F. Graziosi, and E. Panizzi, "Communication control and driving assistance to a platoon of vehicles in heavy traffic and scarce visibility," *IEEE Transactions on Intelligent Transportation Systems* 7(4), 2006.

[22] S. K. Gehrig and F. J. Stein, "Collision avoidance for vehicle-following systems," *IEEE Transactions on Intelligent Transportation Systems* 8(2), 2007.

[23] F. Michaud, P. Lepage, P. Frenette, D. Letourneau, and N. Gaubert, "Coordinated maneuvering of automated vehicles in platoons," *IEEE Transactions on Intelligent Transportation Systems*, 7(4), 2006.

[24] S. Ammoun, F. Nashashibi, and C. Laurgeau, "An analysis of the lane changing manoeuvre on roads: The contribution of inter-vehicle cooperation via communication," In *Proc. of the IEEE Intelligent Vehicles Symposium*, Istanbul, Turkey, 2007.

[25] W. Xiang, P. Richardson, and J. Guo, "Introduction and preliminary experiments of Wireless Access for Vehicular Environments (WAVE) systems," In *Proc. of 3rd IEEE International Conference on Mobile and Ubiquitous Systems*, San Jose, CA, 2007.

[26] DSRC, Standard Specification for Telecomunications and Information Exchange between Roadside and Vehicle Systems—5GHz Band Dedicated Short Range Communications (DSRC) Medium Access Control (MAC) and Physical Layer (PHY), ASTM E2213-03, 2003.

[27] L. Wischhof, A. Ebner, and H. Rohling, "Information dissemination in self-organizing intervehicle networks," *IEEE Transactions on Intelligent Transportation Systems* 6(1), 2005.

[28] T. Nadeem, S. Dashtinezhad, C. Liao, and L. Iftode, "Traffic view: Traffic data dissemination using car-to-car communication," *ACM Mobile Computing and Communcications Review* 8(3), 2004.

[29] J. Bronsted and L. M. Kristensen, "Specification and performance evaluation of two zone dissemination protocols for vehicular ad-hoc networks," In *Proc. of the 39th Annual Simulation Symposium*, Huntsville, KY, 2006.

[30] B-J. Chang, B-J. Huang, and Y-H. Liang, "Wireless sensor network based adaptive vehicle navigation in multihop-relay WiMAX networks," In *Proc. of the 22nd IEEE International Conference on Advanced Information Networking and Applications*, Okinawa, Japan, 2008.

[31] M. Kihl, M. Sichitiu, and H. P. Joshi, "Design and evaluation of two Geocast protocols for vehicular ad-hoc networks," *Journal of Internet Engineering* 2(1), 2008.

[32] A. Benmimoun, J. Chen, D. Neunzig, T. Suzuki, and Y. Kato, "Communication-based intersection assistance," In *Proc. of the IEEE Intelligent Vehicle Symposium*, Las Vegas, NV, 2005.

[33] H. Sawant, J. Tan, Q. Yang, and Q. Wang, "Using Bluetooth and sensor networks for intelligent transportation systems," In *Proc. of the 7th IEEE International Conference on Intelligent Transportation Systems*, Washington, DC, 2004.

[34] C. Sugimoto, Y. Nakamura, and T. Hashimoto, "Development of pedestrian-to-vehicle communication system prototype for pedestrian safety using both wide-area and direct communication," In *Proc. of the IEEE 22nd International Conference on Advanced Information Networking and Applications*, Okinawa, Japan, 2008.

[35] A. Benmimoun, J. Chen, and T. Suzuki, "Design and practical evaluation of an intersection assistant in real-world tests," In *Proc. of the IEEE Intelligent Vehicles Symposium*, Istanbul, Turkey, 2007.

[36] Y. Zang, E. Weiss, L. Stibor, H. Chen, and X. Cheng, "Opportunistic wireless internet access in vehicular environments using enhanced WAVE devices," In *Proc. of the International Conference on Future Generation Communication and Networking*, Jeju, South Korea, 2007.

[37] M. Bechler, S. Jaap, and L. Wolf, "An optimized TCP for Internet access of vehicular ad hoc networks," *Lecture Notes in Computer Science* 3462, 2005.

[38] S. Y. Wang, "The potential of using inter-vehicle communication to extend the coverage area of roadside wireless access points on highways," In *Proc. of the IEEE International Conference on Communications*, Glasgow, UK, 2007.

[39] J. Hubaux, S. Capkun, and J. Luo, "The security and privacy of smart vehicles," *IEEE Security & Privacy Magazine* 2, 2004.

[40] J. Blum and A. Eskandarian, "Fast, robust message forwarding for inter-vehicle communication networks," In *Proc. of the IEEE Intelligent Transportation Systems Conference*, Toronto, Canada, 2006.

[41] C. Laurendeau and M. Barbeau, "Secure anonymous broadcasting in vehicular networks," In *Proc. of the 32nd IEEE Conference on Local Computer Networks*, Dublin, Ireland, 2007.

Chapter 3

Medium Access Control Protocols for Vehicular Networks

Yi Qian and Nader Moayeri

Contents

The main benefit of vehicular communication is seen in active safety systems that increase passenger safety by exchanging warning messages between vehicles. Other applications and private services are also used in order to lower the cost and to encourage vehicular network deployment and adoption. The allocation of 75 MHz in the 5.9-GHz frequency band licensed for dedicated short-range communications (DSRC), which supports seven separate channels, may also enable the future delivery of rich multimedia contents to vehicles at short- to medium range via vehicular communications [1,2]. To effectively use limited wireless spectrum and DSRC channels allocated to vehicular communications, and to provision quality-of-service (QoS) for different vehicular network applications, medium access control (MAC) is one of the major challenges. In this chapter, we give an overview of the recent developments on MAC protocols for vehicular networks, including the standard activities on IEEE WAVE and 802.11p, and the state-of-the-art research on the topic. We also present a priority-based MAC protocol to achieve both QoS and security requirements for vehicular network applications.

3.1 Introduction

Vehicular networks have been developed to improve the safety, security, and efficiency of the transportation system and enable new mobile applications and services. The field of intervehicular communications (IVC), including both vehicle-to-vehicle communications (V2V) and vehicle-to-roadside communications (V2R), is recognized as an important component of the much needed overhaul of the highway information system infrastructure. The immediate impacts include alleviating the vehicular traffic congestions and improving operation management in support of public safety goals, such as collision avoidance. Equipping vehicles with various kinds of on-board sensors and V2V and V2R communication capabilities will allow large-scale sensing and decision or control actions. Communication-based active safety is viewed as the next logical step toward proactive safety systems. These systems provide an extended information horizon to warn the driver or the vehicle of potentially dangerous situations at an early stage.

In spite of the ongoing academic and industrial research efforts on vehicular networks, many research challenges remain. Among them is the design of medium access control (MAC) protocols that can make best use of DSRC multichannel architecture, and schedule application packet transmissions fairly and efficiently in vehicular networks, according to the quality of service (QoS) requirements of the applications. In this chapter, we give an overview on the recent development on MAC protocols for vehicular networks; we also present a priority-based MAC protocol to achieve both QoS and security requirements for vehicular network applications.

In the rest of this chapter, we first give a brief background related to MAC for vehicular communications, including DSRC channel assignment and vehicular network applications. We then present IEEE standards on MAC protocols for vehicular networks, followed by a survey on more research in the area. At the end of the chapter, we propose a secure MAC protocol for vehicular networks with different message priorities for different types of applications to access DSRC channels. We show by simulation and analysis that the proposed MAC protocol can provide secure communications while guaranteeing the reliability and latency requirements of safety-related DSRC applications for vehicular networks.

3.2   DSRC Spectrum and Applications for Vehicular Networks

In a vehicular network, each vehicle is equipped with the technology that allows the vehicle to communicate with others, as well as with the road-side infrastructure, for example, base stations also known as roadside units (RSUs) located in some critical sections of the road such as traffic lights, intersections, or stop signs to improve the driving experience and make driving safer. By using such communication devices, also known as on-board units (OBUs), vehicles can communicate with each other as well as with RSUs. A vehicular network is a self-organized network that enables communications between vehicles and RSUs, and the RSUs can be connected to a backbone network, so that many other network applications and services, including Internet access, can be provided to the vehicles. Figure 3.1 shows an example of a vehicular network.

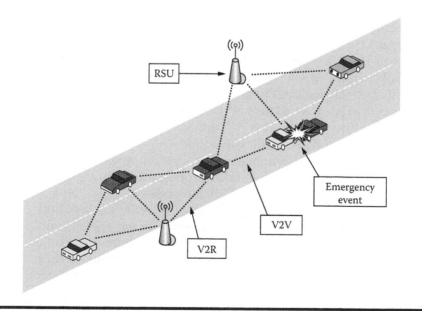

Figure 3.1   An example of a vehicular network.

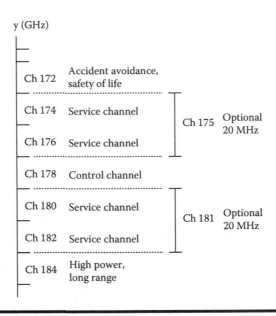

y (GHz)

Ch 172	Accident avoidance, safety of life
Ch 174	Service channel
Ch 176	Service channel
Ch 178	Control channel
Ch 180	Service channel
Ch 182	Service channel
Ch 184	High power, long range

Ch 175 Optional 20 MHz

Ch 181 Optional 20 MHz

Figure 3.2   DSRC Channel assignment in North America.

The U.S. Federal Communications Commission (FCC) recently allocated 75 MHz of DSRC spectrum at 5.9 GHz to be used exclusively for V2V and V2R communications [2]. The primary purpose is to enable public safety applications that save lives and improve vehicular traffic flow. Private services are also permitted to lower the network deployment and maintenance costs to encourage DSRC development and adoption. The DSRC spectrum is divided into seven 10-MHz wide channels, as shown in Figure 3.2. Channel 178 is the control channel, which is generally restricted to safety communications only. The two channels at the edges of the spectrum are reserved for future advanced accident avoidance applications and high-power public safety communication usages. The rest are service channels and are available for both safety and nonsafety applications.

In the following section, we summarize the existing applications and several potential applications that have been proposed for vehicular networks. As studied in "Vehicle-to-vehicle safety messaging in DSRC" [3] and "Design secure and application-oriented VANETs" [4], vehicular networks would support life-critical safety applications, safety warning applications, electronic toll collection, Internet access, group communications, roadside service finder, and the like. We also elaborated on the functions of each application that shall be provided in the MAC layer and the network layer so as to fulfill the requirements of these applications [4].

Table 3.1 lists the characteristics of the vehicular network applications discussed in "Design secure and application-oriented VANETs" [4], with the

Table 3.1 Example Vehicular Network Applications

Applications	Priority	Allowable Latency (ms)	Network Traffic Type	Message Range (m)
Life-critical safety	Class 1	100	Event	300
Safety warning	Class 2	100	Periodic	50–300
Electronic toll collection	Class 3	50	Event	15
Internet access	Class 4	500	Event	300
Group communications	Class 4	500	Event	300
Roadside service finder	Class 4	500	Event	300

priorities of the application message classes, allowable latency as the major QoS requirements of the applications, the network traffic types, and the message transmission ranges.

For safety messages, the amount of information to be transmitted is relatively small, but the transmission reliability as well as the latency and packet dissemination are of great importance.

3.3 IEEE Standards for MAC Protocols for Vehicular Networks

The IEEE has completed the standards IEEE P1609.1, P1609.2, P1609.3, and P1609.4 for vehicular networks and recently released them for trial use [5]. P1609.1 is the standard for the wireless access for vehicular environments (WAVE) resource manager. It defines the services and interfaces of the WAVE resource manager application as well as the message data formats. It provides access for applications to the other architectures. P1609.2 defines security, secure message formatting, processing, and message exchange. P1609.3 defines routing and transport services. It provides an alternative to IPv6. It also defines the management information base for the protocol stack. P1609.4 deals mainly with specification of the multiple channels in the DSRC standard. Figure 3.3 shows the IEEE WAVE standards and Figure 3.4 shows the IEEE protocol architecture for vehicular communications [6].

The WAVE stack uses a modified version of the IEEE 802.11a, known as IEEE 802.11p [6], for its medium access control (MAC) layer protocol. It uses CSMA/CA as the basic medium access scheme for link sharing and uses one control channel to set up transmissions, which then are carried over some transmission channels. The 802.11p PHY layer is expected to work in the 5.850–5.925 GHz DSRC spectrum in North America, which is a licensed radio services band in the United States. By using the OFDM system, it

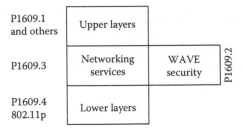

Figure 3.3   IEEE WAVE standards.

provides both V2V and V2R wireless communications over distances up to 1000 m, while taking into account the environment, that is, high absolute and relative velocities (up to 200 km/h), fast multipath fading, and different scenarios (rural, highway, and urban). Operating in 10-MHz channels, it should allow data payload communication rates of 3, 4, 5, 6, 9, 12, 18, 24, and 27 Mb/s. By using the optional 20-MHz channels, it allows data payload capabilities up to 54 Mb/s.

As the original IEEE 802.11 standard is designed only for little mobility, the IEEE 802.11p working group should address important issues such as frequent disconnection and handoff. IEEE 802.11p WAVE defines amendments to IEEE 802.11 to support vehicular network applications. This includes data exchanges between high-speed vehicles and between the vehicles and the roadside infrastructure in DSRC spectrum. The current IEEE 802.11p draft aims at providing the minimum set of specifications

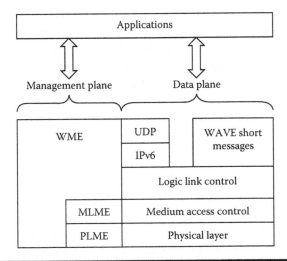

Figure 3.4   IEEE protocol architecture for vehicular communications.

required to ensure interoperability between wireless devices attempting to communicate in potentially rapidly changing communication environments and in situations where transactions must be completed in a timeframe, much shorter than that of infrastructure or ad hoc 802.11 networks. The WAVE mode basic service set (WBSS) in IEEE 802.11p enhances IEEE 802.11 MAC functions for rapidly changing communication environments. Mobile stations in WAVE mode become members of a WBSS in one of two ways, either as a WBSS provider or as a WBSS user. Mobile stations in WAVE mode typically move much faster than legacy 802.11 mobile stations in infrastructure or ad hoc BSS mode. The most important issue is that mobile stations in WAVE join the vehicular network and transmit or receive data as quickly as possible. Due to these rapidly changing communication environments, the WBSS provider and user should be ready for communications as quickly as possible. For this purpose, WBSS do not require MAC sublayer authentication and association prior to being allowed to transmit data. In a WBSS, a WBSS user only needs to receive the WBSS announcement of a WBSS provider before commencing transmissions [7,8].

The WBSS provider first transmits WAVE announcement action frames for which the WBSS users listen. That frame contains all information necessary to join a WBSS. Unlike infrastructure and ad hoc 802.11 BSS types, the WAVE users do not perform authentication and association procedures before participating in the WBSS. To join the WBSS, only configuring according to the WAVE announcement action frame is required [7]. In addition, a mobile station in WAVE mode shall generate a CCA (clear channel assessment) report in response to a CCA request to know the time-varying channel state precisely [8].

As indicated in "A Solicitation-Based IEEE 802.11p MAC Protocol for Roadside to Vehicular Networks" [7], the IEEE 802.11p standard should address important issues such as frequent disconnection, mobility, and the time-varying channel condition, which are the inherent characteristics of vehicular networks. To make an effective IEEE 802.11p standard, the following challenges need to be tackled. Stateless channel access, caching for handoff, and opportunistic frame scheduling. When a WBSS user moves out of a WBSS provider's coverage, the WBSS provider cannot know whether the WBSS user exists in its WBSS or not due to the absence of the deassociation process. Because there are no authentication and no association in a WBSS, it is unlikely that the WBSS provider keeps track of connectivity with WBSS users. IEEE 802.11p needs to support fast handoff among multiple WBSS providers. A WBSS user frequently moves from one WBSS to another while exchanging data because of high-speed mobility. Caching for handoff needs to be addressed, because there is no association process in IEEE 802.11p. When a WBSS user moves within a WBSS, radio link conditions between a WBSS provider and WBSS users are highly dynamic because high-speed mobility causes a large and fast variation of the channel conditions. Data transmissions have to take place depending on the channel state in an opportunistic manner. The bit rate should be determined timely, based on the instantaneous state of the wireless channel to deal with its rapid change.

To address the above challenges, the authors [7] have proposed a new solicitation-based IEEE 802.11p operation mode called WBSS user initiation mode (W-UIM) and a new concept called WBSS-area which is a virtual group of adjacent WBSSs. They assume that WBSS providers share a wired roadside backbone, which provides enough link bandwidth. In W-UIM, a WBSS user solicits data frames destined for itself in an opportunistic manner, by requesting the transmissions of the frames from a WBSS provider by a WAVE-poll frame. Throughput analysis reveals that W-UIM achieves a stable, saturated WBSS throughput, higher than IEEE 802.11 irrespective of the number of contending and moving-away WBSS users.

3.4 More Research on MAC Protocols for Vehicular Networks

Besides the IEEE standardization activities, more research is going on for MAC protocols for vehicular networks. In this section, we survey some the most recent research efforts on MAC protocols for vehicular networks. In particular, we focus on a cluster-based multichannel MAC protocol for vehicular networks [9], and a distributed MAC protocol for safety message dissemination in vehicular networks [10].

3.4.1 A Cluster-Based Multichannel MAC Protocol for Vehicular Networks

Making best use of the DSRC multichannel architecture, the authors have proposed a cluster-based multichannel communication scheme, which integrates the clustering with contention-free and contention-based MAC protocols [9]. In the proposed scheme, the elected cluster-head (CH) vehicle functions as a coordinator to collect and deliver the real-time safety messages within its own cluster and forward the consolidated safety messages to the neighboring CHs. Also, the CH vehicle controls channel-assignments for cluster-member vehicles transmitting or receiving the non–real-time traffic, which makes the wireless channels more efficiently utilized for non–real-time data transmissions. The scheme uses the contention-free MAC (TDMA/broadcast) within a cluster and the IEEE 802.11 MAC among CH vehicles such that the real-time delivery of safety messages can be guaranteed.

The proposed MAC protocol aims at supporting QoS for timely delivery of real-time data, for example, safety-related messages, and increasing

Table 3.2   Definitions of the Seven DSRC Channels for Clustering-Based Multichannel MAC Protocol

Channel Name	Channel Abbreviations	DSRC Channel Numbers
Intercluster control	ICC	178
Intercluster data	ICD	174
Cluster range control	CRC	172
Cluster range data	CRD	176,180,182,184

the throughput for non–real-time traffic over the V2V-based vehicular networks. To achieve these goals, they have developed the cluster-based multichannel communication scheme under the infrastructure-free vehicular network environments. The key of the proposed scheme is to integrate the clustering algorithm with both the contention-free and contention-based MAC protocols under the DSRC architecture. Complying with the seven-channel assignment of DSRC, as shown in Figure 3.2, the authors have defined particular functions for these seven channels in their MAC scheme, which are summarized in Table 3.2 [9]. Specifically, the definitions of these seven channels in the scheme are as follows, Ch 178 is intercluster control (ICC) channel, Ch 174 is intercluster data (ICD) channel, Ch 172 is cluster range control (CRC) channel, and the remaining channels (Ch 176, 180, 182, 184) are cluster range data (CRD) channels.

In the MAC scheme, each vehicle is equipped with two sets of transceivers, denoted by *transceiver 1* and *transceiver 2*, respectively, which can operate simultaneously on three different channels. As shown in Figure 3.5, the proposed scheme works as follows. The vehicles in proximity form a cluster where one of them is elected as the cluster-head based on the given election rules. For each CH vehicle, one transceiver uses contention-free MAC over the CRC channel to collect and deliver safety messages as well as control packets within this cluster, while the other transceiver exchanges consolidated safety messages among CH vehicles through contention-based MAC over the ICC channel. For each cluster-member vehicle, one transceiver is dedicated for communicating with the CH vehicle over the CRC channel within its cluster, and the other transceiver can be used to transmit the non–real-time traffic over one of the ICD/CRD channels assigned by the CH vehicle. As shown in Figure 3.6, the proposed scheme handles the following three tasks: (1) Cluster membership management; (2) Real-time traffic (such as safety messages) delivery; and (3) Non–real-time data communications (such as movies download, etc.).

To accomplish the system functions of the proposed scheme, three different protocols are developed: the cluster configuration protocol, the intercluster communication protocol, and the intracluster coordination and communication protocol. First, the cluster configuration protocol employs

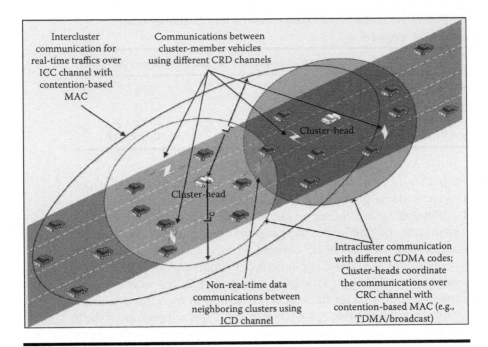

Figure 3.5 The proposed cluster-based multichannel communication architecture (reproduced from [9] by permission of IEEE).

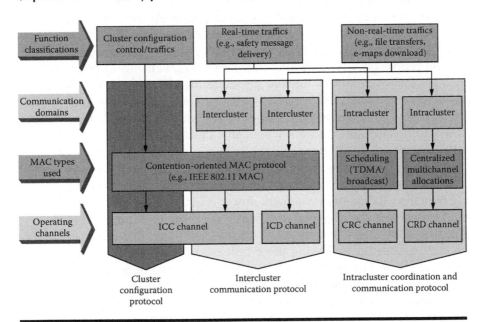

Figure 3.6 Cluster-based multichannel communication scheme structure diagram (reproduced from [9] by permission of IEEE).

contention-based MAC over the ICC channel to perform cluster management tasks (such as joining and leaving a cluster, cluster-head election, etc.). Second, the intercluster communication protocol is responsible for the exchange of safety messages and non–real-time traffic among clusters over ICC and ICD channels, respectively. Third, the intracluster coordination and communication protocol utilizes the multichannel MAC protocol to arbitrate the communication between cluster-head and cluster-member vehicles within a given cluster. Each CH vehicle collects or delivers safety messages and assigns ICD or CRD channels to cluster-members by using contention-free MAC protocol over the CRC channel. Each cluster-member vehicle uses one transceiver to exchange the safety messages with its CH vehicle. Meanwhile, the cluster-member vehicle uses another transceiver to communicate with its peer vehicle within the same cluster over the CRD channel assigned by its CH vehicle [9].

They use a finite-state machine (FSM) to precisely describe the principle and operating process of the proposed scheme. Each vehicle operates under one and only one of the following four states at any given time: (1) cluster-head (CH); (2) quasi–cluster-head (QCH); (3) cluster-member (CM); (4) quasi-cluster-member (QCM). The state transitions of the FSM are controlled by the cluster configuration protocol. There are seven state-transition conditions for the protocol FSM on each vehicle [9]. The intracluster coordination and communication protocol is based on a multichannel MAC protocol, where each CH employs a scheduling scheme over the CRC channel to collect or broadcast safety messages and coordinate the cluster-member vehicles to transfer non–real-time data within or between cluster(s). In the intercluster communication protocol, two types of traffic are served on two separate channels between clusters: (1) the real-time safety messages over the ICC channel; and (2) the non–real-time traffic over the ICD channel. On one hand, cluster-heads, quasi-cluster-heads, and quasi-cluster members use contention-based protocols (e.g., IEEE 802.11) to share the ICC channel. After the CH vehicles collect the safety messages from their own clusters, they use the data-fusion technique to consolidate the safety information, and then contend for the ICC channel to forward the processed information to the neighboring CHs. On the other hand, applying the intracluster coordination and communication protocol, one vehicle is assigned to the ICD channel in each cluster. By employing the contention-based MAC, those vehicles from different clusters contend for the common ICD channel to transmit or receive the non–real-time traffic packets between clusters. They work as gateways to forward the packets for the other CM vehicles.

Through simulations, the authors have compared their proposed scheme with the IEEE 802.11 MAC and V2V-oriented DCA (V2V-DCA) [9]. V2V-DCA, which supports the safety message delivery, is derived from the dynamic channel assignment (DCA) protocol [11]. In addition, they have also developed an analytical model to study the delay and success rate for transmitting

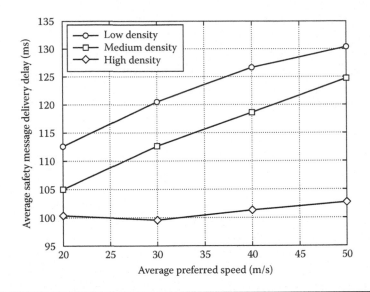

Figure 3.7 **Performance evaluations of the proposed MAC scheme under different traffic densities for safety message delivery delay against the preferred speed (reproduced from [9] by permission of IEEE).**

the consolidated safety messages by the cluster-head vehicles. They have also derived the desirable contention-window sizes for different highway traffic scenarios. Figure 3.7 shows performance evaluations of the proposed MAC scheme under different traffic densities for safety message delivery delay against the preferred speed. Figure 3.8 shows performance evaluations of the proposed MAC scheme under different traffic densities for safety message delivery failure against the preferred speed. Figure 3.9 shows probability of safety message delivery failure against non–real-time traffic arrival rate. The simulation and analysis results obtained show that the proposed MAC scheme for vehicular networks can achieve not only the timely delivery of safety messages, but also the high throughput for the other non–real-time traffic.

3.4.2 A Distributed MAC Protocol for Safety Message Dissemination in Vehicular Networks

MAC for safety message dissemination is a challenging problem in vehicular networks. The reason is that MAC in a typical vehicular network needs to be fully distributed due to the constantly moving and changing nodes in the network. With fully distributed MAC, packets may experience unpredictable delays in media access due to deferrals and backoffs. Long medium access delay is intolerable for safety message dissemination in

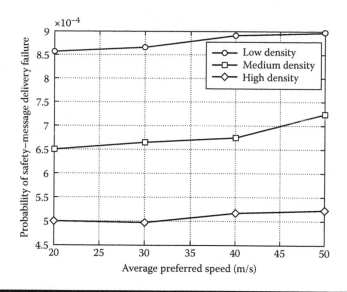

Figure 3.8 Performance evaluations of the proposed MAC scheme under different traffic densities for safety message delivery failure against the preferred speed (reproduced from [9] by permission of IEEE).

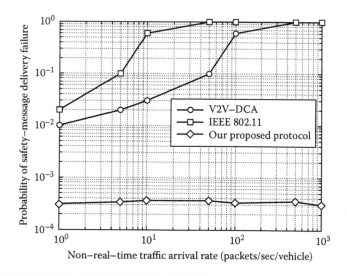

Figure 3.9 Probability of safety message delivery failure against non–real-time traffic arrival rate (reproduced from [9] by permission of IEEE).

vehicular networks because of the short lifetimes of safety messages in such networks. Moreover, different types of safety messages usually have different lifetimes, that is, different levels of emergency such as life-critical safety or safety warning. In a fully distributed way, a MAC scheme for emergency message dissemination must be able to ensure that a message with a longer lifetime yields to other messages with shorter lifetimes. The authors have proposed a new MAC scheme that effectively supports safety message dissemination in vehicular networks [10]. The basic approach of the proposed MAC scheme is the intelligent use of a single control channel for multiple purposes. The proposed MAC scheme realizes low and stable medium access delays for individual safety message packets and does not have the hidden terminal problem. Moreover, the proposed scheme is fully distributed and capable of providing multiple levels of strict priority scheduling for safety message packets.

The basic approach of the proposed MAC scheme is to use pulses in a single control channel to achieve multiple goals [10]. In the proposed scheme, the control channel carries only pulses, and pulses only appear in the control channel. The control channel is monitored by all nodes all the time except when they are transmitting in the channel (note that an antenna usually cannot transmit and receive at the same time). A node that is generating pulses in the control channel still monitors the channel when its pulses pause. The proposed MAC scheme works in the following way [10]. As soon as a safety message packet in a node arrives at the MAC sublayer from the upper layer, the node starts a backoff timer if the control channel has been sensed idle for a specified amount of time. Otherwise, the node keeps monitoring the control channel. The delay of the backoff timer is random but in a range determined by the emergency level of the message to be disseminated. When its backoff timer expires, the node starts to transmit pulses in the control channel. Shortly after starting to transmit pulses, the node starts to broadcast the emergency packet in the data channel. If the node detects a pulse before its backoff timer expires, it cancels its timer and returns to monitor the control channel. The pulses in the control channel are called priopulses. When a node detects a priopulse in the control channel, it aborts its transmissions to release both channels. When a node is generating priopulses but detects a priopulse of another node during one of its own priopulse pauses, the node releases both channels. In addition, a node that is receiving an emergency packet in the data channel relays priopulses in the control channel to suppress hidden terminals. However, the proposed scheme does not require such a node to forward the safety message packet because the proposed scheme is a MAC scheme working at the MAC sublayer.

The above description is for a simplified scenario where there is only a single level of priority for safety messages. In a more realistic scenario, there are multiple levels of priority for safety message packets. In such a case, a safety message source may still contend for the data channel, even if it senses that the control channel is busy.

The pulses in the proposed scheme are basically single-tone waves with pauses of random lengths, which are introduced in details in "A distributed MAC scheme for emergency message dissemination in vehicular ad hoc networks" [10]. In the proposed scheme, the priopulses are relayed by nodes to suppress the hidden terminals of the message source. They use the term relay to distinguish priopulse spreading from packet forwarding. Unlike a packet that is to be forwarded, a priopulse that is to be relayed triggers its relaying nodes to regenerate as soon as it emerges in the control channel; relaying nodes do not wait for the reception of the whole priopulse before regenerating it. Ideally, only neighbors of a safety message source should relay priopulses.

The authors have presented comprehensive simulation results for the proposed MAC scheme [10]. The proposed scheme is named preempprio-MAC because of the preemptive priority service that it provides to emergency packets at the MAC sublayer. To show the performance of the approach that uses in-band control messages, they have also simulated with the IEEE 802.11 and 802.11e MAC protocols. Figure 3.10 shows the average medium access delay for emergency packets versus the size of the packets in the network in the two-ray ground case. The packet size determines the network load in the simulations. As shown in Figure 3.10, preempprio

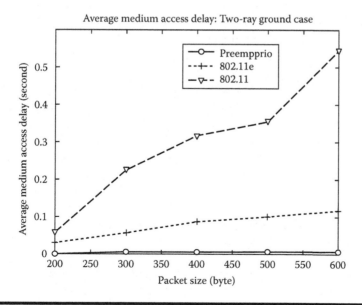

Figure 3.10 **Medium access delay versus network load (two-ray ground) (reproduced from [10] by permission of IEEE).**

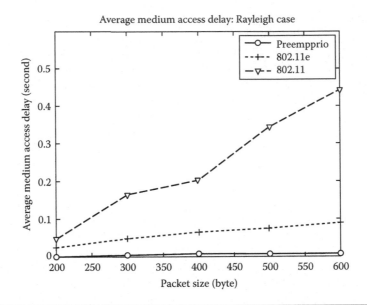

Figure 3.11 Medium access delay versus network load (Rayleigh fading) (reproduced from [10] by permission of IEEE).

introduces negligible medium access delays for emergency packets as compared with the IEEE 802.11 protocols. Meanwhile, IEEE 802.11e has significantly better performance than IEEE 802.11 due to its support of priority medium access. Similar results have been observed in the Rayleigh-fading channel case, as shown in Figure 3.11. Interestingly, when the packet size is the same, the medium access delay for emergency packets is shorter in the Rayleigh-fading case than in the two-ray ground case. This phenomenon occurs because signals attenuate faster over a Rayleigh-fading channel than over a two-ray ground channel, and thus, the actual network load is lower in the Rayleigh-fading channel case. The simulation results show the effectiveness of the proposed distributed MAC scheme in serving safety messages in vehicular networks [10].

3.5 A Priority-Based Secure MAC Protocol for Vehicular Networks

In this section, we propose a secure MAC protocol in consideration of the DSRC channel structures and to accommodate the DSRC applications while providing adequate security for vehicular networks. The proposed secure MAC protocol will use part of the IEEE 1609.2 security infrastructure including PKI and ECC, the secure communication message format for vehicular networks, and the priority-based channel access according to the QoS requirements of the applications.

3.5.1 Message Priorities of Vehicular Communications

As shown in Figure 3.2, the two channels at the edges of the spectrum (Ch 172 and Ch 184) are reserved for future DSRC applications. We assume here that there are four internal queues per OBU for the four different priority message classes, and each message will be queued according to its priority. Class 1 message will always access channel 178 with the highest priority, if channel 178 is full, then it will access either of the channels 174, 176, 180, or 182 with the highest priority; Class 2 message will always access channel 178 with the second highest priority, if channel 178 is full, then it will access either of channels 174, 176, 180, or 182 with the second highest priority; Class 3 and Class 4 messages cannot access channel 178, and it will access channels 174, 176, 180, or 182 with the third or fourth priority, respectively. We assume that there is a scheduler in each OBU, which handles the internal collision. The scheduler will allow higher-priority messages to be transmitted before lower-priority messages. We adopt a preemptive policy, that an arriving high-priority (Class 1 and Class 2) safety-related message will be scheduled to get the channel immediately before the completion of the current low-priority (Class 3 and Class 4) message transmission. Table 3.3 shows the traffic priority classes and the DSRC channels that each class can access.

3.5.2 Secure Protocol

As it is discussed in "Design secure and application-oriented VANETs" [4], vehicular network security requires message authentication and integrity, message nonrepudiation, entity authentication, access control, message confidentiality, availability, privacy and anonymity, and liability identification for the safety-related applications (Class 1 and Class 2).

For nonsafety-related messages (Class 3 and Class 4) different security requirements may be established as compared to those of Class 1 and Class 2. We assume that other security mechanisms will address the security

Table 3.3   Message Priority Classes and the DSRC Channels

Message Priority Classes	DSRC Channels
Class 1	178,174,176,180, and 182
Class 2	178,174,176,180, and 182
Class 3	174,176,180, and 182
Class 4	174,176,180, and 182

requirements of Class 3 and Class 4 messages. We focus our study on the impact of secure safety messages and the priority-based medium access control mechanism for all DSRC applications.

Similar to Raya and Hubaux, Lin et al., and Suthaputchakun and Ganz, we assume that each OBU on a vehicle has a secure database, which stores all cryptographic components used for signing and verifying each message. Each vehicle has to have a valid certificate usually issued by a central trusted party called certification authority (CA). PKI will be used for certificates issued by a CA. For the privacy of a vehicle, such as identity and travel route, a set of anonymous keys can be used to sign each message that will be changed periodically. These keys can be preloaded in the secure database of the OBU for a long period of time, for example, for one year until the next yearly license plate registration. Each key is certified by the issuing CA and has a short lifetime. In case of an accident or other law investigation, the authority can track back to the real identity of the vehicle, using electronic license plate (ELP) [15]. This can also help to prevent nonrepudiation in case of accidents.

For safety-related (Class 1 and Class 2) messages, message authentication and integrity, message nonrepudiation, and privacy and anonymity of the senders are very important. Confidentiality of the safety message itself is not needed, so it can be transmitted in plaintext [12,14]. Under the PKI solution, before an OBU sends a safety message, it signs it with its private key and includes the CA's certificate as follows:

$$V \rightarrow * : M, T, \text{Sig}_{P_r K_V}\{H[M \mid T]\}, \text{Cert}_V \qquad (3.1)$$

where, V is the sender of the safety message, $*$ represents any receivers, M is the safety message sent by plaintext, T is the time stamp to guarantee the freshness of the message (is also sent in plaintext), $\text{Sig}PrKvH[M \mid T]$ is the hash of the message M and time stamp T, signed by the private key of the sender K_V, and Cert_V is the prestored certificate of the sender issued by any CAs. In Eq. (3.1), the total overhead per packet is 140 byte with 56 byte signature and 84 byte certificate.

Note that the attackers cannot alter both message and time stamp, due to digital signature. Because no other OBU knows the private key of the sender, no other OBU can alter the contents of the packet. The certificate of the sender is included in the packet, so that other vehicles can extract the sender's public key and verify the correctness of each message. Once other OBUs receive a message, they retrieve the sender's public key, K_V from Cert_V in order to decrypt the signature to obtain $H[M \mid T]$, hash and time stamp the message, compare the hash with $H[M \mid T]$ and if both of them are the same, the message is verified. Otherwise the message is falsified and will be ignored.

3.5.3   Performance Analysis

In this section, we present the simulation and analysis to show the performance results of the proposed secure MAC protocol. There are two scenarios of the vehicular networks: V2R-based and V2V-based vehicular networks. In V2R-based vehicular networks, we assume that the vehicular communication is controlled by RSUs. Each RSU acts as an access point that broadcasts all the messages received from one vehicle to all others in the range. In V2V-based vehicular networks, on the other hand, we assume no RSU infrastructure exists, each OBU on a vehicle has to rely on its own for communications. It has to broadcast messages to all the nearby nodes. There is no acknowledgment in the V2V-based vehicular network, unlike in the V2R-based vehicular network where acknowledgment is created by the RSU. Figure 3.12 shows a V2R-based vehicular network. In the following text, we show some V2R network simulation results for the proposed secure MAC protocol.

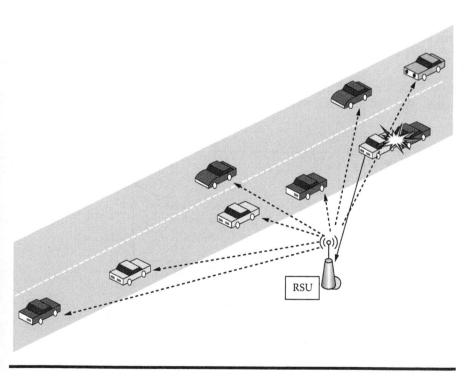

Figure 3.12   A V2R-based vehicular network.

Table 3.4   Simulation Parameters

Parameter	Value
Basic rate	1 Mb/s
Data rate	1 Mb/s
SIFS	10 μs
Slot time	20 μs
DIFS	50 μs
Size of RTS/CTS/ACK	160/112/112 bits
Size of frame header	224 bits
Size of preamble	48 bits
Minimum window size	31
Minimum window size	1023
Retry limits	5

In our simulation, we assume that each vehicle has five interface cards, each of which is operating on a different frequency band. More over, for each channel, we apply the basic parameters of IEEE 802.11. In particular, the main parameters are listed in Table 3.4.

In addition, we assume that the packet arrival of each class of traffic on every node is exponential with an average interval of 100 ms. We also assume that the packet size is fixed to 300 bytes. Because the packet size is rather small, we use the basic access method instead of the RTS/CTS scheme.

In Figure 3.13, we show the throughput performance of the proposed MAC scheme. We can observe that, when the number of nodes in the

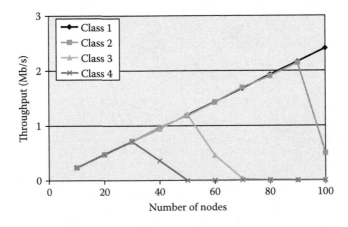

Figure 3.13   Throughput versus the number of nodes.

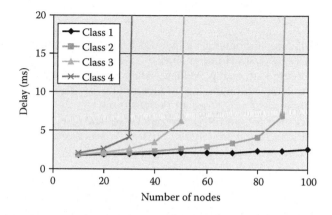

Figure 3.14   Delay versus the number of nodes.

network is small, all traffic will be accepted and be increased linearly with the increasing of the number of nodes. However, if the number of nodes increase to a certain value, then the performance of lower classes will be decreased, while the throughput of Class 1 can still grow.

Figure 3.14 illustrates the corresponding delay performance of the proposed scheme. We can see that the average delay for Class 1 traffic is rather stable with the increase of the number of nodes. The other three classes, on the other hand, will be extremely large at certain thresholds.

The simulations results show that the proposed MAC protocol above can provide secure communications while guaranteeing the QoS requirements of safety-related vehicular network DSRC applications.

3.6   Conclusions

A vehicle that detects an emergency event usually only needs to broadcast its safety message to other vehicles in a short range, such as a range of several hundred meters. A single-hop delivery may therefore be adequate for this type of application, which makes routing for safety message packets less relevant. In such a case, medium access control becomes the most critical segment in the delivery process of a safety message packet. In this chapter, we give a survey on MAC protocols for vehicular networks, including the IEEE standardization activities on MAC protocols for vehicular networks and the current research on the topic. We also present a priority-based MAC protocol to achieve both QoS and security requirements for vehicular network applications.

References

[1] Dedicated Short-Range Communications (DSRC) Home, http://www.leearmstrong.com/DSRC/DSRCHomeset.htm.

[2] Crash Avoidance Metric Partnership,"Vehicle Safety Communication Project Final Report," available through U.S. Department of Transportation. http://www-nrd.nhtsa.dot.gov/pdf/nrd-12/060419-0843.

[3] Qing Xu, Tony Mak, Jeff Ko, and Raja Sengupta, "Vehicle-to-vehicle safety messaging in DSRC," In *Proceedings of the 1st ACM International Workshop on Vehicular Ad Hoc Networks (VANET'04)*, Philadelphia, PA, October 1, 2004.

[4] Yi Qian and Nader Moayeri, "Design Secure and Application-Oriented VANETs," In *Proc. of IEEE VTC'2008-Spring*, Singapore, May 11–14, 2008.

[5] IEEE Draft, "Trial Use Standard for Wireless Access in Vehicular Environments (WAVE)—Architecture," P1609.0/D01, February 2007.

[6] IEEE WG, IEEE 802.11p/D2.01, Draft Amendment to Part 11: Wireless Medium Access Control (MAC) and Physical Layer (PHY) specifications: Wireless Access in Vehicular Environments, March 2007.

[7] Nakjung Choi, Sungjoon, Yongho Seok, Taekyoung Kwon, and Yanghee Choi, "A solicitation-based IEEE 802.11p MAC protocol for roadside to vehicular networks," In *Proc. of 2007 Mobile Networking for Vehicular Environments*, Anchorage, AK, May 11, 2007.

[8] IEEE WG, IEEE 802.11v/D0.07, Draft Amendment to Part 11: Wireless Medium Access Control (MAC) and Physical Layer (PHY) specifications: Wireless Network Management, January 2007.

[9] Hang Su and Xi Zhang, "Clustering-based multichannel MAC protocols for QoS provisionings over vehicular ad hoc networks," *IEEE Transactions on Vehicular Technology* 56(6):3309–3323, November 2007.

[10] Jun Peng and Liang Cheng, "A distributed MAC scheme for emergency message dissemination in vehicular ad hoc networks," *IEEE Transactions on Vehicular Technology* 56(6):3300–3308, Richandson, TX, November 2007.

[11] Shih-Lin Wu, Chih-Yu Lin, Yu-Chee Tseng, and Jang-Laing Sheu, "A new multichannel MAC protocol with on-demand channel assignment for multi-hop mobile ad hoc networks," In *Proc. of ISPAN'00*, December 2000.

[12] Maxim Raya and Jean-Pierre Hubaux, "Securing vehicular ad hoc networks," *Journal of Computer Security* 15(1):39–68, 2007.

[13] Xiaodong Lin, Xiaoting Sun, Pin-Han Ho, and Xuemin Shen, "GSIS: A secure and privacy-preserving protocol for vehicular communications," *IEEE Transactions on Vehicular Technology* 56(6):3442–3456, November 2007.

[14] Chakkaphong Suthaputchakun and Aura Ganz, "Secure priority based inter-vehicle communication MAC protocol for highway safety messaging," In *Proc. of IEEE ISWCS*, Trondheim, Norway, October 16–19, 2007.

[15] Maxim Raya, Panos Papadimitratos, and Jean-Pierre Hubaux, "Securing vehicular communications," *IEEE Wireless Communications*, October 2006, 13(5): 8–15.

Chapter 4

Heterogeneous Wireless Communications for Vehicular Networks

Andrea Conti, Alessandro Bazzi,
*Barbara M. Masini, and Oreste Andrisano**

Contents

---

* "This chapter reflects the research activity made in this field at WiLAB (http://www.
wilab.org/) over the years. Authors would like to acknowledge several collegues with
which a fertile research environment has been created, including M. Chiani, D. Dardari,
A. Giorgetti, G. Leonardi, G. Mazzini, G. Pasolini, V. Tralli, R. Verdone, and A. Zanella."

This chapter introduces and motivates the need for communication vehicular networks (VNs) for infomobility services (Section 4.1) describes main wireless communications technologies enabling (VNs) (Section 4.2), discusses how to evaluate their performance in realistic scenarios (Section 4.3), and presents some case studies (Section 4.4). Results in case studies provide insights on how heterogeneous wireless networks (HWNs) can be exploited in the presence of user mobility and various traffic conditions and how the network will be loaded when several vehicles are sharing information for infomobility, security, and entertainment services. Note that, typically, technologies we are going to describe and discuss have not initially been designed for infomobility applications, thus their performance and behavior are not easily predictable in the new envisioned vehicular scenarios and have to be carefully evaluated.

4.1   Introduction

Road transportation requires new solutions to become more safe and efficient. The safety and efficiency of roads can be substantially improved by the deployment of communication technologies for VNs, which enable new services such as adaptive traffic control, traffic management systems, and acident detection. VNs adopt communications technologies to improve road safety and to enable a set of on-board potential services for drivers and passengers as well as different communication facilities between moving vehicles. These systems can operate either autonomously onboard of vehicles or based on vehicle-to-vehicle (V2V) or vehicle-to-infrastructure (V2I) communications.

The application of wireless communications to VNs started in the late 1980s with the European projects PROMETHEUS and EUREKA, which tried to define short-range communications, among vehicles and to exploit global system for mobile communications (GSM) networks for V2I communications. Even if technologies were not mature, several research activities were devoted, in particular, to short-range V2V communications [1–9]. The advent of new technologies in the last decade enables the adoption of wireless communications in the field of transportation. New projects are, in fact, exploiting, the new communication technologies to allow new transportation services (see, e.g., [E][F] for further details). A significant reduction of congestion and accidents is proved when communication technologies are adopted to provide traffic and roadway information. This increases the interest of today's society in using VN for transportation safety and mobility improvement and productivity enhancement.

Advances with respect to the state-of-the-art, like new and more flexible traffic control devices, software systems, computer hardware, communication and surveillance technologies, and analysis methods, have greatly improved transportation engineering immensly over the last decade.

Many cities are creating traffic management centers with closed-circuit television cameras, traffic and weather sensors, and traffic signals monitor and manage traffic flows on streets and freeways. The information received at the traffic management centers allows them to inform travelers of traffic conditions via radio, television, Internet, electronic variable messages along the roadways and cellular networks. Wireless VNs are intended to enable passenger-oriented services such as (see Figure 4.1) communications between passengers (passenger to passenger, [PASS2PASS]); communications between the single passenger and the hosting car (Passenger to vehicle, PASS2V); data sharing and communications between vehicles (V2V or C2C); streaming and general communications utilizing the external infrastructure (V2I). These services require HWNs.

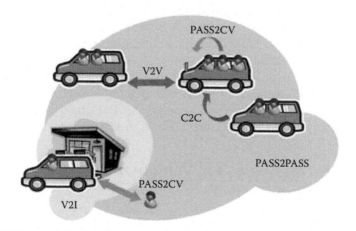

Figure 4.1   Wireless communications enabling passenger-oriented services.

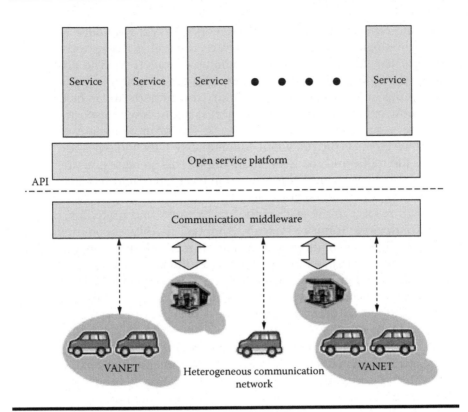

Figure 4.2   A possible functional architecture enabling passenger-oriented services.

In future scenarios, real-time information will be provided to vehicles through HWNs in order to obtain smarter navigation and improved traffic management, as will be discussed in the following sections. The problem of providing services for VNs is typically three folds: (i) to obtain information on the actual state of vehicles and roads, collect them, and send to the management center; (ii) to elaborate the information received at the management center and prepare messages to be sent toward each actor, and (iii) to return to each user the information. In this chapter, we focus on HWNs because they are the enabling technologies to provide real-time information to vehicles and improve traffic conditions. Passenger-oriented services would then require an architecture such as in Figure 4.2.

4.2   Enabling Technologies for Vehicular Communication Networks

There is a variety of wireless communication networks, such as cellular systems from 2G to 3G and evolutions, wireless local area networks (WLANs) IEEE 802.11a/b/g/p, and worldwide interoperability for microwave access

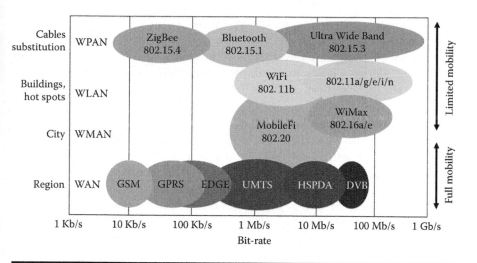

Figure 4.3   Comparison among wireless technologies.

(WiMAX). These can be exploited to enable new services designed for passengers, which can rely on the VN itself: safety applications, infomobility, and entertainment. Some wireless technologies are compared with respect to usage, mobility, and bit rate in Figure 4.3. In addition, because infomobility services are context-aware, it is important to estimate the position of vehicles, which is typically done through satellite navigation systems.

The following sections briefly discusses some characteristics of main wireless technologies enabling VNs, such as (in the order of decreasing coverage): satellite radio navigation systems, cellular systems, WiMAX, WLAN, and wireless sensors and actuators networks (WSANs). In particular, we discuss their main characteristics affecting performance and availability for VNs, such as spectrum usage, coverage, bit rate, physical (PHY) layer techniques, and medium access control (MAC) protocols.

The technologies described in the following sections suffer from wireless channel impairments. To counteract them, they adopt proper PHY level solutions, such as diversity, multiple input multiple output (MIMO), and link adaptation techniques. There is a large literature available on these topics. As examples a reader may refer to [10–13] for diversity techniques, [14,15] for MIMO, and [16,17] for adaptive modulation tracking small-scale or large-scale fading evolutions.

4.2.1   Global Navigation Satellite System

The global positioning system (GPS) is a satellite-based navigation positioning system initially developed in 1973 by the U.S. Department of Defense to meet military needs.[1] This system has been widely used for determining 3D positioning for surveying, scientific, and other applications. In fact GPS has the advantages of being independent of weather conditions, number of users, and time. Hence, GPS is now adopted worldwide for navigation by several applications based on outdoor localization of mobile vehicles and people.

GPS uses the principle of multilateration to find the receiver's position relative to a set of satellites, which orbit the earth at an altitude of about 20.2 km, classified as medium earth orbit (MEO), and broadcast their positions at a consistent time standard called GPS system time. The receiver uses the location of each satellite, the system time, its clock time bias relative to GPS system time, and the time of arrival of the radio message to compute its radial distance from each satellite. The receiver exploits the arrival time of each signal to estimate the distance from each satellite and determine its position using trilateration or multilateration. The satellites also broadcast a secure coded message on a second frequency to improve the accuracy for military use. The resulting coordinates are converted to latitude and longitude or location on a map in order to be displayed to the user. The user position is determined with an accuracy of about a dozen meters.

The GPS carrier frequencies are in the L band.[2] In particular, $L1$ frequency is transmitted on 1575.42 MHz and carries the coarse acquisition code (i.e., the standard positioning signal the GPS satellite transmits to civilian users containing information that the GPS receiver exploits to determine its position and time); P-Code (i.e., the precise code of the GPS signal typically used only by the U.S. military, encrypted and reset every seven days to prevent use from unauthorized persons), and the navigation message (i.e., the message transmitted by each GPS satellite containing system time, clock correction parameters, ionospheric delay model parameters, and the satellite ephemeris data). The information is used to process GPS signals to give the user time, position, and velocity. $L2$ is transmitted on 1227.6 MHz and carries the P-Code and also the navigation message.

Moreover, for military use the transmission is on 1227.6 MHz, for nuclear burst detection at L3 equal to 1381.05 MHz, and for telemetry on 2227.5 MHz. For more technical details a reader may refer to [18].

4.2.2 Cellular Systems: From Second Generation to Third Generation and Further Evolutions

Technologies for mobile radio systems experience a high penetration in the market, covering users in most parts of the world. Thus 2G, 3G, and

[1] Similar satellite navigation systems include the Russian GLONASS, the upcoming European GALILEO positioning system, the proposed COMPASS navigation system of China, and IRNSS of India.

[2] The L Band extends from 390 to 1550 MHz.

evolutions are highly attractive for VNs that can exploit the existent infrastructure, avoiding new setup or expensive installations. Here, we shortly describe some cellular systems without the intention to fully cover this very broad field of technological opportunities.

The GSM represents one of the most popular and adopted standards for cellular phones; it is characterized by a circuit-switched data channel that enables up to 9.6 Kbps data rate in a band of 200 kHz per channel [19, 20]. At the PHY layer, GSM uses a combination of frequency division multiple access (FDMA) and time division multiple access (TDMA). Two frequency bands 45 have been reserved for GSM operation: 890–915 MHz for transmission from the mobile station, that is, uplink, and 935–960 MHz for transmission from the base station, that is, downlink. Each of these bands of 25 MHz width is divided into single carrier channels of width equal to 200 KHz. Some of these frequency channels are allocated to a base station, that is, to a cell. Each of the 200 KHz frequency channels carries eight TDMA channels by dividing each of them into eight time slots forming a TDMA frame. The GSM implements handover, which is necessary to support mobility of users and to enable the interoperability of different network technologies (e.g., between GSM and its evolutions). In fact, the freedom to be able to make and receive calls anywhere, at any time creates a totally new dimension in human communications, which has frequently been advertised as the main advantage of cellular wireless systems. Handover is a key concept in providing this mobility and is generally performed when function cost suggests a change in base station. Handover failures occur when no new resources are available in the target cell or when the radio link quality has decreased below acceptable levels before the call could be handed over.

The first evolution of GSM is represented by GPRS, also known as 2.5G [21, 22]. GPRS supports higher data rate with respect to GSM (up to a theoretical maximum of 140.8 Kbps, though typical rates are closer to 56 Kbps) and it is packet switched rather than connection oriented (circuit switched). Its main purpose is to dynamically and flexibly share physical resources between packet data services and GSM services, thus greatly improving and simplifying wireless access to packet data networks. Users of GPRS benefit from shorter access times and higher data rates. In fact, in conventional GSM, the connection setup takes several seconds and data rates are restricted to 9.6 kbps. GPRS offers session establishment times below 1 second and integrated services digital network (ISDN)-like data rates. The channel allocation in GPRS is different from the original GSM; GPRS allows a single mobile station to transmit on multiple time slots of the same TDMA frame. This results in a very flexible channel allocation: one to eight time slots per TDMA frame can be allocated for one mobile station. Moreover, uplink and downlink are allocated separately, which efficiently supports asymmetric data traffic (e.g., Web browsing). In conventional GSM, a channel is permanently allocated for a particular user during the entire call period (whether data are transmitted or not), while in GPRS the channels are only allocated when data packets are sent or received, then released after the transmission. For bursty traffic, this results in a much more efficient usage of limited radio resources, allowing multiple users to share one physical channel. Note that GPRS can work not only at 890–915 MHz for the uplink, and 935–960 MHz for the downlink, but also at 1710–1785 MHz and 1805–1880 MHz for the uplink and dowlink, respectively.

Enhanced data rates for GSM evolution (EDGE) is a further evolution of GPRS and is often referred to as 2.75G systems. The actual packet data rates can reach around 180 Kbps (effective) and increases the data transmission reliability, thus allowing video services and multimedia applications. Hence, EDGE represents the gateway to enhanced spectrum efficiency and higher data rates. It should be noted that the core of EDGE data capacity improvement is essentially based on link level enhancements: in fact, in addition to binary Gaussian minimum shift keying (GMSK) modulation adopted in GSM and GPRS, octagonal phase shift keying (8-PSK) is also considered, together with adaptive modulation and coding.

In this evolutive path, the universal mobile telecommunications systems (UMTS) is one of the 3G cellular phone technologies standardized by the third generation partnership project (3GPP) [23,24]. UMTS provides a multimedia capability to the user, such as the delivery of data, music, images, and real-time video combining personal communications with universal services.

The main characteristics of UMTS are listed in the following [24]:

- High bit rates theoretically up to 2 Mbps in 3GPP Release 99 and beyond 10 Mbps in 3GPP Release 5. Practical bit rates per user are up to 384 Kbps initially and beyond 2 Mbps with Release 5.
- Increased flexibility in resource usage allowing efficient shared channels for small amount of data or heavy bursty traffic.
- Low delays with packet round trip times below 200 ms.
- Seamless mobility also for packet data applications.
- Quality of service (QoS) differentiation for high efficiency of service delivery with service negotiation at call setup.
- Simultaneous voice and data capability.
- Interworking with existing GSM, GPRS, or EDGE networks.

The air interface of UMTS, also known as universal terrestrial radio access (UTRA), is based on wideband code division multiple access (W-CDMA). Among the performance benefits provided by W-CDMA, we may recall a high user data rate and an increased multipath diversity. In fact, radio propagation is characterized by multiple reflections, diffractions, and attenuation of the signal caused by obstacles such as buildings, hills, and the like. These fading dips make error-free reception of data bits very difficult, especially in mobile environments, but W-CDMA provides some important counter-measures:

■ The various multipath contributions are combined by adopting multiple rake fingers (correlation receivers) allocated to those delay positions on which significant energy arrives.

■ Fast power control, tracking small-scale fading variations, is used to mitigate the problem of fading signal power.

■ Reliable coding, interleaving, and retransmission protocols are used to add redundancy and time diversity to the signal and thus help the receiver in recovering the user bits across fades.

■ Soft and softer handover. (Soft handover enables two simultaneous links between a mobile terminal and two base stations. Softer handover enables a user to have two concurrent connections with one base station.)

The specific frequency bands originally defined by the UMTS standard are 1885–2025 MHz for the uplink and 2110–2200 MHz for the downlink. The access scheme is direct-sequence code division multiple access (DS-CDMA) with information either spread over approximately 5 MHz, 10 MHz, or 1.6 MHz depending on the operating mode. In fact, UTRA supports three basic modes of operation: frequency division duplex (FDD), time division duplex (TDD), and low chip rate. In the FDD mode two separate bands of 5 MHz are used for the uplink and downlink, respectively, whereas in TDD only one 5 MHz band is time shared between the uplink and downlink. The TDD mode was added in order to leverage the basic W-CDMA system also for the unpaired spectrum allocations. The low chip rate TDD with 1.28 Mcps, namely [time division-synchronous code division multiple access (TD-SCDMA)] is adopted in China and includes suitability for IP services, ability to support asymmetric services in up/down link, and flexibility to incorporate new technologies.

On top of spreading, the transmitter performs the scrambling operation, which does not change the signal bandwidth but only makes the signals from different sources separable from each other. The spreading codes are based on the orthogonal variable spreading factor (OVSF) technique. The use of OVSF codes allows the spreading factor to be changed while maintaining orthogonality between different spreading codes of different lengths for interference rejection. The information rate varies with the symbol rate, which is derived from the chip rate and the spreading factor.[3] Thus the respective modulation symbol rates vary from 1920 Ksymb/s to 15 Ksymb/s (7.5 Ksymb/s) for FDD uplink (downlink), and for TDD the modulation symbol rates vary from 3.84 Msymb/s to 240 Ksymb/s.

To improve reliability and quality of service (QoS), UTRA adopts channel coding, with two options supported for FDD and three options supported for TDD: (i) convolutional coding; (ii) turbo coding; and (iii) no coding (for TDD, only) [24]. To randomize transmission errors, bit interleaving is also performed. The UTRA modulation scheme is quadrature phase shift keying (QPSK).

In order to efficiently transmit the same content to several users simultaneously, the 3G standard has been enhanced with multimedia broadcast or multicast system (MBMS) [25,26]. MBMS point-to-multipoint transmissions are intended for multiple users, hence, it is not possible to dynamically adapt the transmission parameters according to the user reception conditions, thus constant transmit power and bearer data rate are employed. The transmission has to be configured statistically to serve the worst case user contemplated. In MBMS, multimedia content is delivered either as a streaming service or as a file download service to the end user. For streaming services a continuous data flow of audio, video, and subtitling is transmitted to the terminals; for file download services, a finite amount of data is delivered and stored into the terminals as a file.

The evolution of UMTS is represented by high-speed downlink packet access (HSDPA) in downlink and high-speed uplink packet access (HSUPA) in uplink. They improve the performance by introducing higher-level modulations, new adaptations and control mechanisms to enhance peak data rates, spectral efficiency, and QoS control for packet services. In some years, the long-term evolution (LTE) project will get data rates up to 100 Mbps in the downlink and 50 Mbps in the uplink. Hence, the high achievable data rate allows UMTS to compete with other network technologies such as WLAN which offers access to the Internet and other data services on mobile devices but in small areas covered by the access points (APs).

Overall, cellular systems answer to new lifestyles and customer needs and represent technologies spread worldwide. For this reason, they can be highly considered for infomobility applications. In particular, the GPRS system is already adopted to transmit position and speed information both from public (e.g., buses) and private vehicles to the network for monitoring applications and to transmit congestion or weather information from the network to the vehicles for smart navigation. For future applications, real-time information could be provided to vehicles in order to manage

---

[3] Spreading factors are from 256 to 2 with FDD uplink, from 512 to 4 with FDD downlink, and from 16 to 1 for TDD uplink and downlink.

urban and suburban road traffic and to drive smart (dynamic) navigation choices. These kind of applications could require the transmission of high amount of data, thus requiring high bandwidth and critical cellular design. For example, it will be a critical choice whether to transmit personalized information following specific requests or to broadcast all available information to all the users leaving the on-board navigator to selectively use them. In the latter case, MBMS over UMTS can be adopted.

4.2.3 Wireless Metropolitan Area Networks

One appealing solution for wireless metropolitan area networks (WMANs) is given by the family of IEEE 802.16 standard, which enables the convergence of fixed and mobile broadband networks through a common wide area broadband access technology and a flexible network architecture. The IEEE 802.16-2004 air interface standard [27,28], which is the basis of the WiMAX technology, is one of the most promising solutions for the provision of fixed broadband wireless services in a wide geographical scale and proved to be an effective solution for the establishment of wireless metropolitan area networks (WMANs). On February 2006, the IEEE 802.16e-2005 amendment [29] to the IEEE 802.16-2004 standard, which introduced a number of features aimed at supporting users mobility has been released; it originated the so-called mobile-WiMAX profile. The result of the IEEE 802.16 standardization activity is a complete standard family that specifies the air interface for both fixed and mobile broadband wireless access systems, thus enabling the convergence of mobile and fixed networks through a common wide area radio access technology. Four PHY modes

- WirelessMAN-SC, which has been mainly developed for back-hauling in line-of-sight (LOS) conditions and operates in the 10–66 GHz frequency range adopting a single carrier (SC) modulation scheme
- WirelessMAN-SCa, which has the same characteristics of Wireless-MAN-SC but operates even in non line-of-sight (NLOS) conditions in frequency bands below 11 GHz, adopting SCa
- WirelessMAN-OFDM, which has been developed for fixed wireless access in NLOS conditions and adopts the orthogonal frequency division multiplexing (OFDM) modulation scheme in frequency bands below 11 GHz
- WirelessMAN-OFDMA, which has been conceived for mobile access and adopts the orthogonal frequency division multiple access (OFDMA) scheme in the 2–6 GHz frequency range

Because VNs require mobility, we will focus our attention on WirelessMAN-OFDMA provided by IEEE 802.16e amendment. At the PHY level, the

OFDMA technique is based on the OFDM modulation scheme [30–32] (with a number of subcarriers equal to 128, 512, 1024, or 2048). OFDMA systems provide transmission resources both in the time domain, by means of groups of OFDM symbols, and in the frequency domain, by means of groups of pseudorandomly chosen subcarriers (each group defining a subchannel). The IEEE 802.16e-2005 supports TDD, FDD, and half duplex FDD operations, but the initial release of mobile WiMAX certification profiles only includes TDD, because it enables adjustments of the downlink to uplink ratio to support asymmetric traffic. Both FDD and TDD duplexing schemes are based on a time-frequency frame structure to accommodate uplink or downlink data flows.

Data to be transmitted are subjected to a channel-coding process, which includes randomization, forward error correction (FEC) coding, and bit interleaving. The mandatory channel coding is performed by means of a bit-tail punctured convolutional code with variable code rate R_c, and provides seven fixed combinations of modulation and coding rate, from QPSK with $R_c = 1/2$ up to 64-QAM with $R_c = 3/4$. Mobile WiMAX improves NLOS coverage by adopting advanced antenna diversity schemes and hybrid automatic repeat request (ARQ). Dense subchannelization improves system gain and indoor penetration, enabling better coverage and capacity trade-off. It also uses adaptive antennas and MIMO technologies to improve the coverage.

Like UMTS, IEEE 802.16e-2005 is designed to efficiently manage services with different QoS requirements. In particular, it divides all possible services into five classes: unsolicited grant service (UGS), real-time polling service (rtPS), extended real-time polling service (ErtPS), non–real-time polling service (nrtPS), and best effort (BE). Each of them is associated with a set of QoS parameters that quantify characteristics such as (a) maximum sustained rate, (b) minimum reserved rate, (c) maximum latency tolerance, (d) jitter tolerance, and (e) traffic priority. These quantities are the basic inputs for the service scheduler placed in the base station. The design and implementation of service scheduler are left to the manufacturer, which is aimed at fulfilling service-specific QoS requirements.

As a performance example for mobile WiMAX, in [33] it is shown that a single user can obtain a transmission control protocol (TCP) throughput of almost 13 Mbps in downlink with UGS service adopting the TDD version and an asymmetry of 1/3 between uplink and downlink traffic, assuming a total bandwidth of 7 MHz; adopting the (realistic) settings that are not fully reported here (please refer to [33]), a maximum coverage distance of 1.4 km is estimated using a 120 degrees antenna in the suburban environment with no other cells interfering (at this distance the throughput decreases to about 2 Mbps).

Mobile WiMAX also supports multicast and broadcast service (MBS). Through flexible radio resource allocation it enables full or partial allocation of radio resources to MBS dynamically. The MBS can be supported by either constructing a separate MBS zone in the downlink frame with unicast service, or the entire frame can be dedicated to MBS for broadcast service (downlink only). The MBS zone supports a multibase station MBS mode using a single frequency network (SFN) operation, and flexible duration of MBS zones permits the scalable assignment of radio resources to MBS traffic. Note that multiple MBS zones are feasible. The multibase station MBS does not require a mobile terminal to be registered with a base station.

Multimedia stream services provided by broadband wireless networks are recently attracting much attention, with particular remark to multimedia applications for mobile devices. From this point of view, mobile WiMAX provides users seamless broadband connectivity, hence representing a viable and efficient technology for wireless multimedia services (e.g., voice over Internet protocol (VoIP), mobile TV, etc.), due to the offered flexibility in both radio access and network architecture. In the future, the possibility of mobile WiMAX to efficiently support a broad range of applications and services for VNs, such as traffic information distribution (both personalized or broadcasted) and multimedia contents transmission (e.g. advertisements or tourist information) will allow this technology to compete or integrate with cellular systems. It must be noted, however, that huge investments are required to reach that target, aside from the fact that WiMAX was initially conceived for fixed use or at most nomadic use (i.e., the user may connect anywhere, but mobility is not supported during an active session), rather than for mobile use; moreover, in many parts of the world, the adoption of high frequencies (more than 3 GHz) makes it difficult to support coverage with full mobility in urban or suburban areas. In any case, WiMAX may be a key technology for rapid deployment of backhaul connecting roadside devices to control centers. In this case, efficient and flexible routing protocols over hybrid networks will be one of the most challenging issues for researchers.

4.2.4 Wireless Local Area Networks

Starting from the last years of the 1990s, IEEE 802.11 WLAN technologies have been developed to provide wireless connectivity integrated with existing wired networks, aiming to have features such as easy installation, flexibility, and user mobility.

The family IEEE 802.11 is a set of standards developed by the IEEE LAN/MAN Standards Committee (IEEE 802), and its first version [34] was approved in 1997 to work in the unlicensed band at 2.4 GHz with a bit rate of 1 or 2 Mbps. The IEEE specifications define the lower levels of the protocol pillar, involving all PHY and data link control aspects. At the PHY level, different options were provided, such as direct sequence spread spectrum

(DS-SS), frequency hopping spread spectrum (FH-SS), and infrared, while at the MAC level a decentralized approach was defined through a carrier sensing multiple access with collision avoidance (CSMA-CA) protocol. This choice at the MAC level also allows peer-to-peer communication, but has critical problems when nodes communicating to the same destination are not in radio visibility of each other (the so-called hidden terminal problem [35,36]). A centralized protocol is also considered, but it reduces the advantages of the technology itself; thus, it is scarcely adopted in practice. After the first version of the standard, a number of amendments were defined or are still under definition, in order to solve a number of weaknesses and to further increment capacity and coverage.

In 1999 the amendments IEEE 802.11a [37] and IEEE 802.11b [38,39] significantly increased the nominal throughput, with modifications limited to the PHY level. In fact, the IEEE 802.11a version works at 5 GHz and allows rates from 6–54 Mbps through four different modulation schemes and three possible coding rates (eight combinations of modulation and coding rate are defined); IEEE 802.11b works at 2.4 GHz and provides up to 11 Mbps by adopting the complementary code keying (CCK) modulation. Due to the backward compatibility and the fact that an unlicensed band at 5 GHz is not available in most countries, the IEEE 802.11b version spread earlier; its massive widespread adoption underlined the interest for these systems and highlighted their limits, requiring the IEEE 802.11 working group to define new task groups for specific improvements.

Among all the versions of the standard (a complete list can be found at [40]), the major enhancements provided are the IEEE 802.11g [41] and IEEE 802.11e [42] versions. The IEEE 802.11g, ratified in 2003, defines a PHY level at 2.4 GHz with backward compatibility to the "b" version, but adopting the same solutions as the "a" version (including the OFDM modulation scheme); IEEE 802.11e, released in 2007, has the scope to provide an enhanced MAC level protocol in order to introduce the concept of QoS in the frame delivery procedure of wireless local area networks (WLANs); the adoption of the "e" version at the MAC level is compatible with non-"e" devices and remains independent to the choice at the PHY level (e.g., either IEEE 802.11a or g).

To properly compare IEEE 802.11a to IEEE 802.11g, it is important to remark that although they allow almost the same throughput, the different frequencies adopted make their coverage planning different; in particular, at 5 GHz the path loss due to propagation and obstacles is higher (thus a higher power is required for the same average performance at the same distance); moreover, only three nonoverlapping frequency bands are allowed for the "g" version, while more then 10 nonoverlapping frequency bands are allowed for the "a" version. For both standards the 6–54 Mbps nominal throughput corresponds to the range from about 4–5 to 20–25 Mbps effective user data rate, respectively, depending on the protocols adopted at higher levels and on some choices at PHY and MAC layers; this is due to the heavy overhead introduced (in particular by the CSMA/CA protocol). Please note that this throughput can be achieved by a single active connection, whereas more active connections would share a total throughput that slightly decreases with the number of active connections.

To conclude this excursus related to the main IEEE 802.11, it is important to mention two other versions still under development: IEEE 802.11n and IEEE 802.11p. In particular, IEEE 802.11n will introduce MIMO techniques and other modifications in order to increase the throughput up to 100 Mbps.

For VNs, great attention has to be paid also for IEEE 802.11p [43, 44] that has the scope to become the main standard for vehicular communications (both V2V and V2I). Its aim is, on one hand, to define rules for fast network recognition and setup and, on the other, to allow differentiation between normal use and emergency use. Distance coverage up to 1000 m and devices speed of 200 km/h are in the scope of the task group. Higher transmitted power will be defined for special uses (e.g., connection among firemen or medical vehicles).[4]

Due to the low expense and large diffusion of this technology, IEEE 802.11 (in particular in the "p" version) will probably become the main standard for vehicular communication at least for V2V communication. In order to understand the main applications for infomobility services over WLANs, it must be remarked that this technology allows ad hoc networking and multihop communication. Once the problem of connection setup delay is solved (the IEEE 802.11p aims at overcoming this problem), it allows a high throughput communication with low delay among vehicles. This will lead to efficient emergency communications: for example, in case of accident, an alerting message transmitted among vehicles can be faster and, thus, well-timed, rather than a communication sent through an infrastructure network (such as cellular systems). Moreover, the exchange of information between vehicles can ease the respect of a safe distance, especially when travelling at high speed.

4.2.5 Wireless Sensor Networks

A wireless sensor network (WSN), in its simplest form, can be defined as a network of (possibly, low-size and less-complex) devices, denoted as *nodes* that can sense the environment and communicate the information

---

[4] Higher power levels defined for this version will overcome coverage difficulties encountered when dealing with actual standards, but new regulations will be needed to allow its adoption in all countries. At this time, a frequency of 5.9 GHz is envisioned for this use, at least in the United States.

gathered from the monitored field (e.g., an area or volume) through wireless links; the data is forwarded, possibly via multiple hops relaying, to a *sink* (sometimes denoted as controller or monitor) that can use it locally or is connected to other networks (e.g., the Internet) through a gateway. The nodes can be stationary or moving, aware of their location or not, they can be homogeneous or not [45–49].

In the presence of actuators, (i.e., devices able to manipulate the environment, rather than observe), we have a WSAN. From the communication protocol point of view, the inclusion of actuators does not represent a simple extension of a WSN. In fact, the information flow must be reversed in this case: the protocols should be able to manage many-to-one communications when sensors provide data, and one-to-many flows when the actuators need to be addressed, or even one-to-one links if a specific actuator has to be reached. The complexity of the protocols in this case is even larger.

Given the very large number of nodes that can constitute a WSAN (more than hundreds sometimes), it is clear that the MAC and the network layers are very relevant parts of the protocol stack. Tens of proposals specifically designed for WSANs have been made in the past few years. The communication protocols of a WSAN should also allow an easy deployment of nodes; the network must be able to self-organize and self-heal when some local failures are encountered even in the presence of mobile nodes.

The traditional architecture of a sensor node is constituted by a microprocessor that manages all tasks; one or more sensors used to take data from the environment; a memory included over the board used to store temporary data, or during its processing; and a radio transceiver (with the antenna). All these devices are powered by a battery that should be parsimoniously used for the whole duration of the network lifetime by all these devices. In some cases, energy scavenging techniques can be introduced to increase the lifetime of the nodes, but in few applications this can really be considered as a viable technique. As a result of the need to have energy-efficient techniques implemented over the board, all data processing tasks are normally distributed over the network; therefore, the nodes cooperate to provide the data to the sinks. This is also because of the low complexity that is accepted for the architecture of such nodes. In conclusion, a WSAN can be generally described as a network of nodes that cooperatively sense the environment and may control it, enabling interaction between people or computers and the surrounding environment.

The density of nodes and sinks is a very relevant parameter for WSANs. The density of sensor nodes defines the level of coverage of the monitored space (i.e., what percentage is such that if an event happens inside, it is detected by at least one node); however, it also defines the degree of connectivity or reachability, which is a relevant issue. On the other hand, the density of sinks plays a significant role in defining the performance of the network in terms of success rate of data transmissions, and the like.

Although a standard explicitly and specifically devoted to WSANs does not exists, the enabling technologies whose standardization process has already provided or is currently producing steady releases are discussed; namely, ZigBee, with two options as regards to PHY layer (IEEE 802.15.4 and ultrawide bandwidth [UWB] IEEE 802.15.4a), and Bluetooth (IEEE 802.15.1).

ZigBee wireless technology is a short-range communication system intended to provide applications with relaxed throughput and latency requirements in wireless personal area networks (WPANs). The key features of ZigBee wireless technology are low complexity, low cost, low power consumption, and low data rate transmissions, supported by cheap, fixed or moving devices. The IEEE 802.15.4 working group [50] focuses on the standardization of the bottom two layers of protocol stack. The upper layers are normally specified by industrial consortia such as the ZigBee Alliance [51]. The ZigBee core system consists of a radio frequency (RF) transceiver and the protocol stack. The ZigBee PHY layer operates in three different unlicensed bands (and with different modalities) according to the geographical area where the system is deployed. DS-SS is adopted and mandatory to reduce the interference level in shared unlicensed bands. Because the interface is provided with a physical medium, the PHY is in charge of radio transceiver activation and deactivation, energy detection, link quality, clear channel assessment, channel selection, and transmission and reception of the message packets. Moreover, it is responsible for establishment of the RF link between two devices, bit modulation and demodulation, synchronization between the transmitter and the receiver, and, finally, for packet level synchronization. IEEE 802.15.4 specifies a total of 27 half-duplex channels across the three frequency bands organized as follows:[5]

- The 868 MHz band mode, ranging from 868.0 to 868.6 MHz is used in the European area. It adopts a raised-cosine–shaped binary PSK modulation format, with DS-SS at chip rate 300 Kcps (a pseudorandom sequence of 15 chips transmitted in a 50 μs symbol period). Only a single channel with data rate 20 Kbps is available and the ideal transmission range (i.e., without considering wave reflection, diffraction, and scattering) is approximatively 1 km, with a required minimum −92 dBm RF sensitivity.

---

[5] The ideal transmission range reported is evaluated considering that, although any legally acceptable power is permitted, IEEE 802.15.4 compliant devices should be capable of transmitting at −3 dBm.

■ The 915 MHz band mode, ranging between 902 and 928 MHz is used in the North American and Pacific area. It adopts a raised-cosine–shaped binary PSK modulation format, with DS-SS at chip rate of 600 Kcps (a pseudorandom sequence of 15 chips is transmitted in a 25 μs symbol period). Ten channels with rate 40 Kbps are available and the ideal transmission range is approximatively 1 km, with a required minimum −92 dBm RF sensitivity.

■ The 2.4-GHz industrial, scientific, and medical (ISM) band mode, which extends from 2400 to 2483.5 MHz is used worldwide. It adopts a half-sine–shaped offset quadrature PSK modulation format, with DS-SS at 2 Kcps (a pseudorandom sequence of 32 chips is transmitted in a 16-μs symbol period). Sixteen channels with data rate 250 Kbps are available and the ideal transmission range is approximatively 220 m, with minimum −85 dBm RF sensitivity required.

Because power consumption is a primary concern in WSN to achieve long battery life, the energy must be drained continuously at an extremely low rate, or in small amounts at a low power duty cycle. This means that IEEE 802.15.4-compliant devices are active only during a short time. The standard allows some devices to operate with both the transmitter and the receiver inactive for over 99% of the time. Thus, the instantaneous link data rates supported (i.e., 20 Kbps, 40 Kbps, and 250 Kbps) are high with respect to the data throughput in order to minimize device duty cycle. According to the IEEE 802.15.4 standard, transmission is organized in frames, which can differ according to the relevant purpose. In particular, there are four frame structures: a beacon frame, a data frame, an acknowledgment frame, and a MAC command frame. The meaning of the four possible frame structures depends on possible network topologies and MAC channel access strategies.

UWB radio [52,53] is a fast emerging technology with uniquely attractive features for academia, industry, and global standardization bodies. The UWB technology has been around the 1960s, when it was mainly used for radar and military applications, whereas nowadays it is a very promising technology for localization systems and advanced wireless communications, networking, radar, imaging, positioning systems, and, in particular, WSNs. The most widely accepted definition of a UWB signal is given by instantaneous spectral occupancy in excess of 500 MHz or a fractional bandwidth of more than 20%. In 2002, the U.S. Federal Communications Commission (FCC) issued the first Report & Order, which permitted unlicensed UWB operation and commercial deployment of UWB devices. There are three classes of devices defined in the document: (i) imaging systems (e.g., ground penetrating radar systems, wall imaging systems, through-wall imaging systems, surveillance systems, and medical systems); (ii) vehicular radar systems; and (iii) communications and measurement systems. The FCC allocated a block of unlicensed radio spectrum from 3.1 to 10.6 GHz at the noise floor of -43 dBm/MHz for the above applications where each category was allocated a specific spectral mask and UWB radios overlaying coexistent RF systems can operate. With similar regulatory processes currently under way in many countries worldwide, government agencies responded to this FCC ruling. Such UWB systems can be realized through conventional modulation schemes by stressing the bandwidth to be larger than 500 MHz, for example, by adopting OFDM signaling leading to multiband ultrawide bandwidth (MB-UWB). This approach has been first followed by the IEEE 802.15.3a Task Group then included in the WiMedia alliance standard for high-speed applications, which involve imaging and multimedia. Very promising, especially for WSN applications, is the impulse radio ultrawide bandwidth (IR-UWB) technique, which relies on ultrashort (nanosecond scale) waveforms that can be free of sine-wave carriers and do not require intermediate frequency (IF) processing because they can operate at baseband. As information-bearing pulses with ultrashort duration have UWB spectral occupancy, UWB radios come with unique advantages that have long been appreciated by the radar and communication communities, such as the enhanced capability to penetrate through obstacles, the ultrahigh precision ranging at the centimeter level, the very high data rates in harsh multipath environments along with a commensurate increase in user capacity, and potentially small size and processing power. The IR-UWB technique has been selected as the PHY layer of the IEEE 802.15.4a Task Group for WPAN low rate alternative PHY layer [54], thus also suitable for VNs. The IEEE 802.15.4a is based on two optional PHYs. The one based on IR-UWB (operating in unlicensed UWB spectrum) will be able to deliver communications and high-precision ranging. The UWB PHY layer supports an over-the-air mandatory data rate of 851 Kbps, with optional data rates of 110 Kbps, 6.81 Mbps, and 27.24 Mbps. The choice of the PHY depends on the local regulations, application, and user preferences. Frequency bands foreseen by the standard (some of them are optional) are 250–750 MHz for UWB Sub-GHz, 2400–2483.5 MHz for 2450 CSS, 3244–4742 MHz for UWB low-band, and 5944–10234 MHz for UWB high-band. The adopted modulation format combines both binary PSK and pulse position modulation (PPM) signaling so that both coherent and low-complexity noncoherent receivers can be used to demodulate the signal. Through measurements of two-way ranging, applications can derive the node position information with high accuracy (UWB is intended to provide submeter localization [55]).

Bluetooth wireless technology is a short-range communication system intended to replace the cables in WPANs [56,57]. The key features of Bluetooth wireless technology are robustness, low power, and low cost and many features of the core specification are optional, allowing product differentiation. The IEEE 802.15.1 group has derived a WPAN standard based on the Bluetooth v1.1 Foundation specifications. The Bluetooth RF PHY operates in the unlicensed ISM band, for the majority of countries around 2.4 GHz in 2400–2483.5 MHz. The system employs a FH-SS transceiver (the nominal hop rate is 1600 hops/s) to combat interference and fading. RF operation uses a binary Gaussian frequency shift keying (GFSK) modulation to minimize transceiver complexity and a FEC coding technique. The symbol rate is 1 Msymb/s supporting the bit rate of 1 Mbps. With enhanced data rate of Bluetooth 2.0, a gross air bit rate of 2 or 3 Mbps can be reached changing the modulation format. Between the devices connected, one provides the synchronization reference and is known as the master; the others are known as slaves. A group of devices synchronized in this fashion form a piconet. This is the fundamental form of communication for Bluetooth wireless technology. Devices in a piconet use a specific frequency hopping pattern which is algorithmically determined by certain fields in the Bluetooth specification address and clock of the master. The basic hopping pattern is a pseudorandom ordering of the 79 frequencies[6] with channel spacing of 1 MHz in the ISM band (e.g., $f = 2402 + k$ MHz, with $k = 0, \ldots, 78$). To comply with out-of-band regulations in each country, a guard band is used at the lower and upper band edge, respectively, of 2 MHz and 3.5 MHz. The hopping pattern may be adapted to exclude a portion of the frequencies that are used by interfering devices. The adaptive hopping technique improves the coexistence of Bluetooth technology with static (nonhopping) ISM systems when these are colocated.[7] The physical channel is subdivided into time units known as slots with duration 625 μs. Data is transmitted between Bluetooth-enabled devices in packets that are positioned in these slots. When circumstances permit, a number of consecutive slots may be allocated to a single packet. Frequency hopping takes place between the transmission or reception of packets. Bluetooth technology provides the effect of full duplex transmission through the use of a TDD scheme. Above the physical channel there is a layering of links and channels and associated control protocols [59].

4.2.6 Heterogeneous Networks for Infomobility Services

In the preceding subsections, a brief description of all main enabling technologies has been given and some potential applications for each of them were depicted. It must be remarked, however, that the joint adoption of

---

[6] In some countries, like France, the number of frequencies is 23.
[7] Coexistence of Bluetooth and WLAN sharing the same frequency band is studied, e.g., in [58].

all or some of the proposed technologies in heterogeneous communication networks is necessary to allow the infomobility services envisioned nowadays. As an application example, automatic emergency calls from a vehicle to ambulance, through a GPS positioning and a GPRS call, is already available on some vehicles. In the future, the addition of a WLAN connection would allow a faster distribution of the warning message among vehicles. A WiMAX transmitter could also be adopted as a backup option to increase the system reliability or collect information from the roadside (e.g., information from WSNs) and retransmit them to control centers.

4.3 A Platform for the Design and Simulation of Heterogeneous Wireless Networks

4.3.1 Introduction to the Platform SHINE

In Section 4.2, an overview of the enabling technologies was given, highlighting the complexity and the challenges of future VNs scenarios. It can be easily understood that although specific problems dealing with one or two technologies in defined scenarios can be investigated analytically or through testbeds, it is almost impossible to investigate a scenario that includes more technologies, vehicles with realistic mobility, and data flows for realistic services without the adoption of a complex and advanced simulation platform.

In this section we describe the simulation platform for heterogeneous interworking networks (SHINE) developed over the years at WiLAB within several projects also involving manufacturers and operators [60–62]. When possible it has been tested through measurements in order to investigate single wireless communication networks in realistic scenarios as well as interoperating networks.

The SHINE platform has been designed to investigate the integration or interworking of different access technologies toward fourth-generation (4G) communication networks which perform, for instance, intersystem handovers within overlapped coverage areas. This is the next step in the evolution of actual access-networks toward 4G to provide always the best Qos to users on the basis of their needs and network status (according to the always best connected paradigm). The availability of more than one access network to support multimedia applications and user requests, with a common pool of resources provided by each single cooperating networks, requires the adoption of efficient common radio resource management (CRRM) strategies in order to make the best of the "global" access network.

This platform aims at taking care of all aspects related to every single protocol level of each access technology affecting the achieved performance following an integrated approach [63]. SHINE realization was planned in order to overcome the limitations of off-the-shelf network simulation tools such as Opnet [64] or ns-2 [65], in terms of accuracy in wireless link modelling jointly with capability to simulate large scenarios (hundreds of terminals) where users perform realistic sessions (statistically distributed beginnings, realistic traffic exchanges, realistic durations) and have realistic mobility (also including, obviously, heterogeneous networks operating simultaneously).

Actually SHINE includes the following access technologies, whose performance have been in some cases already investigated (see references):

- WLAN IEEE 802.11a/g/e (see [66–68])
- UMTS TD-SCDMA (see [69,70] and for UMTS-WLAN interworking see [71,72])
- WiMAX IEEE 802.16e (see [33])
- UMTS FDD
- GSM-GPRS-EDGE
- Multicarrier code division multiple access (MC-CDMA) [73–75]

The platform is open for evolutions and other standards; in fact the different technologies at lower layers are used as plug-ins for SHINE.

4.3.2 Structure of SHINE

The SHINE platform is constituted by one server simulator, hereafter called upper layers simulator (ULS), and a client simulator for each access technology, called lower layers simulators (LLSs) (see Figure 4.4 where, e.g., two LLSs are depicted: UMTS and WLAN). The ULS takes care of all the information related to users' mobility and end-to-end aspects of each connection, such as the generation of the application-level traffic and the transport protocols (TCP, UDP, etc.) dynamics. It is worth noting that the ULS structure, being related to the end-to-end aspects of communications, is independent of any particular access technology (UMTS, WLAN, WiMAX, etc.) adopted to establish the user connection, whereas its performance is strongly dependent on this.

All aspects related to the access technologies, hence related to the data link and physical layers, are managed by the LLSs, which are the client simulators and are specific for each access technology, such that the simulation platform provides the presence of all LLSs as technologies adopted in the investigated scenario. Let us emphasize that, within the whole simulation platform, ULS and LLS are distinct executables; nonetheless during simulations the ULS communicates run time with LLSs through the TCP sockets of the computer operating system, thus simulating both vertical communications among the protocol layers of the single access technology and,

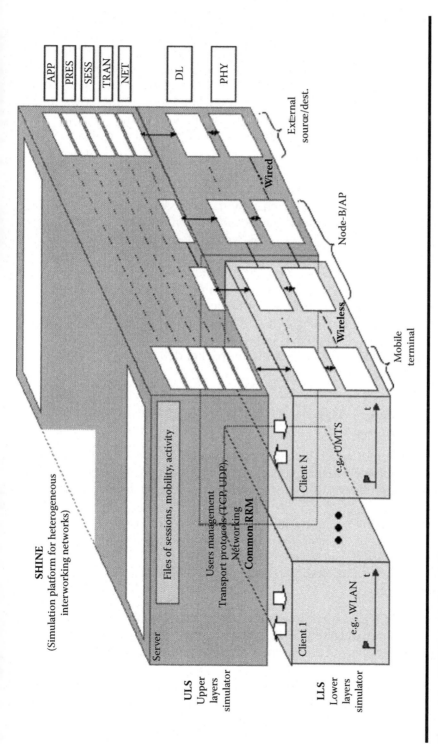

Figure 4.4   SHINE simulation platform.

since ULS is on top of all LLSs and manages all end-to-end communications aspects, the access networks interworking.

As ULS manages end-to-end aspects of each connection (no matter the access technology supporting them at the physical and data-link levels). Its tasks are mainly concerned with communications management (e.g., connections setup and closure, management of application level traffic flows, etc.), simulation of transport level protocols (TCP and UDP), and processing of simulation outcomes to provide application level performance. In particular, the main tasks of ULS are

- To manage connection setup and closure procedures
- To set the starting instant of each new traffic session originated by users according to the arrival statistics of the traffic class it belongs to (http, e-mail, voice calls, etc.), as well as users' positions within the investigated scenario
- To generate the bit flows uploaded and downloaded by users in each session according to the statistics of their class of traffic
- To reproduce the transport protocol behavior
- To perform packet segmentation and reassembly
- To collect, finally, all simulation outcomes and to generate the outputs (user satisfaction rate, throughput, packet delivery delays, etc.) from an end-to-end point of view

Common radio resource management (CRRM) peformed by ULS is also of great importance. Thus at ULS

1. It selects through which technology (i.e., through which LLS) each user should connect to the network on the basis of user- or network-defined rules and available information on the current status of each network.
2. It moves ongoing communication sessions from an LLS to another (i.e., from a given technology to another) when required, thus simulating vertical (i.e., intersystem) handovers.
3. It rejects connections when, according to the network policies, they cannot be served.

With reqard to the LLSs, because they are specific for the particular access technologies investigated, their tasks are mainly concerned with data link and PHY level aspects of communications and, in particular:

- To perform the call admission control (CAC) specific to the technology it simulates and all technology-specific radio resource management (RRM) (such as the link adaptation).
- To manage the transmission scheduling at the data-link level

- To perform MAC and RLC fragmentation and reassembly of network level packets
- To simulate MAC and RLC behavior of the given technology
- To reproduce all PHY layer procedures related to each communication: power control, handover, radio frequency measurements, channel coding, modulation, information detection, decoding, and the like.
- To collect, finally, all simulation outcomes and to generate the outputs (user satisfaction rate, throughput, packet delivery delays, etc.) from the wireless links point of view (i.e., at data link and PHY level).

The implementation of the technologies through different executables allows a careful and specific implementation of all aspects related to propagation, including path loss, small-scale and large-scale fading, and interference. In fact, when wireless communication is investigated, these aspects significantly affect the QoS perceived by the user, and they must be reproduced with proper time and frequency correlation. It is also important to highlight that how to model these aspects depends on the specific technology.

In Figure 4.5, a block scheme shows some insights on ULS and LLS functionalities. Note the presence of additional simulators, on top of SHINE,

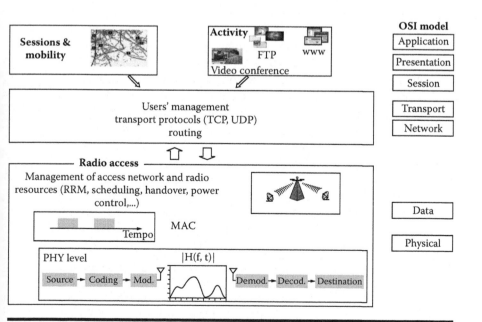

Figure 4.5   Communication flows within SHINE.

which generate input files (for ULS) reproducing users' mobility (according to the street layout, for instance) and users' traffic characteristics (according to the kind of service provided). Also note that SHINE enables the investigation of the performance in scenarios (mobility, traffic, topology) given as input as well as the evaluation of several metrics and parameters affecting the QoS at different layers, thus allowing the assessment of proper statistics. Examples of usage in some case studies are reported in the following section.

4.4 Case Studies

To give the reader some examples on how some of the described technologies operate in VNs, we propose two case studies: the former is related to the exploitation of GPRS in providing alerting message for emergency situations by focusing, in particular, on the delays affecting the network response in real-time applications; the latter refers to the broadcasting of traffic information to vehicles through MBMS via UMTS. Without being exhaustive, in this section we would like to highlight the impact that the introduction of infomobility services would have on VNs [76–78].*

4.4.1 Emergency Warning through GPRS

We describe here how a GPRS system can provide road safety applications, exploiting its capabilities for data applications, radio localization, and telemetry. In fact, in case of accidents on a highway, the main danger is caused by vehicles obstructing the road (cars involved in the accident, emergency vehicles, etc.). Successive accidents, which are often more serious than the original one, could be avoided if the accident was detected immediately. Here we aim at evaluating if a prompt radio notice, exploiting the GPRS infrastructure, can help avoid risks situations [76]. Because GPRS provides a radio infrastructure, it allows the prompt transmission of the alerting message, potentially also providing traffic information.

As a case study, we consider the highway scenario shown in Figure 4.6, and we assume that the whole highway trunk is covered by the GPRS network; thus vehicles can take advantage of the GPRS ability to support

---

* Portions reprinted, with permission, from B. M. Masini, L. Zuliani, and O. Andrisano, "On the effectiveness of a GPRS based intelligent transportation system in a realistic scenario," VTC 2006-Spring. IEEE 63rd, 6:2997–3001, 2006. ©2006 IEEE, and from A. Bazzi, B. M. Masini, A. Conti, O. Andrisano, "Infomobility provision through MBMS/UMTS in realistic scenarios," ITSC 2008, 12–15 October, Bejing, China, ©2008 IEEE.

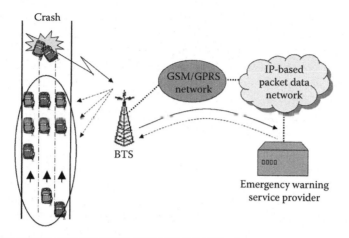

Figure 4.6   GPRS infrastructure on a highway.

packet-switched services and priority-based scheduling. Note that the presence of a GPRS device on board does not change either the vehicle architecture or the energy consumption; hence, it is fully compliant with normal driving conditions. When an accident occurs, a GPRS-equipped vehicle identifies itself as crashed (e.g., by sensors detecting airbag ignition) and alerts the approaching vehicles generating a warning message through the GPRS network. In particular, an emergency warning service provider (EWSP) retransmits the message in multicast mode to all the incoming vehicles as shown in Figure 4.6 [77].

To provide a warning service, it is important to reduce the time spent by the packet in the network which depends on the time spent to obtain the radio access, the delays introduced by the core network, and the delays in the Internet protocol (IP) network. The longer these delays, the longer is the interval time for accident notification to approaching vehicles and less effective would be the emergency warning system.

When a vehicle receives a warning message, the driver is informed in advance about the accident occurrence rather than by his own perception. Hence, in order to evaluate the importance of a well-timed warning message to avoid accidents, we exploit SHINE simulation platform to compare the crash rate in case of GPRS-equipped vehicles or not. We make the following assumptions:

- Each vehicle moves at a variable speed v measured in km/h.
- We assume that the speed, a Gaussian random variable with mean value v_m and its standard deviation σ_v, expressed in Km/h. The standard deviation of velocity quantifies the rapidity of the velocity variations.

■ We define the mean interval time between two consecutive speed changes: the instants to update the vehicles speed follow a negative exponential distribution with mean value equals to the mean interval time between two consecutive velocity variations.

■ We take into account the possibility of lane changes and car overtaking (every 170 m).

■ The mobility model takes into account the driver's perception of risk by means of a suitable variable denoted by the visibility v.

We also assume that when a driver receives a warning message he reacts and slows down until the car stops. To avoid hitting against the obstacle, the distance covered by the vehicle after the reception of the warning message has to be lower than the distance from the last queued car. Hence, the parameters that really affect the number of collisions are the distance between a vehicle and the queue; the vehicle speed (and, as a consequence the distance to stop); the human reaction time; the road conditions, pointed out by the grip coefficient f; the presence or not of the GPRS on board. Hence, a vehicle has to stop in the distance D_{stop} given by

$$D_{stop} = RS + BS \qquad (4.1)$$

where RS is the reaction space and BS the braking space. In particular, RS represents the distance, in meters, covered during the time spent by the packet in the network, round trip time (RTT), and during the driver's reaction time (estimated at about 0.7 seconds), while BS is the distance covered since the vehicle begins to slow down until it stops. Hence,

$$RS = (0.7 + RTT)u \qquad (4.2)$$

and

$$BS = \frac{v^2}{250 \cdot f}, \qquad (4.3)$$

where u is the speed measured in m/s, ($u = \frac{v}{3.6}$) and f is the grip coefficient (some typical values of f are reported in Table 4.1). Since a fast warning message can help in reducing D_{stop}, network delays assume a fundamental importance. For this reason, we measure the RTT in the laboratory on a suitable test bed [79]. In particular, the RTT delay represents the interval time that an Internet control message protocol (ICMP) packet takes to come back to the sending host, after being processed by the receiving one. Hence, the time required has been measured for sending ICMP packets from a GPRS-equipped laptop toward a local host or a remote host (a ping operation). We measured the RTT both in unloaded and loaded

Table 4.1 Typical values of *f*

f	*road type*
0.1	iced road
0.3	mud road
0.4	wet asphalted road
0.5	smooth asphalted road
0.6	coarse asphalted road
0.7	dry asphalted road
0.8	dry asphalted road with uniform granular surface

networks (the network was loaded with one or two file transfers simultaneously executed during the ping operation). Through the measurements, we can derive the RTT frequency, shown in Figure 4.7, corresponding to the case in case of a network loaded with one FTP transfer (e.g., a map download), which the condition we assume hereafter.

It is worth noting that in the proposed scenario, the RTT is the time that a warning message takes to come back to the hosts close to the sending one (and not back to the sending host), after being processed by the GPRS network and the EWSP. However, assuming that the uplink scheduling is independent from that in downlink, we can consider that the measured RTT is a good approximation also for the simulated scenario.

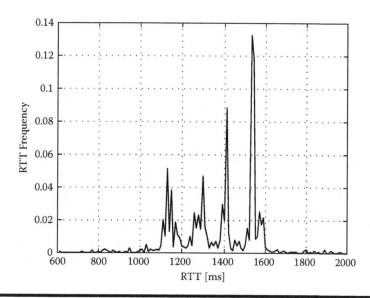

Figure 4.7   RTT frequency in case of network loaded with one FTP transfer.

Table 4.2   Parameters of the scenario

description	value
length of the highway trunk	2.4 km
lanes in each direction	3
speed's standard deviation	10 km/h
low traffic density	40 veh/km
high traffic density	70 veh/km

We compare situations with or without GPRS on board in a highway trunk of length 2.4 Km with σ_v equal to 10 Km/h.[8] We propose the results for two different traffic densities: a low traffic density (40 vehicles/km) and a high traffic density (70 vehicles/km) distributed in all the considered lanes. Table 4.2 summarizes the parameters of the considered scenario.

In Figure 4.8 the crash rate as a function of the grip factor f is plotted when a mean speed of 100 km/h and 200 m of visibility are considered with high traffic density. The advantage of a GPRS warning message can be observed; in particular, when $f = 0.4$, (i.e., wet asphalted road) a vehicle running at 100 km/h without alerting system cannot avoid the accident, while it is involved with a probability of 10% with a GPRS driver assistance system. When $f = 0.8$, (i.e., dry asphalted road) the percentages decrease to 40% and 2%, respectively.

In Figure 4.9 the number of crashed vehicles as a function of the mean speed v_m for different values of f can be observed with or without an alerting system on board. High traffic density and $v = 150$ m are assumed. The improvement in avoiding an accident provided by GPRS on board is quite evident for any kind of road asphalt.

4.4.2   MBMS over UMTS for Infomobility

This case study envisions to transmit a relatively high amount of traffic information to all users in order to dynamically help in defining the best route on a real-time basis. At this scope we consider the UMTS (FDD version) as the enabling technology, exploiting the potentialities given by MBMS. As already discussed in Section 4.2.2, the advantage of considering an existing and largely deployed technology, such as UMTS, is the spreading of its radio coverage in most countries and the existent infrastructures. It is also worth noting that an infrastructure-based system, such as UMTS, typically makes security and privacy protocol implementation easier. The objective of this section is to show the feasibility of the proposed service,

---

[8] $\sigma_v = 10$ Km/h means that the most users run with speed near to the average, but also guarantees a not negligible number of users driving with different speeds.

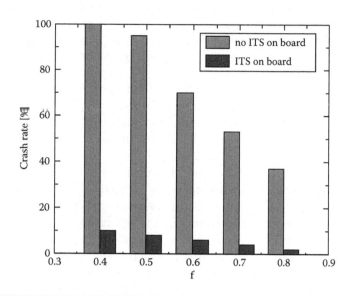

Figure 4.8 **Crash rate versus the grip factor,** f**, with mean speed** $v_m = 100$ **km/h and high traffic conditions. Comparison between GPRS on board and no alerting system.**

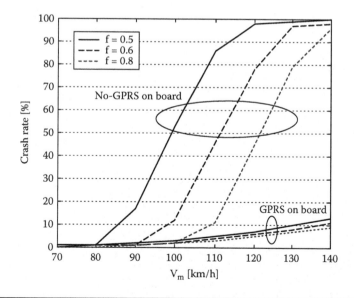

Figure 4.9 **Crash rate vs. the mean speed,** v_m**, for different values of the grip factor,** f**, and high traffic conditions. Comparison between GPRS on board and no alerting system.**

and the impact that it would have on the performance of concurrent traffic sessions, contemporarily active on the same network. We want to evaluate if the adoption of MBMS offered via UMTS can provide downlink traffic information and navigation support without affecting typical traffic already present in the network.

The reference scenario is shown in Figure 4.10. We assume that a number of pedestrians are performing voice calls through the UMTS network and that a multicast channel is also transmitted in each cell; cars moving along the road are equipped with a smart device able to listen to that channel. In particular, by adopting multicast radio channels, we assume to transmit up-to-date real-time information, such as the mean speed (and its variance) of all urban surrounding roads and those of the main regional roads and national highways; main suggestions for alternative routes (e.g., to avoid the stadium area while a football match is about to conclude), and the like. As reference scenario, we consider the city of Bologna, Italy, which is a densely populated medium-size city affected by traffic and congestions in many areas, thus representing a suitable environment to evaluate the effectiveness of the proposed service. In particular, we consider a realistic number of UMTS base stations distributed in the city center, and we

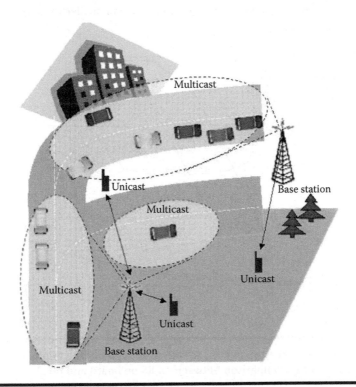

Figure 4.10   Scenario and technologies.

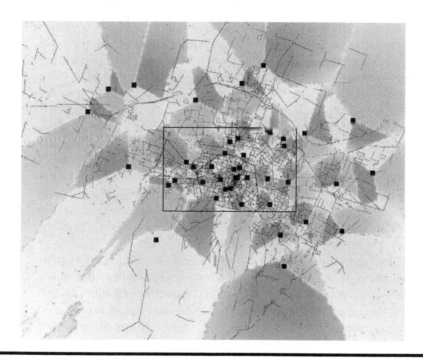

Figure 4.11 **City map of Bologna and frequency planning. Filled black squares correspond to UMTS Nodes-B locations. Three frequency bands are considered, represented by three different gray gradiations; higher brightness corresponds to lower signal to noise and interference ratio. The black rectangle indicates the area considered for numerical results.**

perform a frequency planning of the scenario in order to grant a good coverage and interference mitigation. The situation is depicted in Figure 4.11, [78] where different colors corresponding to different frequencies and color degradation indicates the decreasing of the estimated mean signal-to-noise ratio. We assume to have a background traffic constituted by UMTS voice users and additional MBMS users receiving infomobility services. The objective is to evaluate if the network can support the new load introduced by infomobility information and its impact on the performance perceived by the classical UMTS users.

Typical ITS traffic flows are assumed to be made up of small packets often directed to many users in different moments [80]. Common UMTS networks are, instead, designed and optimized to serve relatively large traffic flows directed to a single user by assigning dedicated channels. Here, we assume to send infomobility information to many users adopting the MBMS multicast channel. Note that MBMS uses part of the available power at the base station, thus limiting the number of dedicated channels that can be established for other applications. Obviously, the broadcast or multicast nature of the channel does not allow fast power control, a feature that is of main importance for an interference-limited system like UMTS; the base station preassigns a certain amount of power to MBMS services depending on the coverage planning and the desired bit rate (according to [81]).

To estimate the amount of resources that the proposed service can require, we make some assumptions. First of all we note that more than 8 bytes are needed in order to uniquely determine a street segment; thus a 16 byte payload should be supposed in order to include velocity measurements and other parameters. Second, we count more than 7000 street segments in the simulated area; although they are not all subject to measurements, the main regional streets and national highways should be also considered, as stated earlier; thus information on 10,000 streets could be updated. A refresh time of about 10 seconds is considered. A 129.2 Kbps one is assumed among possible bearers of MBMS.

In order to simulate this scenario we adopt SHINE as simulation platform, thus we take into account both the territory map with road traffic and the UMTS network architecture from PHY up to application layers. As mentioned in Section 4.3, SHINE simulation platform has been realized to simulate heterogeneous wireless networks, according to a client–server structure. In this case, only the UMTS-FDD LLS is considered.

4.4.2.1 Scenario

For the scenario considered, we reproduced the entire road network of Bologna, made up of 7201 roads with real street distribution. We considered 43 UMTS sites made up of 119 cells and we performed a frequency planning adopting three different frequencies. Figure 4.11 shows the UMTS base stations (filled black squares) and the different cells of the frequency planning; road map can also be observed as given from the simulation output.

4.4.2.2 Mobility Model

Users belong to two classes: pedestrian or vehicular. Pedestrians move on the entire scenario without constrains on roads and perform normal voice traffic, while vehicular users can move only on road and receive infomobility information. Each vehicle moves at a speed modeled as a Gaussian random variable truncated between a minimum and a maximum speed value and characterized by its mean value and its standard deviation (the standard deviation quantifies the rapidity of the velocity variations). In Table 4.3 values related to the main mobility parameters are indicated.

4.4.2.3 Traffic

In average, 15 Erlangs of voice traffic and 15 Erlangs of infomobility traffic per cell are considered. Please note that the traffic is not uniformly distributed over the territory, being the center of the city more crowded than

Table 4.3   Users' classes and characteristics for urban traffic

Class	Pedestrian	Vehicular
Mean session duration [sec]	90	90
Minimum speed [km/h]	0	0
Maximum speed [km/h]	1.5	30
Speed standard deviation [km/h]	0.25	0.25
Direction/speed changing [sec]	30	10
Maximum direction variation [degree]	180	150

suburbs. The infomobility traffic is constituted by a continuous file download. Both voice calls and infomobility sessions start following Poissonian statistics with an exponentially distributed duration having mean equal to 90 seconds.

4.4.2.4   Bearers

The channels and bearers adopted to follow 3GPP specifications: A 12.2 Kbps bearer is considered for the voice traffic while a 129.2 Kbps bearer is considered for the infomobility traffic. Convolutional coding with 1/3 rate and soft decision is adopted both for voice and MBMS, with quaternary PSK modulation.

4.4.2.5   Radio Interface

- The power transmitted at the base stations is within the range of 4–13 dBW (a lower power is set for the smaller cells) and 120 degree antennas are adopted with maximum gain of 15 dB, while 0 dB gain is assumed at the mobile terminals.
- A three-fingers rake receiver is assumed.
- Hard handover is considered.
- For the channel model, we assume a power attenuation due to propagation according to the Walfish-Ikegami model (i.e. with parameters settings in [82] the path-loss model results given by $PL(d) = 15.3 + 37.6 \cdot \log_{10}(d)$) log-normal, shadowing attenuation (in dB it is random variables with zero mean and variance equal to 5), and fast fading following the International Telecommunication Union (ITU) pedestrian A channel (pedestrian voice users) or ITU vehicular A channel (infomobility vehicular users) [83]. Note that the channel behavior is reproduced in our simulation platform taking carefully into account time and frequency correlation; this is of main importance to effectively simulate the real dynamic and its effects on the performance of each single user.

- When calculating the intracell interference power, the following orthogonality factors are assumed: 90% in downlink (due to the adoption of OVSF codes[9]) and 70% in uplink (due to the adoption of pseudorandom codes and multiuser detection at the base stations).
- A CAC algorithm is performed. A new radio link is successfully setup provided that the necessary orthogonal variable speeding factor (OVSF) codes are available, the initial power required is available at both the base station (downlink) and the terminal (uplink), and the estimated interference are less than a given threshold.
- Calls under critical conditions are dropped, following a "leaky bucket" algorithm acting on a transmission time interval (TTI) scale.

4.4.2.6 Quality Requirements

To evaluate the network performance in terms of served users, we define some reference values related to the connection quality. In particular, for voice traffic: (i) one user (i.e., a voice call) is in outage when the mean bit error rate (BER) of that TTI is greater than 0.02 (uplink and downlink are evaluated independently of each other); (ii) an ended voice call is considered in outage when either in downlink or in uplink, the TTI outage intervals exceed a threshold of 5% and (iii) if the quality of the signal is not sufficient either in downlink or in uplink the voice call is forced to end, following a "leaky bucket" algorithm acting on a TTI time scale.

For what concern multicast traffic, the application level flow is fragmented into data link level blocks. For each block, the block error rate (BLER), that is the probability to loose a block, is evaluated taking into account the channel conditions. Then, a random value determines if that block is correctly received or lost (no retransmissions are possible for multicast service). The following rules apply: (i) an ended infomobility session is considered in outage when it is not correctly received by at least 95% of blocks; and (ii) if during the call, the quality of the signal is not sufficient, the infomobility connection is forced to end, following a "leaky bucket" algorithm acting on a TTI time scale.

4.4.2.7 Performance Indexes

Having defined these reference values, we then simulated voice traffic together with multicast transmissions in the entire city of Bologna depicted in Figure 4.11. To avoid border effects deriving from the edges of the scenario,

---

[9] Orthogonality cannot be 100% due to multipath.

where adjacent cells are not present, we focus the attention and analyze our results only inside the rectangular area plotted at the center of Figure 4.11. The performance indexes we will analyze are defined in the following:

■ *Call setup success rate (CSSR)* : It regards the *service access* and describes the ratio between the number of calls successfully connected to the system and the overall number of call attempts:

$$\text{CSSR} \triangleq \frac{N_{\text{attempt}} - N_{\text{block}}}{N_{\text{attempt}}} \tag{4.4}$$

where N_{attempt} is the number of users attempting to access the network and N_{block} is the number of blocked calls (i.e., the number of users that try to access but are not admitted to the network, hence cannot perform their session).

■ *Drop call rate (DCR)*: It regards the *service retainability* and is defined as

$$\text{DCR} \triangleq \frac{N_{\text{drop}}}{N_{\text{release}}} \tag{4.5}$$

where N_{drop} is the number of dropped calls (i.e., abnormal releases for poor radio conditions or congestions) and N_{release} is the number of users admitted to the system among the N_{attempt} trying to access (i.e., both the normal and the abnormal call releases). Hence, DCR describes the termination of services against the willingness of the user.

■ *Outage rate (OutR)*: It is only for speech calls and regards the *service integrity*. It describes the ratio between the number of calls, which, although normally released, perceived an unacceptable outage time when exceeding 5% of the duration of the call, and the overall number of calls released (that is both the normal and the abnormal released):

$$\text{OutR} \triangleq \frac{N_{\text{outage}}}{N_{\text{release}}} \tag{4.6}$$

where N_{outage} is the number of calls with unacceptable outage time.

4.4.2.8 Numerical Results

To obtain results related to the above described performance indexes, simulations are performed varying the transmitted power used to broadcast infomobility information via the MBMS channel; this means that a constant

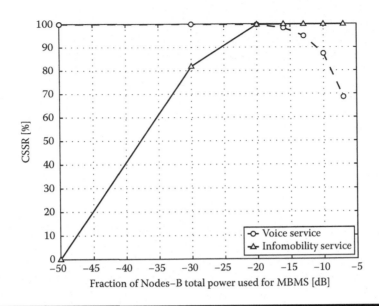

Figure 4.12   Call setup success rate [%] versus fraction of Nodes-B total power used for MBMS [dB]. Voice and infomobility services compared.

fraction of the maximum available power at the base station is reserved for this use. This parameter is given in dB; for example, if −20 dB is considered for the multicast or broadcast channel in a cell where the maximum power is 10 dBW, it means that the voice is transmitted with −10 dBW.

Figure 4.12 shows the CSSR that is depicted both for voice and infomobility services. It can be noted that increasing the power used for MBMS, the CSSR increases for the infomobility users while it decreases for voice users. Please note that a fraction of −50 dB dedicated to the infomobility service corresponds to no setup succeeded for that class of service; due to the low interference introduced to other channels and the absence of MBMS users, this case corresponds to the UMTS network in the absence of MBMS service. These curves would suggest a −20 dB as an optimal case, since neither voice nor MBMS users are rejected by the CAC algorithm; however, as we will see in the rest of the section, the impact on the quality of service perceived by the final users should be deeply investigated.

Figure 4.13 shows the DCR. As expected, the DCR of voice increases when a higher power generates a heavier interference, while it decreases for MBMS users. Please note that a block event would be significantly better than a drop event, since in the latter case the user would not be satisfied for a service he paid for. Also consider that a different CAC could reduce the DCR at the expense of a lower CSSR, because less admitted users means

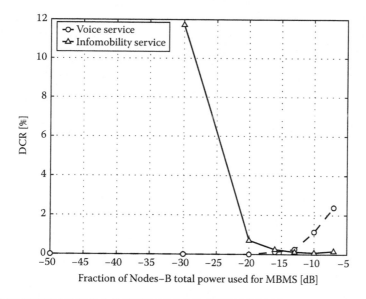

Figure 4.13 **Drop call rate [%] versus fraction of Nodes-B total power used for MBMS [dB]. Voice and infomobility services compared.**

lower interference. If a maximum DCR of 2% is considered as acceptable by the operator, a fraction ranging between −20 dB and −10 dB seems to be acceptable.

Finally, in Figure 4.14 the outage rate, out R, is given. This performance index allows us to understand the QoS that is perceived by the final user. In particular, a heavy degradation in the quality of the voice calls is determined by the presence of the infomobility service; in fact, a voice call transmitted with a fractional power of −20 dB (i.e., the minimum acceptable level, as deducted by Figures 4.12 and 4.13), causes a degradation of the voice quality, making the outage rate to exceed 10%. Still, the outage rate for the infomobility service is quite high, 30%.

Looking at these results in their completeness, we can state that the adoption of an MBMS channel of relatively high capacity (129.2 Kbps) heavily impacts on the performance of the network. Please remember that here soft handover for voice calls and soft combining for infomobility sessions have not been considered (it is expected to increase the QoS). We would also like to remark that results are obtained in a realistic scenario and, although not including optimized algorithms and planning, highlight both the feasibility and the impact that the introduction of such infomobility service would have on real cellular networks.

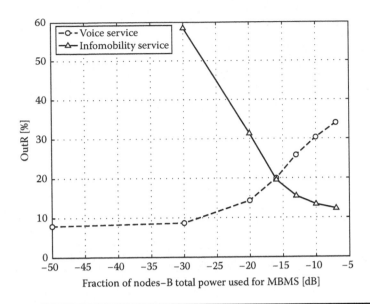

Figure 4.14 Outage rate [%] versus fraction of Nodes-B total power used for MBMS [dB]. Voice and infomobility services compared.

References

[1] O. Andrisano, M. Chiani, M. Missiroli, C. Moss, and V. Tralli, "Short range 60 GHz communication systems for vehicle-to-vehicle applications," *App. to Proc. of 5th PROMETHEUS Workshop*, Munchen (D) and *10th PROCOM European Meeting*, Aachen (D), Tecnoprint, October 1991.

[2] O. Andrisano, M. Chiani, M. Frullone, C. Moss, and V. Tralli, "Propagation effects and countermeasure analysis in vehicle-to-vehicle communication at millimeter waves." *Vehicular Technology Conference*, CO, Denver, May 1992.

[3] O. Andrisano, M. Chiani, V. Tralli, and R. Verdone, "Impact of cochannel interference on vehicle-to-vehicle communication at millimeter waves," *Proc. of ICCS/ISITA '92*, Singapore, November 16–20, 1992.

[4] O. Andrisano, M. Chiani, M. Frullone, C. Moss, and V. Tralli, "Millimetre wave short range communications for advanced transport telematics," *European Transactions on Telecommunications, Luglio-Agosto 1993*. DRIVE, CEE.

[5] O. Andrisano, D. Dardari, and R. Verdone, "Code division and time division multiple access networks for vehicle-to-vehicle communications at 60 GHz," *IEEE Vehicular Technology Conference (VTC'94)*, Stockholm, Sweden, June 7–10, 1994.

[6] O. Andrisano, P. Daniele, M. Frullone, and R. Verdone, "Roadside to vehicle communications at 60 GHz," Workshop on Short Range Radio Communications, *24th European Microwave Conference*, Cannes, France, September 9, 1994.

[7] O. Andrisano, D. Dardari, and R. Verdone, "Short range communication systems for vehicle-to-roadside in urban environment," *Rova'95 International*, Bolton Institute, Bolton, U.K., September 11–13, 1995.

[8] http://www.ertico.com/.

[9] http://www.cvisproject.org/.

[10] M. K. Simon and M.-S. Alouini, *Digital Communication over Fading Channels*, 2nd ed. New York: John Wiley & Sons, Inc., 2004.

[11] M. Z. Win and J. H. Winters, "Virtual branch analysis of symbol error probability for hybridselection/maximal-ratio combining in Rayleigh fading," *IEEE Transactions on, Communications*, November 2001.

[12] A. Conti, M. Z. Win and M. Chiani, "On the inverse symbol-error probability for diversity reception," *IEEE Transactions on Communications*, 51, (5): 753–756 (May 2003).

[13] W. M. Gifford, M. Z. Win, and M. Chiani, "Diversity with practical channel estimation," *IEEE Transactions on, Wireless Communications*, July 2005.

[14] M. Chiani, M. Z. Win, and A. Zanella, "On the capacity of spatially correlated MIMO channels," *IEEE Trans. on Inf. Theory* 49(10) (October 2003):2363–2371.

[15] A. Conti, V. Tralli, and M. Chiani, "Pragmatic space-time codes for cooperative relaying in block fading channels," *EURASIP Journal on Advances in Signaling, Processessing*, vol. 2008 (2008), Article ID 872151, 11 pages doi: 10.1115/2008/872151. vol. 2008:1–11.

[16] S. T. Chung and A. J. Goldsmith, "Degrees of freedom in adaptive modulation: A unified view," *IEEE Transactions on, Communications*, September 2001.

[17] A. Conti, M. Z. Win, and M. Chiani, "Slow adaptive M-QAM with diversity in fast fading and shadowing," *IEEE Transactions on, Communications*, 55, no. 5 (May 2007):895–905.

[18] G. French, *Understanding the GPS. An introduction to the Global Positioning System. What It Is and How It Works*, Bethesda, MD: GeoResearch Inc., 1996.

[19] M. Mouly and M.-B. Pautet, "The GSM system for mobile communications," ed. by Cell and Sys. Correspondence, 1992.

[20] A. Mehrotra, "GSM system engeneering," Mobile Communication Series, Edited by Artech House, Inc., 1997.

[21] G. Heine and H. Sagkob, "GPRS: gatway to third generation mobile networks," Edited by Artech House, Inc. 2003.

[22] C. Bettstetter, H. J. Vögel, and J. Eberspächer, "Gsm phase 2+ general packet radio service GPRS: Architecture, protocols, and air interface," *IEEE Communication Survey*, 1999.

[23] 3Gpp, http://www.3gpp.org

[24] H. Holma and A. Toskala, *WCDMA for UMTS—Radio Access for Third Generation Mobile Communications*, 3rd ed. New York: Wiley, 2004.

[25] 3GPP TS 23.246, "Multimedia Broadcast/Multicast Services (MBMS): Architecture and functional description."

[26] Correia, A. M. C. Silva, J. C. M. Souto, N. M. B. Silva, L. A. C. Boal and A. B. Soares, "Multiresolution broadcast/multicast systems for MBMS broadcasting," *IEEE Transactions on communications*, 53, no. 1 (Pt. 2, March 2007):224–234.

[27] IEEE, Std 802.16-2004 IEEE standard for local and metropolitan area networks, Part 16: Air interface for fixed broadband wireless access systems, October 2004.

[28] Yang Xiao, *WiMAX/MobileFi: Advanced Research and Technology*, ed. Boca Raton, FL: Auerbach Publications, Taylor & Francis Group, 2008.

[29] IEEE Std 802.16e-2005 and IEEE Std 802.16-2004/Cor1-2005 IEEE standard for local and metropolitan area networks, "Part 16: Air interface for fixed and mobile broadband wireless access systems amendment 2: Physical and medium access control layers for combined fixed and mobile operation in licensed bands and corrigendum 1," February 2006.

[30] J. Cimini, "Analysis and simulation of a digital mobile channel using orthogonal frequency division multiplexing," *IEEE Tranactions on communications*, COM-33, no. 7, (June 1985):665–675.

[31] R. Van Nee and R. Prasad, *OFDM for Wireless Multimedia Communications*, Boston: Artech House, 2000.

[32] B. J. Choi, T. Keller, L. Hanzo, M. Munster, *OFDM and MC-CDMA for Broadband Multi-user Communication, WLANs and Broadcasting*, New York: John Wiley & Sons, Inc., September 2003.

[33] G. Leonardi, A. Bazzi, G. Pasolini and O. Andrisano, IEEE802.16e Best Effort Performance Investigation, IEEE ICC 2007, Glasgow, Scotland, June 24–27, 2007.

[34] IEEE, IEEE Std. 802.11—1997, "Part 11: Wireless LAN medium access control (MAC) and physical layer (PHY) specifications."

[35] F. Tobagi and L. Kleinrock, "Packet switching in radio channels: Part II–the hidden terminal problem in carrier sense multiple-access and the busy-tone solution," *IEEE Transactions on Communications*, 23, no. 12 (December 1975):1417–1433.

[36] A. Giovanardi, G. Mazzini, "Theoretical throughput evaluation in CSMA based WLAN systems considering hidden terminals and capture effects," IST 2003, July 2003, Eshafan, Iran.

[37] IEEE, IEEE Std 802.11a—1999 "Part 11: Wireless LAN medium access control (MAC) and physical layer (PHY) specifications high-speed physical layer in the 5 GHz band."

[38] IEEE, IEEE Std 802.11b-1999 "Part 11: Wireless LAN medium access control (MAC) and physical layer (PHY) specifications: Higher-speed physical layer extension in the 2.4 GHz band."

[39] IEEE, IEEE Std 802.11b-1999/Cor 1-2001 "Part 11: Wireless LAN medium access control (MAC) and physical layer (PHY) specifications amendment 2: Higher-speed physical layer (PHY) extension in the 2.4 GHz band - Corrigendum 1."

[40] IEEE 802 LAN/MAN Standards Committee http://www.ieee802.org/11/QuickGuide_IEEE_802_WG_and_Activities.htm

[41] IEEE Std 802.11g.-2003 "Part 11: Wireless LAN medium access control (MAC) and physical layer (PHY) specifications amendment 4: Further higher data rate extension in the 2.4-GHz Band."

[42] IEEE 802.11, "Part 11: Wireless medium access control (MAC) and physical layer (PHY) specifications: Medium access control (MAC) enhancements for quality of service (QoS)," IEEE 802.11e/D3.0, May 2002.

[43] IEEE P802.11p/D0.26, Std. 802.11 "Part 11: Wireless LAN medium access control (MAC) and physical layer (PHY) specifications: Amendment 3: Wireless access in vehicular environment (WAVE), 2006."

[44] Alesander Paier et al., Smart Antennas 2008. WSA 2008. International Workshop on 26–27 Feb. 2008, 9–15. Non-WSSUS vehicular channel characterization in highway and urban scenarios at 5.2 GHZ using the local scattering function.

[45] I. F. Akyildiz, W, Su, Y Sankarasubramaniam and E. Cayirci, "A survey on sensor networks," *Communications Magazine*, IEEE, August 2002. Digital Object Identifier: 10.1109/MCOM.2002.1024422.

[46] C.-Y. Chong and S. P. Kumar, "Sensor networks: evolution, opportunities, and challenges," *Proc. of the IEEE*, 91, no. 8 (August 2003).

[47] D. Culler, D. Estrin, and M. Srivastava, "Overview of Sensor Networks," *IEEE Computer*, August 2004.

[48] D. Dardari, A. Conti, C. Buratti, and R. Verdone, "Mathematical evaluation of environmental monitoring estimation error through energy-efficient wireless sensor networks," *IEEE Transactions on Mobile Computing*, 6, no. 7, (July 2007), Digital Object Identifier: 790–802, 10.1109/TMC.2007. 1041.

[49] R. Verdone, D. Dardari, G. Mazzini, and A. Conti, "Wireless sensor and actuator networks: Technologies, analysis, and design," Oxford, UK: Elsevier.

[50] IEEE 802.15 WPAN TG4, http://www.ieee802.org/15/pub/TG4.html.

[51] ZigBee Alliance, https://www.zigbee.org.

[52] M. Z. Win and R. A. Scholtz, "Impulse radio: How it works," *IEEE Commun. Lett.*, 2, no. 2, (February 1998):36–38.

[53] M. Z. Win and R. A. Scholtz, "Ultrawide bandwidth time-hopping spread-spectrum impulse radio for wireless multiple-access communications," *IEEE Trans. Commun.*, 48, no. 4, (April 2000): 679-691.

[54] IEEE 802.15 WPAN Low Rate Alternative PHY TG4a, http://ieee802.org/15/pub/TG4a.html.

[55] D. Dardari, A. Conti, J. Lien, and M. Z. Win, "The effect of cooperation on UWB based localization systems using experimental data," EURASIP *J. Appl. Signal Processing* (Special issue on wireless cooperative networks), 2008:1–11

[56] Bluetooth, http://www.bluetooth.com.

[57] IEEE 802.15 WPAN TG1 http://www.ieee802.org/15/pub/TG1.html.

[58] A. Conti, D. Dardari, G. Pasolini, and O. Andrisano, "Bluetooth and IEEE 802.11b coexistence: Analytical performance evaluation in fading channels," *IEEE Journal on, Selected Areas in Communications*, 21, no. 2, (February 2003):259–269.

[59] B. M. Masini, A. Conti, G. Pasolini, and D. Dardari, "On the benefits of diversity schemes for Bluetooth coverage extension in the presence of IEEE802.11g interference," *Wiley's Journal Wireless Communications and Mobile Computing*, 8, no. 5:585–595, Special Issue: ISWCS'2006, doi: 10.1002/wcm.571, December 2007.

[60] O. Andrisano, M. Dell'Acqua, G. Mazzini, R. Verdone, and A. Zanella, "On the parameters optimization in handover algorithms," *IEEE Vehicular Technology Conference 1998 (VTC'98)*, Ottawa, Canada, May, 18–21, 1998.

[61] A. Bazzi, G. Pasolini, and C. Gambetti, "SHINE: Simulation platform for heterogeneous interworking networks," *IEEE International Conference on Communications*, (June 12, 2006):5534–5539.

[62] L. Zuliani, A. Zanella, A. Marazzi, R. Moretti, E. Agrati, R. Verdone, and O. Andrisano, "A simulation tool for performance assessment of realistic mobile radio networks," *PIMRC* 2004, Barcelona (Spain).

[63] O. Andrisano, D. Dardari, and G. Mazzini, "An integrated approach for the design of wide-band wireless LANs," In *Proc. of IEEE Int. Conf. Telecommunications*, Porto Carras, Greece, June 1998, 121–126.

[64] OPNET Technologies, Inc., http://www.opnet.com.

[65] Information Sciences Institute, http://www.isi.edu/nsnam/ns/.

[66] A. Bazzi, M. Diolaiti, and G. Pasolini, "Measurement based call admission control strategies in infrastructured IEEE802.11 WLANs," *IEEE PIMRC*, 2005.

[67] A. Bazzi, M. Diolaiti, and G. Pasolini, "Link adaptation algorithms over IEEE802.11 WLANs in collision prone channels," *IEEE VTC Spring 2006*, Melbourne, Australia, May 7–11, 2006.

[68] A. Bazzi, N. Dimitriou, and A. Conti, "Adaptive cross-layer techniques for cellular systems and WLANs: Simulative results within NEWCom Proj.C," *IEEE VTC-2007 Spring*, Dublin, Ireland, April 22–25, 2007.

[69] C. Gambetti and A. Zanella, "Trade-off between data throughput and voice user satisfaction in TD-SCDMA networks: The impact of power control," *Wireless Pers. Multim. Comm.* (WPMC 2004), Abano Terme, Italy, September 12–15, 2004.

[70] C. Gambetti, A. Zanella, R. Verdone, and O. Andrisano, "Performance of a TD-SCDMA cellular system in the presence of circuit and packet switched services," *IEEE Vehic. Tech. Conf. (VTC 2004 Spring)*, Milan, Italy, May 17–19, 2004.

[71] O. Andrisano, A. Bazzi, M. Diolaiti, C. Gambetti, and G. Pasolini, "UMTS and WLAN integration: Architectural solution and performance," *IEEE PIMRC 2005*, Berlin, Germany, September 11–14, 2005.

[72] A. Bazzi, M. Diolaiti, C. Gambetti, and G. Pasolini, "WLAN call admission control strategies for voice traffic over integrated 3G/WLAN networks," *IEEE Consumer Communications and Networking Conference 2006* (CCNC 2006), Las Vegas, U.S., 2:1234–1238, January 8–10, 2006.

[73] A. Conti, "MC-CDMA bit error probability and outage minimization through partial combining," *Communications Letters, IEEE* 9, no. 12 (December 2005):1055–1057.

[74] A. Conti, B. Masini, F. Zabini, and O. Andrisano, "On the down-link performance of multicarrier CDMA systems with partial equalization," *IEEE Transactions on Wireless Communications*, 6, no. 1, (January 2007):230–239. Digital Object Identifier: 10.1109/TWC.2007.05112.

[75] B. M. Masini, G. Leonardi, A. Conti, G. Pasolini, A. Bazzi, D. Dardari, and O. Andrisano, "How equalization techniques affect the TCP performance of MC-CDMA systems in correlated fading channels," *EURASIP Journal on Wireless Communications and Networking*, (2008), Article ID 286351.

[76] B. M. Masini, L. Zuliani, and O. Andrisano, "On the effectiveness of a GPRS based intelligent transportation system in a realistic scenario," VTC 2006-Spring. *IEEE 63rd*, 6:2997–3001, 2006.

[77] B. M. Masini, C. Fontana, and R. Verdone, "Provision of an emergency warning service through GPRS: Performance evaluation," *Proc. of the 7th International IEEE Conference on Intelligent Transportation Systems*, (October 3–6, 2004):1098–1102.

[78] A. Bazzi, B. M. Masini, A. Conti, O. Andrisano, "*Infomobility provision through MBMS/UMTS in realistic scenarios*," ITSC 2008, 12–15 October, Bejing, China.

[79] http://www.bo.ieiit.cnr.it/5percento.php.

[80] D. Valerio, F. Ricciato, P. Belanovic, and T. Zemen, "UMTS on the road: Broadcasting intelligent road safety information via MBMS," VTC Spring 2008. *IEEE*, 11–14:3026–3030. May 2008.

[81] 3GPP TR 25.993 v7.6.0.

[82] 3GPP TR 25.942, "Radio frequency (RF) system scenarios."

[83] ITU-R Recommendation M.1225, "Guidelines for Evaluation of Radio Transmission Technologies for IMT-2000," recommendation M.1034-1.

Chapter 5

Routing in Vehicular Networks

*Francisco J. Ros, Victor Cabrera, Juan A. Sanchez,
Juan A. Martinez, and Pedro M. Ruiz*

Contents

Recent advances in wireless communication technologies are enabling new vehicular networking scenarios. In these networks, also known as vehicular ad hoc networks (VANETs), vehicles can communicate wirelessly through multihop paths. That is, vehicles use intermediate vehicles as relays to reach other vehicles or even nodes in the fixed network infrastructure and roadside units. These communications, which are usually known as vehicle-to-vehicle (V2V) and vehicle-to-infrastructure (V2I), respectively, are gaining a lot of momentum within both industry and academic.

The increasing importance of vehicular networks is highly recognized by governments, industries, standardization bodies, funding agencies, and the academic community in general. In particular, governments have allocated spectrum for short-range vehicular communications, and standards such as IEEE 802.11p are being produced. We can also see an important increase in the number of research projects and initiatives being funded in the United States and Europe, where car manufacturers are playing a highly active role. Also a number of conferences and workshops have been created along the lines of VANET communications such as VANET, MoVeNET, V2VCOM, and the like.

VANETs propose a new kind of scenario with a great projection of a future full of possibilities. Their multihop nature enables numerous applications and services that are unique to the vehicular context. They include among others safety applications to avoid accidents, roadside information services (e.g., free parking slots and gas prices), infotainment and so on. These applications are certainly not possible with the communication capabilities in existing cars. Much research is still required to be able to deploy all those services efficiently.

Routing is one of the key research issues in vehicular networks as long as it supports most emerging applications. Recent research showed that existing routing solutions for mobile ad hoc networks (MANETs) are not able to meet the unique requirements of vehicular networks. Thus, a lot of effort has been devoted during the last years to design VANET-specific routing protocols being able to exploit additional information available in VANET nodes (e.g., trajectories of nodes, city maps, traffic densities, constrained mobility, etc.). In this chapter, we perform a comparative study among the most relevant routing categories and explain their advantages and drawbacks. We analyze the main technical challenges for VANET routing protocol designers and describe the operation of existing solutions detailing the main advantages and issues of current proposals.

In this chapter, the reader will be introduced to the main challenges and requirement specifics for the routing protocols in VANETs as a first step to know the special conditions of VANET scenarios. In the following section, we classify the main routing protocols for VANETs depending on the type of information they use. We start with the protocols that only use information about the neighboring cars. Then, we move on to protocols that use traffic information, maps, or even the trajectories of the vehicles to decide how to deliver the packets to their destination. These protocols are described in detail in the last sections of the chapter. Finally, we discuss open issues and research opportunities.

5.1 Challenges and Requirements for Routing Protocols in VANETs

Routing is one of the key research issues in vehicular networks as long as many applications require multihop communications among vehicles. VANET routing protocols are designed to deliver data packets to their intended destinations using available vehicles as relays. Vehicular networks exhibit unique properties such as a possibly large number of vehicles, high dynamics, and fast-changing vehicle densities and partitions. Even if a large population of vehicles are present, constrained mobility, traffic lights, intersections and the liked lead to frequent network partitions and an uneven network density. These properties represent a real challenge for routing protocol designers. On the other hand, some characteristics of VANETs provide excellent support for routing protocols. For instance, mobility constraints, predictable mobility, and access to additional information (e.g., geographic coordinates, city maps, etc.) can be exploited by VANET-specific routing solutions to increase their performance. This means that VANET routing protocol designers are required to explore a distinctive set of technical challenges and design alternatives. We describe the most important ones below.

> **Localized operation.** One of the key requirements for VANET routing is scalability. This means that the performance of the protocol must scale with the number of vehicles in the network. Localized algorithms are those in which a node takes routing decisions based solely on information locally available in the close vicinity of that node. Protocols with localized operation are highly desirable for VANETs because its control overhead can be greatly reduced by not requiring nodes to know the topology of other parts of the network.
>
> **Neighborhood discovery.** Discovering neighbors is a fundamental part of routing protocols. It can be performed as part of the route establishment or using dedicated one-hop control messages called beacons. Most routing protocols assume that nodes send periodic beacon messages informing neighboring nodes about their identifier, position, and other relevant information. However, the selection of a proper beaconing interval becomes really important to find a good trade-off between control overhead and updated

neighborhood information. Some protocols may use adaptive beaconing intervals depending on the mobility of the network. Another recent approach is the reactive discovery of neighbors in a per packet basis as part of the forwarding of data packets. These protocols are usually called beaconless.

Identification of the destination. In traditional MANET routing protocols, routing is based on the identifier of the nodes. However, in VANETs most routing protocols usually route messages to a particular area or position. In these geographic routing solutions the source node needs to know the position of the destination. To do that, some routing protocols rely on broadcasting query messages including the identifier of the destination. When such a message is received by the destination it answers with a response that includes its current position. Some other protocols just assume that the location of the destination can be obtained using any other external protocol. The external protocol is the one that would then broadcast the query. Another alternative is to use a distributed location service in which destinations update their positions periodically. Data sources can query these location servers to obtain the position associated to that particular identifier. If the destination is mobile, it can also include its intended trajectory or even velocity vectors.

Trajectory precomputation. For VANET routing protocols it may be advantageous to precompute a desired trajectory for data packets to follow (e.g., list of streets to follow). The advantage of this approach is that the trajectory does not refer to individual nodes, and it is not affected by mobility. Data packets are then forwarded (as we shall explain in the following section) by nodes that are close to the trajectory at that particular instant. However, it may happen that some parts of the trajectory cannot be followed if there are not enough vehicles in the selected streets. Also, traffic conditions may have changed because the source precomputed the trajectory. Hence, allowing intermediate nodes to reevaluate and recompute a new trajectory when needed may be helpful. A trade-off between the advantages of using a precomputed trajectory and the cost of doing it should be considered.

Data forwarding. Many ad hoc routing protocols create routing tables containing the next hop to reach the destination based on a given metric. In that case, next hop selection is quite simple. However, this can be inefficient in highly dynamic scenarios due to the need to repair routes after link breakages. A more appealing solution for VANET networks could be to route in a per packet basis. This is very common in geographic routing solutions in which the data packet is routed according to the current neighborhood of the forwarding node in the very moment of forwarding the message. There are multiple options for a node to select its next hop toward the destination or the next vehicle in the precomputed trajectory. Traditional geographic routing protocols usually select the neighbor that provides a greater advance. That is, the one that is closer to the destination. Other metrics taking into account wireless link properties and probabilities of correct reception by neighbors would be more effective for real deployments.

Dealing with network partitions. VANET networks are characterized by an uneven vehicle density. Even if traffic is dense, crossings, traffic lights, and the like produce frequent network partitions. This means that some data packets may eventually reach a vehicle that is not able to continue routing the message as planned. Some protocols just neglect this issue by assuming that there is always enough vehicle density. There are better alternatives to deal with those situations. For instance, the node detecting the situation may try to find a different path. Some protocols use void-avoidance ideas from geographic routing over the street map. Finally, another option is to store the message until a new forwarding opportunity (i.e., new neighbor) appears. This is interesting for delay-tolerant information. However, in heavily congested streets it is not very likely that a new neighbor is discovered within a reasonable amount of time. Probably the best option would be an adaptive scheme that varies the operation mode depending upon the network conditions.

Prediction of future events. Some salient features of VANETs such as the constrained mobility and the knowledge of odometry and position information allow vehicles to predict future positions. When that information is exchanged with neighboring vehicles, it also allows routing protocols to take more informed decisions. While using prediction seems like a good approach, it must be carefully considered. Inaccurate information or predictions may be more harmful than its absence.

Use of additional information. Vehicles have access to lots of information about their context. They are expected to be able to use navigation software and even access external information services providing information about traffic rates and so on. Some of this information may provide additional advantages to routing solutions. They help routing protocols to take informed decisions when selecting paths, neighbors, and the like. Being able to take advantage of that information is a winning strategy for VANET routing protocol designers.

In the following section we classify the most relevant VANET routing protocols into categories, describe their operation, and explain how each protocol relates to these design issues.

5.2 Classification of Routing Protocols for VANETs

VANET routing protocols can be classified according to a number of factors. Figure 5.1, shows our classification based on the type of information that they use. Each category corresponds to protocols using that particular type of information in addition to that of previous categories.

- *Basic solutions* are those that are able to work only with control messages from neighbors. Inside this group we can find CAR [1] and GPCR [2] that follow a position-based routing scheme. While the first one during a route discovery process saves the coordinates of the points where the direction of the packets sent has changed, the second one includes special roles for nodes depending on their location, making different routing decisions depending on the role taken.
- *Map-based solutions* include those protocols that use a street map. This map helps the routing protocol in setting the junctions needed to get to the destination. GSR [3] is probably the best example of that kind of operation. The street map is used to establish the whole path that the packets must follow to get to the destination. Then, geographic routing is applied to traverse each street of the path, until the message is eventually delivered to the destination.
- *Based-on trajectory solutions* include those protocols that work with information about the planned trajectory of the vehicle. Therefore, nodes follow a store-carry-forward scheme deciding the next node in charge of the delivery of the packet to the destination. For instance,

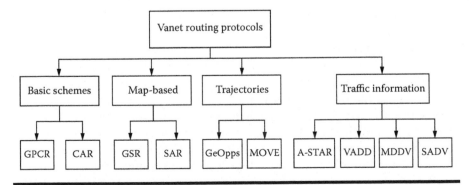

Figure 5.1   Classification of VANET routing protocols.

in GeOpps [10] a node sends a message to a neighbor if its trajectory is closer to the destination than the current one. MoVe [9] is aware of the velocity of the nodes to select the most appropriate next hop toward the destination.

■ *Traffic information solutions* include those protocols that, in addition to being equipped with digital maps, assume that vechicles can obtain information about the traffic conditions. For instance, A-STAR [5] assumes the knowledge of the density of vechicles at each streets. In this way, the protocol can take information into account when computing the best path from a source to a destination (dense streets are preferred in order to avoid partitions). Other protocols like VADD [6], employ this information to decide whether a data packet must be immediately forwarded or stored until a more connected street in reached.

At first glance one would expect that the more information is available the better the performance of the routing protocol. However, even if perfect information is known the way in which the protocol is designed still has a lot of influence in the overall performance of the protocol. Below we describe the operation, advantages, and shortcomings of the protocols belonging to each category.

5.3   Basic Solutions

We consider in this category those protocols working without using information specific to VANET scenarios such as a map of the city, statistics about traffic density on the different roads, number of lanes per road, speed limits, information about trajectory estimations, and the like.

The two protocols analyzed here are CAR [1] and GPCR [2]. The first one can be considered an extension of the proactive routing protocols widely used in MANETs. That is, it establishes paths between source and destinations by means of an initial flooding and then a path maintenance algorithm is applied to keep the path connected in spite of topology changes. The second one, GPCR [2] can be seen as an application of the classic geographic routing algorithms used in wireless sensor networks.

5.3.1   *Connectivity-Aware Routing*

Connectivity-aware routing (CAR) [1] is a position-based routing scheme. The protocol is aimed at solving the problem of determining connected paths between source and destination nodes. VANETs' nodes present a high degree of mobility, and nodes cannot know the position of the rest of the vehicles due to several well-known scalability problems. This lack of information makes it impossible to determine, a priori, which streets have enough vehicles to allow messages to be routed through them.

CAR's algorithm is designed to deal with these problems, and to do that it is divided into three stages: (i) finding the location of the destination as well as a connected path to reach it from the source node, (ii) using that path to relay messages, and (iii) maintaining the connectivity of the path in spite of the changes in the topology due to the mobility of vehicles.

In the first stage, the source node broadcasts a route request message. The idea behind this initial broadcast is the following. The reception of, at least, one of these route request messages at the destination means that, at least one connected path exists. The destination node answers the route request message with a response message including its current location so that the first problem is solved. But the source node also needs to know the path to reach the destination.

CAR's authors propose to include in the header of every route request message the list of junctions (called anchor points) traversed by that message in its way toward the destination. Thus, adding that list to the response message issued by the destination solves the second problem. Besides, nodes periodically transmit short messages including the issuer's identifier, location, and current velocity vector. These short messages (called beacons) keep the neighbor's tables updated. Moreover, the determination of junctions is made by means of comparing the direction of the vehicles. That is, a node determines if it is currently located at a junction when the angle between its velocity vector and the one of some neighbor are not parallel.

Additionally, to select not only a connected path between the source and the destination, but also a short one, the destination does not respond immediately. Instead, it waits a predefined amount of time and then the path selected is the shortest one among those included in the different route request messages received. CAR uses the preferred group broadcast (PGB) [11] protocol to reduce as much as possible the overhead of flooding.

Once the source node has determined a path to reach the destination, data messages are routed geographically from an anchor node to the next one until the destination is reached. To do that, the source node uses a source routing approach. The full list of anchor nodes is included as a header in every data message transmitted. CAR uses the advance greedy forwarding (AGF) [11] algorithm to deliver messages between each pair of anchor nodes. In AGF, relay nodes select as next hop the neighbor located closest to the destination. In this case, it is the vehicle located closest to the next anchor point.

CAR defines the concept of "guards" (see Figure 5.2) to help nodes determine if a message has reached a certain anchor point. A guard is a set of information tied to a geographical area. That area is defined by the location of the anchor point and a radius. Thus, guards contains both the location and the radius. Nodes create guards when they identify a new anchor point. A node creating a guard is the first one including it on its beacon messages.

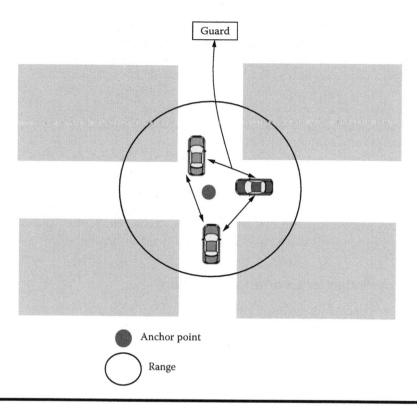

Figure 5.2   Three vehicles interchanging a guard about an anchor point.

Nodes store the guards received during beaconing periods, but only nodes located inside the area defined in the guard retransmit it on their beacons.

A node receiving a message being routed toward the anchor point a can determine that the message has reached a by checking if it has already received a guard for a. Moreover, as we have already said, the mobility of vehicles cause constant topology changes. Thus, the connected path found at the beginning of a data transmission can become disconnected over time. To overcome this issue, CAR's authors propose to use the guards to help maintain the connectivity of the path, or at least to dynamically auto-adjust it on the fly without resorting to a new route discovery process.

Concretely, authors assume that there cannot be disconnection problems between anchor points, so that only the movement of the destination node represents an issue. Therefore, when a destination node changes its direction, then a new guard is generated including also the new velocity vector of the destination node. When a data message arrives to the old destination node's location, the guarding nodes (those interchanging that guard) can retransmit the packet toward the new estimated location of the destination. Of course, this assumption may not hold valid in general, which means that the protocol may fail to maintain the path connected.

Finally, as CAR makes extensive use of beacons, an adaptive beaconing mechanism is proposed to reduce control overhead while keeping neighbor tables as accurate as possible, specially when the number of neighbors or their mobility makes them very unstable. The idea is to adapt the beaconing rate to the nodes density, so that the fewer the number of neighbors, the higher is the beaconing frequency.

Unfortunately, there are some flaws in the protocol. For instance, the use of broadcast as a discovery process makes the protocol unscalable. Besides, message's header includes a whole list of anchor points that can be too much information to store in a single message. Additionally, the idea of guards is only valid assuming that there is at least one node near the junctions at every moment, which cannot be guaranteed in general.

5.3.2 Greedy Perimeter Coordinator Routing

Greedy perimeter coordinator routing (GPCR) [2] does not use map information or densities in the streets, nor does it use the idea of a list of junctions that a packet must pass until it reaches the destination. It aims to avoid overhead. GPCR proposes a position-based algorithm. The main idea is to take into account the fact that the streets and junctions form a natural planar graph and, hence, it is possible to apply geographic routing directly. GPCR consists of two parts:

> **Restricted greedy routing.** Data packets should be routed along streets because they cannot get through buildings. The junctions are the only places where actual routing decisions are taken. Therefore, packets should always be forwarded to a node on a junction rather than being forwarded across a junction. This is illustrated in Figure 5.3, where node b would forward the packet beyond the junction to node c if greedy forwarding is used. But by forwarding the packet to any of the nodes in the corner it finds an alternative path to the destination without getting stuck in a local optimum. A local optimum is produced when a forwarding node does not find a neighbor closer to the destination than itself. A node on a junction usually has more available options to route a message. Nodes that are located close to a junction are called *coordinators*. A *coordinator* broadcasts its role into its beacon packets. Thus, its neighbors will know its role when they have to forward these beacons. As a node must know whether it is a coordinator or not, the authors propose two methods to learn that.
>
> The first one consists of including neighbors' locations and identifiers in beacons, so that each node can have information about

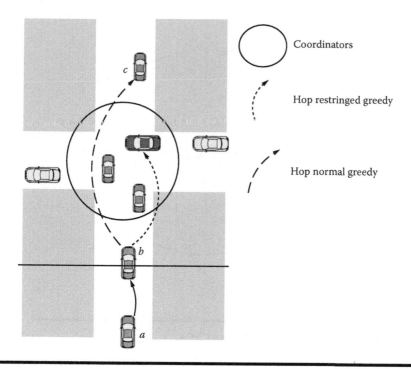

Figure 5.3 Preference coordinator nodes in GPCR.

two-hops distance. Then, a node is considered to be in a junction when it has two neighbors that are within transmission range to each other but do not list each other as neighbors. We can see in Figure 5.4 how v and w are neighbors of node u but they do not list each other as neighbors. This method might have problems. In Figure 5.5 the requirement is correct but node u is not really a coordinator because it is located in a curve.

The second one consists of calculating the correlation coefficient with respect to the position of the neighbors. In this method, it is not necessary to include additional information into beacon messages. Let x_i and y_i be the (x, y) coordinates of a node i. Also let $\bar{x}$ and $\bar{y}$ be the mean of x-coordinates and y-coordinates, respectively. σ_{xy} indicates covariance of x and y. σ_x and σ_y indicates the standard deviation of x and y, respectively. Finally, the correlation coefficient is defined in Equation (5.1):

$$\rho_{xy} = \left| \frac{\sigma_{xy}}{\sigma_x \sigma_y} \right| = \left| \frac{\sum_{i=1}^{n}(x_i - \bar{x})(y_i - \bar{y})}{\sqrt{\left(\sum_{i=1}^{n}(x_i - \bar{x})^2\right)\left(\sum_{i=1}^{n}(y_i - \bar{y})^2\right)}} \right| \tag{5.1}$$

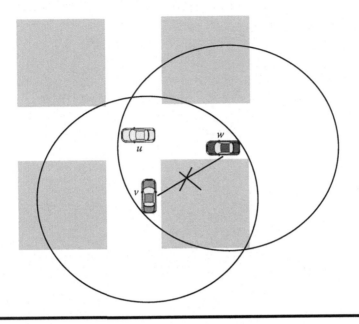

Figure 5.4   Discovery method of coordinators in GPCR.

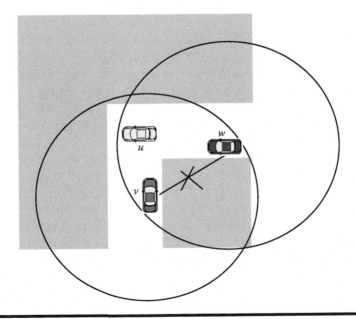

Figure 5.5   Failed discovery coordinator in GPCR.

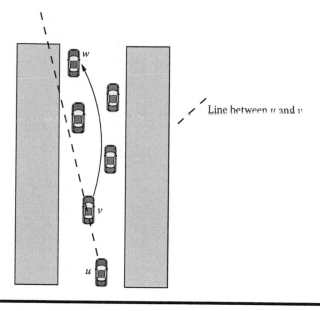

Line between *u* and *v*

Figure 5.6   Restricted Greedy in a street.

with $\rho_{xy} \in [0,1]$. If the value is close to 1 it indicates a linear coherence that is normally found when a vehicle is located in the middle of a street (Figure 5.6). On the other hand, a value close to 0 might indicate a linear coherence and, hence, we can determine that a vehicle is located in a junction. By adjusting a threshold ϵ a node can evaluate the correlation coefficient and assume with $\rho_{xy} \geq \epsilon$ that it is located on a street and then the node is a coordinator. But if $\rho_{xy} < \epsilon$ we can conclude that the node is close to a junction. The authors consider that a good value for ϵ is 0.9. They think that this value provides a good behavior, but it may be arbitrary.

If the forwarding node is located on a street and not on a junction the packet is forwarded along the street toward the next junction. To achieve this, the forwarding node draws a line between the forwarding node's predecessor and the forwarding node itself. The neighbors that approximate the extension of the line will be candidates. The farthest candidate node is selected. We can see in Figure 5.6 that node *w* is selected based on the line between u and v. In the event that some candidate node is a coordinator it will be selected before any noncoordinator node. If there were more coordinators, one of them is randomly selected (Figure 5.3). This prevents a packet from crossing a junction. Once a packet reaches a coordinator a decision has to be made about the street that the packet should follow. This is done in a greedy mode: the

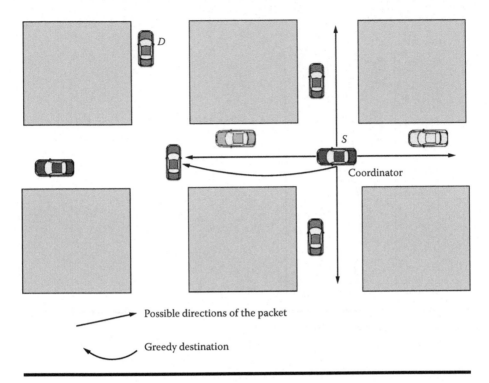

Figure 5.7   Coordinator applying greedy in GPCR.

neighbor with the largest progress toward the destination is chosen. This implies a decision on the street that the packet should follow (Figure 5.7).

Repair strategy. Despite the new greedy routing model, there exists the risk that a packet gets stuck in a local optimum. Hence a repair strategy is required. The vehicle tries to infer the topology of the roads by applying the recovery strategy over the set of neighbors. If the forwarding node is a coordinator and the packet is in repair mode, then the node needs to determine which street the packet should follow next. To this end the right-hand rule [13] is applied (Figure 5.8). Using the right-hand rule it chooses the street that is the next one counterclock-wise from the street the packet has arrived on. But if the forwarding node with a packet in repair mode is not a coordinator, then the node applies restricted greedy routing.

In GPCR there exists a risk that a packet could be forwarded back over the same street from which the packet has arrived. When a packet is being forwarded in repair mode and reaches a coordinator node, it applies perimeter routing. In Figure 5.9, we can see how if the node u applies the

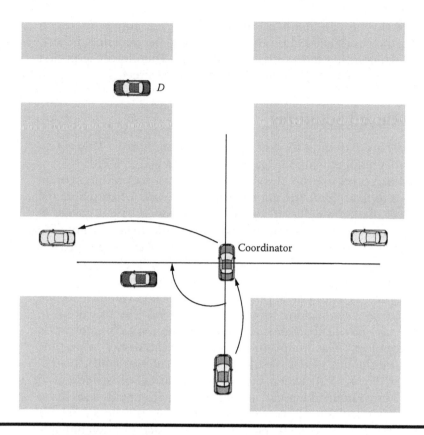

Figure 5.8   Coordinator applying perimeter in GPCR.

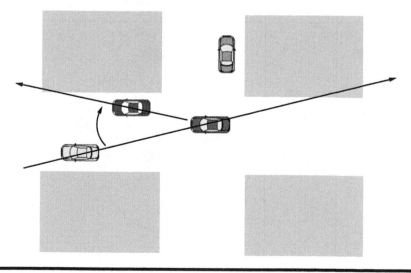

Figure 5.9   Problem of perimeter mode in GPCR.

right-hand rule (in this case left-hand rule) from the line formed between nodes u and v, the coordinator chooses the node w as the next hop, instead of the node x which is located along the next street. Therefore, the packet does not turn the junction, but it remains on the same street.

5.4 Map-Based Solutions

These protocols are based on the knowledge of a street map. This is reasonable for cars equipped with navigation capabilities (e.g., GPS).

Map-based protocols use the idea of spatially aware forwarding. For an explanation we look at the following example. In Figure 5.10a, where the circle represents the radio range, if source node is forwarding a packet in greedy mode it chooses the neighbor closest to the destination, the node B. But if we look at Figure 5.10b, taking into account the street map, is not advisable to choose node B because the packet will take a path that goes away from the destination. Instead, choosing node A turns out to be the right choice. Hence, it is important to have such awareness of the space environment when making forwarding decisions. The mistake is to assume that nodes physically closest to the destination are also closest on the network topology. This is not true in vehicular networks, where the geographic distribution is heavily restricted by road infrastructure.

Returning to Figure 5.10, when the source node chooses node B as the best neighbor without taking into account the environment, it seems to have taken the best decision. However the road map reveals the error. If we opt for vehicle B, the packet will be potentially forwarded through multiple

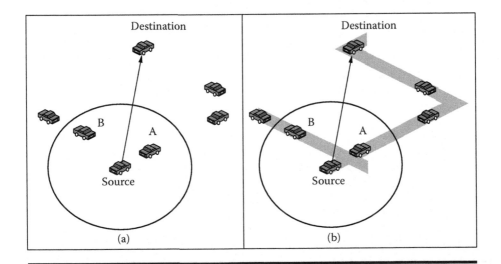

Figure 5.10   Spatial awareness in greedy forwarding.

hops (if there is any neighbor closest to destination). If the only path to the destination is on the road segment to the right, the packet has to be forwarded back and goes through vehicle B. Because a greedy failure will not be memorized in stateless routing, the forwarding of each subsequent packet may fail in the same way and has to be recovered each time. However, with spatial awareness, the source can avoid the forwarding failure in this situation by forwarding packets to the more suitable neighboring vehicle A instead of B. Below we study some examples of map-based routing protocols for VANETs.

5.4.1 Geographic Source Routing

Geographic source routing (GSR) [3] is a position-based routing scheme supported by the map of the city.

Using the location of the destination, the map of the city and the location of the source node, GSR computes a sequence of junctions the packet has to traverse to reach the destination. The protocol aims to calculate the shortest route between origin and destination applying Dijkstra's algorithm over the street map. The calculated path is a list of junctions that the packet should go through (Figure 5.11). From here, it applies greedy forwarding, where the greedy destination is the position of the next junction of the list. That is, a node forwards the packet to one that is the closest to next junction. Once a junction of the path is reached, the greedy destination is changed

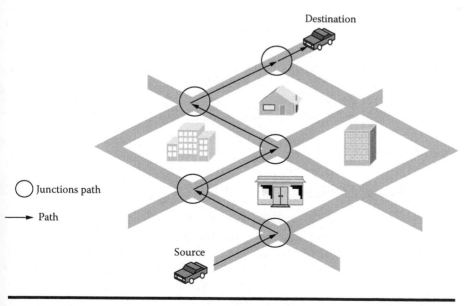

Figure 5.11   List of junctions that determine the path of the packet.

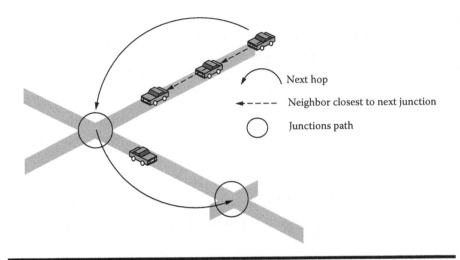

Figure 5.12 Greedy forwarding example.

to the next junction and greedy forwarding is applied again. The protocol works in this way until that packet eventually reaches the destination node (Figure 5.12).

The authors propose two different approaches to take advantage of the list of junctions: (1) put the whole list in the header of the message and (2) recompute it at each forwarding node. The first option reduces the computation workload at nodes at the cost of increasing the bandwidth consumption (messages are bigger). The second option represents the contrary situation, a low bandwidth usage at the cost of higher computation cost.

There are several problems related to GSR. GSR's authors do not mention some implementation issues inherent to the protocol such as how nodes decide if a certain junction has been reached or not. Also, another GSR problem is that if the connectivity between vehicles is low many packets could be dropped. If a forwarding node cannot find any neighbor closer to the next junction, the packet gets stuck in a local optimum, and it drops the packet.

5.4.2 Spatially Aware Routing

Spatially aware routing (SAR) [4] is also a position-based routing scheme supported by the map of the city. SAR employs the idea of using information about the underlying road topology. Nodes decide the whole list of junctions that messages must traverse to reach the destination. These paths (called geographic source route) are computed using Dijkstra's algorithm over the graph of roads and junctions. Then, the path is included in the header of every message sent by the source node.

Intermediate hops forward the message to the neighbor that is located closest to the first junction in the GSR. This first junction is removed from the GSR when the forwarding node finds the junction's coordinates to be located within its radio range. Hence, the message will be forwarded toward the next junction in the path getting successively closer to the destination. So SAR is based on the GSR protocol.

SAR also defines a mechanism to deal with situations where the forwarding node cannot find any neighbor closer to the next junction in the GSR. Three strategies are proposed:

Suspend the packet. This strategy consists of storing the packet in a buffer with limited storage space and periodically trying to forward it. If it is not possible the packet remains in the buffer until it exceeds the limit of time or storage. If the buffer is full and a new packet arrives, the oldest packet in the buffer is discarded.

Switch to greedy forwarding. In this case, the protocol applies standard greedy routing, that is, it forwards the packet toward the destination instead of the next junction.

Recompute path. The last one consists of discarding the path calculated by the source and recalculating the path toward the destination without taking into account the current street.

Nevertheless, the second option is not suitable for vehicular networks (as seen at the beginning of this section) because the street topology heavily restricts the potential locations of vehicles.

5.5 Based on Trajectories

Because vehicles are equipped with multiple sensors such as odometers and speedometers, it is easy to obtain the current velocity vector of a moving node. Furthermore, navigation systems are currently very popular and are able to suggest routes to the drivers. Therefore, not only the current trajectory but also the whole route up to a given destination can be estimated.

With this information, trajectory-based routing protocols forward messages to those neighbors that are expected to get closer (or quicker) to the destination. In this section, we discuss two of those protocols.

5.5.1 MoVe

Motion vector scheme (MoVe)[9] is an opportunistic delay tolerant forwarding protocol that makes use of velocity information. It is targeted at routing packets from a source to a destination by means of a network of

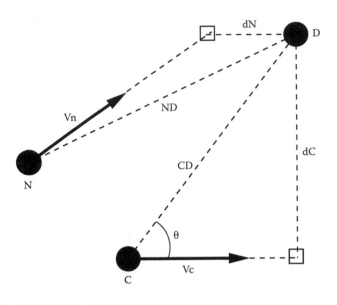

Figure 5.13   Example of the MoVe protocol operation.

mobile vehicles whose movements are not under control. The main problem that arises is the decision process, by which a message is either forwarded or stored by the holding vehicle. Such decision is taken based on the closest distance that the vehicles are predicted to get to the destination, following their current velocity vectors. Therefore, nodes are assumed to know their own position, velocity vector, and the destination's position.

Vehicles that hold packets to be routed, issue periodic *hello* messages. Neighbor vehicles receive these announcements and reply with a *response* message to be taken into account in the forwarding decision. Both *hello* and *response* messages piggyback the sender's predicted closest distance to the destination, according to its current velocity vector. This computation is easy to obtain by applying basic trigonometric operations. Eventually, the packet is forwarded to the neighbor that is predicted to get closest to the destination, or borne by the current vehicle if it follows the most promising trajectory. The process is repeated until the destination is reached or the packet expires.

In the example in Figure 5.13, D is the destination, vehicle C is currently bearing the packet, N is a neighbor, and $\vec{v}_C$ and $\vec{v}_N$ are their velocity vectors, respectively. Let us concentrate on vehicle C, that is interested in computing dC, which is the closest distance following its current trajectory from D. Applying the definition of sine first, and dot product thereafter, dC

is computed as in Equation (5.2). N applies the same equation to obtain dN. Since $dN < dC$, C forwards the packet to N.

$$dC = |\vec{CD}| \sin(\theta) = |\vec{CD}| \sin \left[\arccos \left(\frac{\vec{v}_C \cdot \vec{CD}}{|\vec{v}_C||\vec{CD}|} \right) \right] \qquad (5.2)$$

The main problem of the MoVe approach is that the current velocity vector is likely to be very different in the near future. That is, vehicles' movements are constrained by the streets, topology and other vehicles' movements. Therefore, the forwarding decisions that are taken might be erroneous or really far from optimal.

5.5.2   GeOpps

GeOpps (geographical opportunistic routing for vehicular networks)[10] is a trajectory-based protocol that exploits both the opportunistic nature of vehicular mobility patterns and the geographic information provided by navigation systems. Each vehicle is assumed to know its complete trajectory. The more deviation there is from the suggested route of the navigator system to the actual route followed by the driver, the worse performance is expected for the protocol. It also employs a delay-tolerant approach, hence a vehicle may store data packets until a suitable next hop (maybe the destination itself) for them is found later on.

To choose the next hop, vehicles compute the closest point in their trajectory with respect to the destination of the packet. Figure 5.14 shows an example of nearest points NP of routes P toward a packet's destination D. Given this point and the map of the related area, it is possible to estimate the minimum time required for the packet to reach the destination. Among all the neighbor vehicles and the holding one, the node that can deliver the packet quicker is chosen as the carrier.

To estimate the time required for a vehicle to deliver a packet, it makes use of the following *utility function*. Once the vehicle knows its NP, it uses the map to compute the estimated time of arrival (ETA) at point NP (this is performed by considering the maximum allowed speeds at each street). In addition, it must estimate the time that the packet would take from NP to D. The sum of these two values indicates how much time the packet would take if it were carried by this vehicle. This is called the minimum estimated time of delivery (METD) for the packet and is shown in Equation (5.3).

$$METD = ETA \text{ to } NP + ETA \text{ from } NP \text{ to } D \qquad (5.3)$$

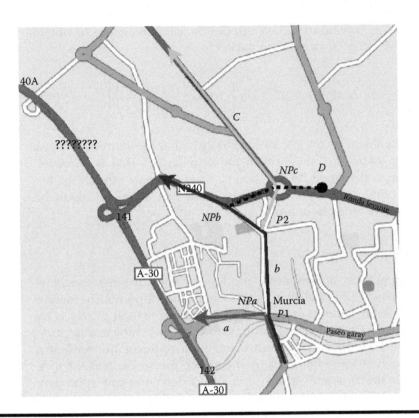

Figure 5.14   Example of calculation of the nearest point in GeOpps.

The GeOpps protocol consists of looking for neighbors with lower METD than the packet holder. To accomplish this, the following algorithm is employed:

1. Vehicles issue periodic broadcast messages containing the destinations of those packets that are buffered.
2. One-hop neighbors that receive those messages compute their METD for each destination and send a reply to the vehicle that holds the packets.
3. The vehicle with the packets keeps on carrying them if it has the lowest METD for the destination(s) under evaluation. Otherwise, the corresponding packets are forwarded to the neighbor with lowest METD.
4. This process is repeated until the packet reaches the destination or the packet expires.

Returning to Figure 5.14, at point P1 the packet holder encounters vehicles a and b. Vehicle b computes the lowest METD, gets the packet and holds it until point P2. There it finds a vehicle c whose trajectory gets closer to D than b's. Therefore, c has a lower METD than b, and the former obtains the packet and holds it until it is forwarded to D.

GeOpps is a suitable protocol for delay-tolerant data, but its performance depends very much on how accurate the trajectory information is. Thus, if the driver does not follow the route suggested by the navigation system, the routing decision taken might be erroneous.

5.6 Based on Traffic Information

These protocols assume that nodes have updated information on the state of the different roads and streets. This information includes traffic density and the like.

5.6.1 *Anchor-based Street- and Traffic-Aware Routing*

Anchor-based street- and traffic-aware routing (A-STAR) [5] follows the approach of anchor-based routing with street awareness. The idea of street awareness is the same as spatial awareness, discussed in Section 5.4.2 for SAR. This is having consciousness of the physical environment around the vehicles, the protocol can take wiser routing decisions. On the other hand, the use of anchor-based routing is not novel either. It consists of including within the packet header the list of junctions (anchors) that the packet must traverse. This approach has been employed in the GSR protocol (see Section 5.4.1). In fact, A-STAR relies on GSR to perform the routing task.

However, one novelty provided by A-STAR is the inclusion of traffic density information to weigh the streets of the scenario. This contribution modifies the behavior when computing the route of junctions that a packet must go through. In this way, every streets' weights are defined as a function of their traffic density, and the Dijkstra's algorithm is employed to compute the shortest route between source and destination. With this improvement, data packets are expected to be routed through those streets with more vehicles and, therefore, higher connectivity among nodes.

To obtain density information, two different mechanisms are proposed. One option is to use statistic information. That is, rely on the traffic history to estimate densities. This information is not exact, but it is relatively easy to obtain. The other alternative consists of using dynamic information, obtaining real-time traffic density status. This approach is very costly because it would require the deployment of sensors by the streets to collect the required information and the development of a query service to provide the vehicles with it.

Probably the best contribution of A-STAR is its recovery strategy when a packet reaches a local optimum. When a node n is located in a street without neighbors which provide progress up to the next junction, it computes

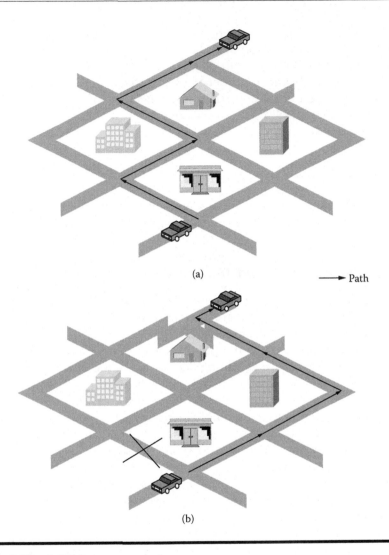

Figure 5.15 A-STAR recovery strategy.

the route from that point to the destination. Thus, it temporarily removes the street where the local optimum happened from the map and applies Dijkstra's algorithm again. Such street is marked as "out of service" and is included within the packet headers generated by node n. Every node that receives one of those packets would mark that same street as "out of service," preventing other packets from being attempted to be routed through it. A given packet can only be salvaged a finite number of times.

As you can see in Figure 5.15a, there is an existing planned route at the left side. However, the current packet holder does not find any neighbors in that trajectory and computes a new one, after removing the street that caused the local optimum, as shown in Figure 5.15b.

However, this strategy presents some drawbacks. In Figure 5.16a, the arrows indicate the original path computed for the packet. Vehicle *S* holds the packet. A local optimum occurs, because *S* has no neighbors to forward the packet to. So, it executes the recovery strategy by computing a new route for the packet. *S* is next to junction *a* (Figure 5.16b), so the new route calculation is performed by starting in that junction. This means that the same route is obtained again because the removed street does not participate in the new shortest path toward destination *D*. Because of this problem, *S* has no neighbors again as the same route is trying to be followed. When the packet has tried to be salvaged a maximum allowed number of recoveries, it is discarded. Nonetheless, Figure 5.16c shows what had happened if *S* would have been closer to junction *b* when entering in recovery mode. Applying Dijkstra's algorithm from junction *b* to destination *D*, the computed route is a true alternative where there are neighbor vehicles to forward the packet to.

5.6.2 *Vehicle-Assisted Data Delivery*

Zhao and Cao [7] proposed several vehicle-assisted data delivery (VADD) protocols. All of them share the idea of storing and forwarding data packets. That is, nodes can decide to keep the message until a more promising neighbor appears on their coverage range, but trying always to forward them as soon as possible. Additionally, decisions about which streets must be followed by the packet are made using vehicle and road information such as current speed, distance to the next junction, and maximum speed allowed. These routing decisions are dynamically taken at junctions because the authors state that precomputed optimal paths used by other protocols might rapidly lose their optimality due to the unpredictable nature of VANETs.

In VADD the main goal is to select the path with the smallest packet delivery delay. The behavior of the protocol depends on the location of the node holding the message. Two cases are considered: when nodes routing the message are located in the middle of a road and when they are located in a junction. The first case (also called routing in straight way) presents less alternatives: forwarding the packet toward the next junction or to the previous one. However, the second case (also called routing in intersections) is much more complicated because at junctions, the routing decision must consider the different roads, so that the number of options is higher.

Both cases use the same approach, determining the next road the message must follow, and then selecting the next relay among the current neighbors. VADD's authors propose a common way of determining the next road while the determination of the next hop remains different. Concretely, the outgoing road with the lowest estimated delay would be selected. In the "straight way" case, there are two possible outgoing roads, the two

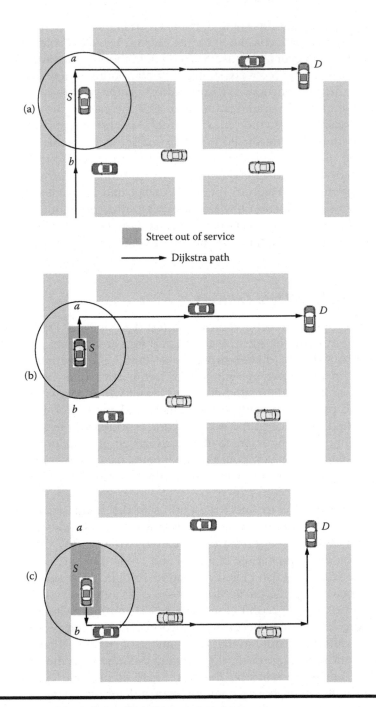

Figure 5.16 Problem of the A-STAR recovery strategy.

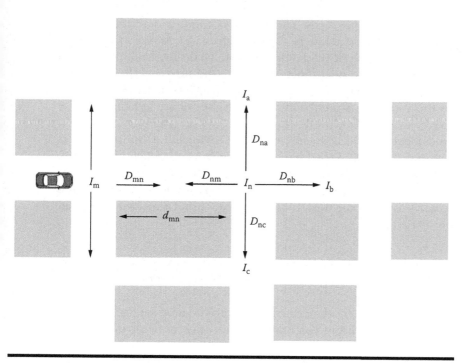

Figure 5.17 Example of VADD model.

segments of the roads in which the node divides the current road. In the "intersection" case, each road starting in that junction represents a different option.

The authors call roads to the street segments delimited by two consecutive junctions. The estimation of the delay of routing a message through a certain road takes into account the road's length, its maximum speed, the mean traffic density, and other traffic specific parameters. But, to estimate the delay to the destination the authors also incorporate the estimated delay of the next possible roads along with the probability of choosing them. In Figure 5.17 we depict the delay model used in VADD. A car located near intersection I_m computes the delay for the road between I_m and I_n (D_{mn}) accounting also the estimated delay of choosing the road between I_n and I_a, the one between I_n and I_b or the one between I_n and I_c.

To estimate message delays for the different roads, the authors propose to solve a $n \times n$ linear equation system using the Gaussian elimination algorithm [$\Theta(n^3)$], n being the number of junctions. To limit the complexity of this computation, a boundary area around the current location is defined, so that only the junctions inside that area are considered in the equation system.

Once the next road has been selected, it is time to determine which neighbor must be the next relay. In the "straight way" the decision is simple, the one located closest to the next junction according to the next road selected. In this case, the next junction can be the next one in the direction of the current vehicle or the one the vehicle has just passed by. In both cases the packet is stored only if no neighbors are available at the moment.

The "intersection" case is more complex. Obviously, if no neighbor is available, or every outgoing road has a longer estimated delay than the the current one, the decision taken consists of storing the message waiting for the next forwarding opportunity. Additionally, the authors propose three alternative ways to select the next forwarder when more than one candidate neighbor is available:

■ Location first: The node located closest to the next selected junction in the most promising road is chosen. This scheme presents routing loops.

■ Direction first: The node located closest to the next selected junction in the most promising road among the ones moving in the right direction.

■ Hybrid: Location first is applied unless a cycle is detected, in that case direction first scheme is used. This scheme seems a little bit unrealistic due to the difficulty of detecting routing loops.

VADD's main drawbacks are its complexity and difficulty of parametrization. The size of the bounding area is by far the most important parameter and, at the same time it is responsible for the complexity of the computation needed at every node. Determining a value for this parameter to achieve a good trade-off between computational complexity and accuracy can be a hard task. Additionally, the authors claim that their hybrid scheme achieves the best performance, however it is not clear how to implement this scheme due to the difficulty of detecting cycles.

5.6.3 Mobility-Centric Data Dissemination Algorithm

Mobility-centric data dissemination algorithm (MDDV) [7] combines the ideas of opportunistic forwarding, trajectory-based forwarding, and geographic forwarding. In MDDV source nodes compute the forwarding trajectory (the list of junctions in the path toward the destination) taking into account the road distance and traffic conditions. However, the authors only take advantage of static network topology information because vehicles cannot have access to real-time traffic information in all scenarios.

Concretely, each road in the map is assigned a weight related to the number of lanes on that road. The authors state that network connectivity is tightly coupled to the density of vehicles, and they assume that the more lanes the higher the vehicles' density. This decision is justified by the fact that, usually, road networks are engineered to match transportation demand.

Once the path is established, the data packet is forwarded along it. Forwarding between junctions is made in a greedy way. Additionally, to increase the delivery ratio, the authors propose an opportunistic approach. Upon receiving a new message, every node moving toward the destination assumes the role of next forwarder for a given time. Moreover, messages include a field indicating its generation time, so that only newer messages must be retransmitted.

Though the ideas of MDDV seems appealing, the major problem of this protocol is its implementation. The authors define five different parameters related to timers nodes have to take into account. All of them should be estimated locally and dynamically by individual vehicles based on their knowledge about the vehicle traffic condition around the location of the node holding the message. From our point of view, computing such estimation seems to be very hard in practice.

5.6.4 Static Node-Assisted Adaptive Routing Protocol

Static node-assisted adaptive routing protocol (SADV) [8] is a multihop routing protocol for VANETs. The authors propose to use static nodes located at junctions to improve data delivery when the node density is low. These gateways store data packets until a vehicle on the best delivery path appears. Data packets are greedily forwarded by mobile nodes traveling between gateways, and each gateway recomputes the best delivery path to reach the destination for every data packet received.

Gateways measure the delay of all their incoming links, that is, the time required by messages coming from static nodes located at adjacent junctions. The computation of the best delivery path is performed by using that link-state information, and it is used to select the next gateway for data packets.

Link-state information is broadcast to all gateways to improve the accuracy of the information used by all static nodes when computing delivery paths. Additionally, when the buffers of a static node are full due to lack of connectivity through the best delivery path, instead of dropping messages, they are sent over the next best delivery path. Moreover, the authors propose to apply a multipath strategy when the packet load is low, that is, forwarding data messages through additional links, but only at intersections.

5.7 Open Issues and Research Opportunities

Throughout this chapter, we have classified and reviewed the most salient routing protocols for VANETs in the literature. Although a lot of work has been already done, this is not but the underpinnings of the emerging vehicular communication technologies that will certainly dominate the market for the oncoming years. Therefore, researchers from both industry and academia have a great opportunity to concentrate their efforts on developing new routing algorithms for this challenging environment.

In fact, current state-of-the-art solutions have settled down the basis for VANET routing, but do not cover many open issues yet. In the following, we show the main research opportunities that we have detected.

Forwarding strategy. Most reviewed protocols rely on traditional geographic forwarding to select the next hop in a route, that is, when routing toward a destination (either the actual packet's destination or a predetermined junction), protocols choose as next hop the vehicle that provides a greater advance onto the desired destination. If we think of real wireless networks, where signal fluctuations, noise, and interferences make the communication so challenging, this approach is not the most suitable because the probability of reception is lower as the distance increases. New forwarding strategies that take into account real physical conditions are mandatory for the success of vehicular communication technologies.

Related to the former, there is another drawback with the criteria employed in maximum progress forwading that shows up with the unique characteristics of vehicular mobility patterns. Even considering an ideal communication range, the farthest known neighbor (which is the one that provides more advance) is likely to be out of range when the forwarding actually occurs (either because the driver is running faster or traffic lights and crossings make them be farther away). This happens due to not fully fresh neighborhood information at each vehicle, which is largely related to the beacon interval and the time that the beacon information is considered useful. The lower interval, the more up-to-date information is acquired. However, this also increases the overhead and generates interferences and more collision likelihood for user transmissions, which are the ones that matter. This trade-off deserves being studied to obtain high-throughput and low-congestion solutions.

Multiple interfaces and technologies. The state-of-the-art VANET multihop routing has generally assumed that all nodes are equipped with one network interface of a same technology. Nonetheless, the on-board units equipped within next-generation vehicles will surely incorporate several interfaces of different technologies. The latter is clear when observing the work developed at some standardization bodies such as the ISO TC 204 Working Group 16 (CALM) [12].

Thus, routing protocol designers must consider this scenario and assume that multiple communication capabilities will be present at the same time. Then the routing protocol must evaluate the different available options and choose the optimum interface at that instant.

Higher-layer traffic awareness. Some of the protocols that have been reviewed throughout this chapter are concerned with delay-tolerant traffic, while others just try to deliver data packets as soon as possible. Actually, there is a large variety of requirements from data traffic depending on the service/application that generated it. Therefore, the routing protocol should act in consequence of the application's expectations and provide different treatments to different requisites. In this way, a safety message should be given the highest priority and be routed through the best possible route, while an infotainment flow could be directed toward a less reliable path if it relieves congestion in the critical one.

Network conditions adaptability. One of the characteristics of vehicular scenarios is their great variability. Traffic density and mobility patterns vary a lot depending on the type of road (urban, city, or highway), region under consideration, timeframe, meteorological conditions, and the like. Routing solutions must be aware that they can be run on many different situations, from high congestion to really sparse connectivity. Then a good protocol can try to reduce congestion in the first case while take the most possible advantage of a communication opportunity in the latter.

Infrastructure exploitation. Another situation that many protocols neglect is the existence of previous infrastructure along the roads. Such infrastructure consists of devices deployed by road operators and private telecommunications companies. Routing protocols could benefit a lot from those devices, which could act as relays, buffers, and so on. Moreover, useful information about the traffic state could be obtained from them, helping algorithms to make more intelligent decisions.

To sum up, in the authors' opinion there is still a long way to go for efficient VANET routing useful to commercial applications. The current gaps to be filled are interesting research lines to follow. Specifically, a unified routing approach based on the current state, but solving the aforementioned issues, is an effort that would be worth the effort.

5.8 Conclusions

In the chapter, the reader has been introduced extensively to specific protocols for vehicular networks.

The beginning of the chapter is devoted to make the reader aware of the main challenges for routing in vehicular networks. We have exposed several considerations that must be taken into account by routing protocol designers, due to the unique characteristics of VANETs.

Afterward, we provided a classification of the different routing protocols that have been specifically proposed for VANETs. Such classification considers the information that is assumed to be known by the protocol to perform its task. Moreover, full descriptions of the main protocols in each of the resulting categories have been presented, outlining their main advantages and drawbacks.

Finally, we have stated those open issues, which, in our opinion, lead to research opportunities targeted at filling the gaps that current proposals do not cover.

We hope that this chapter provides insight on the current state of VANET routing and motivates further research on this subject.

References

[1] V. Naumov and T. R. Gross, "Connectivity-aware routing (CAR) in vehicular ad-hoc networks," In *Proc. of 26th IEEE International Conference on Computer Communications (INFOCOM '07)*, Anchorage, Alaska, May 2007:1919–1927.

[2] C. Lochert, M. Mauve, H. Fusler, and H. Hartenstein. "Geographic routing in city scenarios," *ACM SIGMOBILE Mobile Computing and Communications Review*, 2005:69–72.

[3] C. Lochert, H. Hartenstein, J. Tian, H. Füßler, D. Hermann, and M Mauve, "A routing strategy for vehicular ad hoc networks in city environments," In *Proc. of the IEEE Intelligent Vehicles Symposium 2003*. Columbus, OH, June 2003:156–161.

[4] J. Tian, L. Han, K. Rothermel, and C. Cseh, "Spatially aware packet routing for mobile ad hoc inter-vehicle radio networks," In *Proc. of IEEE Intelligent Transportation System Conference (ITSC '03)*, Shanghai, China, October 2003:1546–1551.

[5] B. C. Seet, G. Liu, B. S. Lee, C. H. Foh, K. J. Wong, and K. K. Lee, "A-STAR: A mobile ad hoc routing strategy for metropolis vehicular communications," In *Proc. of 3rd International Networking Conference IFIP-TC6 (IFIP '04)*, Athens, Greece, Dec 2004. *Lecture Notes in Computer Science* 3042:989–999.

[6] J. Zhao and G. Cao, "VADD: Vehicle-assisted data delivery in vehicular ad hoc networks," In *Proc. of 25th IEEE International Conference on Computer Communications (INFOCOM '06)*, Barcelona, Spain, April 2006.

[7] H. Wu, R. Fujimoto, R. Guensler, and M. Hunter, "MDDV: A mobility-centric data dissemination algorithm for vehicular networks," In *Proc. of First ACM International Workshop on Vehicular Ad Hoc Networks (VANET '04)*, Philadelphia, PA, October 2004:47–56.

[8] Y. Ding, C. Wang, and L. Xiao. "A static-node assisted adaptive routing protocol in vehicular networks," In *Proc. of the Fourth ACM International Workshop on Vehicular Ad Hoc Networks*, 2007:59–68, ACM Press.

[9] J. Lebrun, C.-N. Chuah, and D. Ghosal, "Knowledge-based opportunistic forwarding in vehicular wireless ad hoc networks," In *Proc. of 61st Vehicular Technology Conference* 4:2289–2293, May 2005.

[10] I. Leontiadis and C. Mascolo, "GeOpps: Opportunistic geographical routing for vehicular networks," In *Proc. of the IEEE Workshop on Autonomic and Opportunistic Communications*, Helsinki, Finland, 2007.

[11] V. Naumov, R. Baumann, and T. Gross, "An evaluation of inter-vehicle ad hoc networks based on realistic vehicular traces," In *Proc. of 6th ACM International Symposium on Mobile Ad Hoc Networking and Computing (MOBI-HOC'06)*, Florence, Italy, May 2006:108–119.

[12] ISO TC 204 Working Group 16, "CALM—Communications Access for Land Mobiles," http://www.calm.hu/.

[13] H. Frey and I. Stojmenovic, "On delivery guarantees of face and combined greedy face routing in ad hoc and sensor networks," In *Proc. of the ACM Annual International Conference on Mobile Computing and Networking (Mobicom)*, Los Angeles, CA, 2006.

Chapter 6

Routing in Vehicular Networks: A User's Perspective

Bertrand Ducourthial and Yacine Khaled

Contents

This chapter presents an analysis of the routing problem in vehicular networks centered on the users' (or applications') needs. VANET routing protocol families are first analyzed. The applications' requirements and the traffic conditions are then studied, and a new taxonomy based on usage and context is introduced. Following these considerations, a new approach for routing in VANET is discussed, called *conditional transmissions*. It relies on conditions instead of addresses. Its purpose is to better fit the applications' needs and to better adapt to a variety of road situations. Finally, a rigorous performance comparison of some of the main protocols is presented. The new strategy offers very good performance in many road situations and for different application requirements.

6.1   Introduction

Routing in vehicular ad hoc network (VANET) has attracted much attention during the last few years. VANET is characterized by strong mobility and high dynamics of the nodes, as well as some specific topology patterns. Moreover these networks encounter a significant loss rate and a very short duration of communication. These properties affect the performance of the routing solutions usually applicable to traditional MANET (creating routing tables, discovering routes, detecting stable structures, and so on). Many studies have tried to overcome these difficulties and proposed new protocols.

Routing messages in vehicular networks remain a great challenge. The applications in such networks have specific requirements. In our opinion, such high dynamic networks should be studied with a new perspective, different from other networks, including MANET. An analysis of the requirements should help in the design of efficient adapted protocols.

This chapter presents a user's perspective of routing in VANET. First the routing protocols are briefly presented. Then they are analyzed with an application and road traffic point of view. A new taxonomy based on usage and context is then sketched. Such taxonomy helps in the design and evaluation of efficient solutions according to the targeted applications and traffic conditions.

Based on these considerations, a new approach for routing in VANET is discussed, and the conditional transmissions technique is introduced. Instead of transporting addresses or positions, a message is sent with some conditions used for retransmission or reception. Thanks to the dynamic evaluation of the conditions, the conditional transmissions can efficiently support the high dynamics of the VANET. Moreover, it is versatile and can then be adapted to many applications and contexts.

Finally, some performance comparisons are presented and analyzed. Concluding remarks end the chapter.

6.2 Brief Review of Routing Protocols in Vehicular Networks

In this section, a brief overview of routing protocols in ad hoc networks is presented, with a special focus on the vehicular networks. Routing solutions are presented according to their design, without consideration of the applications' requirements. Then, we summarize previous works relying on network topologies, the geographical positioning, the clusters of nodes, and the nodes' movements. We end by discussing the broadcasting approach.

6.2.1 Topological Routing Protocols

Principle. *Topological* routing algorithms rely on the network topology, composed of nodes and communication links. The *proactive protocols* (e.g., OLSR [15]) build routing tables, even if there is no message to route. On the contrary, the *reactive ones* (e.g., AODV [13]) determine a route to a destination only on request.

Several enhancements have been brought to these protocols to support high network dynamics. In fast optimized link state routing (OLSR) [9], the frequency of beaconing messages is adapted to the network dynamics. Fukuhara et al. proposed to add geographical information in the routing request packets of ad hoc on-demand distant vector (AODV) [23]. Menouar, Lenardi and Filali proposed [48] another improvement of AODV and OLSR protocol for highly dynamic networks; future positions are predicted on the basis of current positions, speed, and direction. Note that other prediction approaches [24, 28] can also be considered as topological routing enhancement.

Analysis. The proactive routing algorithms, such as OLSR, avoid flooding the entire network. However, the control messages consume an important part of the bandwidth. On the other hand, the reactive algorithms such as AODV avoid a delay at the beginning of each communication in order to discover the route. In proactive algorithms, the traffic overhead needed to maintain the routing tables increases with the dynamics of the network, because the neighborhood is always changing. Hence, the more the network dynamics increases, the more the construction overhead increases, in order to try to keep the routing table as accurate as possible. Similarly, for reactive routing algorithms, the flooding needed to discover the routes is less efficient when the network dynamics increases. More control messages will be necessary to reconstruct broken routes while route durations decrease.

6.2.2 Geographical Routing Protocols

Principle. *Geographical routing protocol* mainly rely on positioning data information given by GPS receivers (one per node). The *geocast protocols* (e.g., GAMER, LBM [44]) address all the mobiles belonging to an area. On the contrary, the *position-based protocols* (e.g., LAR, DREAM, GPSR [46]) address a single node.

Geocast approach. LBM protocol [44] avoids to flood the entire network by defining a forwarding area that includes at least the destination region and a path between the sender and the destination region. GAMER protocol [44] adapts dynamically the size of the forwarding area according to the current network environment.

Several works deal with improvements or adaptations of geocast protocols for VANET. In "Stored geocast" [45], after receiving the geocast message inside the destination region, nodes start an election process. The elected node stores the message and delivers it periodically or on request. Legner uses the digital map and the mobility of vehicles to improve the geocast approach in VANET [38]. Harshvardhan I. used [54] a distance-based approach to define relay node and an angle-based algorithm to determine implicit acknowledgment. The VTRADE protocol [58] uses velocity vectors and last positions to sort neighbor vehicles in different groups and select the most appropriate one for the message retransmission. The UMB [35] protocol adapts RTS/CTS mechanism to make a directional broadcasting.

Position-based approach. The position of the destination is known either from a location management service or by flooding in the expected destination area. In case of a full-duplex communication, the receiver can inform the sender of its new position.

The unicast approach can be divided in two main families: *greedy forwarding* (GPSR [30]) and *directional flooding* (LAR and DREAM [6,34]). In the greedy forwarding, a node selects the closest to the destination as the next hop. On the other hand, in the directional flooding approach, the messages are sent to the nodes situated in the same direction as the destination. MURU protocol is based on the path quality prediction [49]. Lochert et al. proposed a GSR protocol that combines geographical routing and digital map information to build an adapted knowledge of the road environment [41]. The same authors also propose GPCR protocol, which appears as an enhancement of GPSR by using a digital map [42]. Some adaptations of positions-based solutions are proposed in several texts [25,49–51]. The authors take into account the position and the direction of movement of vehicles.

Analysis. While many works rely on geographical information, it is important to notice that the more the dynamics increase, the more the positions are unstable. As a consequence, the receiver position will no longer be valid when the message arrives. The position inaccuracy leads generally to performances limitations. To deal with the mobility, the geographical area defining the destination can be increased, but this also increases the number of involved nodes and leads to bandwidth waste.

Nevertheless, some studies have shown that the routing algorithms using location information are useful for routing in VANET compared to topology-based routing protocols [22].

6.2.3   Hierarchical Routing Protocols

Principle. With the *hierarchical routing protocols*, the network is composed of several *clusters*. A cluster is defined as a connected set of nodes sharing some characteristics, supposed to remain stable for a given period. The message propagation is done from cluster to cluster by means of gateway nodes. These nodes are in the transmission range of several cluster heads. Several levels of clusters can be defined (e.g., HSR [26]).

Analysis. The hierarchical approach has been adapted to the topological routing protocols (e.g., CBRP [27], HSR [26]) or to the geographical routing protocol (e.g., ZHLS [29], GeoGRID [44]).

The hierarchical protocols try to optimize the resource usage. However, the overhead needed to build such clusters increases with the network dynamics, while the clusters' compositions are less and less stable. For VANET, the algorithms may take into account road traffic characteristics in order to increase the clusters' lifetime [3,7,40,47,64]. Nevertheless, a routing solution should not rely entirely on clusters, because in some situations, their size is too small, they have a very short lifetime, or they do not exist at all. Instead, clusters should be used (when possible) to enhance routing strategies by limiting the number of retransmitters for instance.

6.2.4   Movement-Based Routing Protocols

Principle. In the *movement-based routing protocols*, the messages move forward to the destination by means of a node's movements. Hence, a node may carry messages until they meet their destinations [3,36]. The message propagation can either rely on epidemic or random schemes [59] or on subgraphs of the entire network. Such a structure is similar to a backbone in fixed networks, but needs to evolve in dynamic networks. The evolutions of the structure can be controlled [39] [14] with support-based routing. Some optimizations can be performed when the context is known [16]. For instance, Zhao and Cao exploited the predictable vehicles' mobility [65].

Analysis. In VANET networks, the vehicles cannot be deviated from the aim of ensuring routing. Hence, the movement-based approach is neither sufficient nor practical in all the situations. As a consequence, a routing protocol should not rely entirely on the movement of the nodes. Instead, the movement-based approach could be used to improve the performances of more general routing strategies.

6.2.5 Broadcasting Approach

Principle. In the broadcasting approach, each node receiving a message retransmits it to the neighboring nodes. This ensures that as many nodes as possible receive the message. Generally, upon receiving a message, each node makes a decision whether it will forward the packet or not, on the basis of some neighborhood-related parameters.

In the probability-based broadcast algorithms [4], the decision relies on some random polling involving the neighborhood. In the location-based broadcast algorithms, the decision relies on the node's positions [4,8]. Whenever a node retransmits a message, it adds its own location in the message. It then computes the additional coverage area it would cover itself by retransmitting the message. The improvements for the location-based protocol have been proposed by using the direction-based broadcast algorithms. The broadcast decision is made according to the trajectory of the nodes or from a digital map [35,58]. The broadcast decision can also rely only on the cluster heads [40]. The nodes can also estimate the message utility to decide which message should be retransmitted first [61,62].

Analysis. All these protocols have been designed for vehicular ad hoc networks. Most of them rely on information related to the neighborhood. Some control messages are required to gather information on the neighborhood. When the dynamics increase, the neighborhood information is less precise. In order to keep accuracy, the more the dynamics increase, the more the number of control messages should increase. This leads to bandwidth consumption.

6.3 A Use-Case Analysis of the Protocols

In this section, we analyze the previously presented routing protocols regarding the application perspective. We then refine the description with traffic density consideration.

6.3.1 VANET Applications and Services

Both academic and industrial teams began to design applications based on vehicular ad hoc networks. Road safety applications are certainly the main motivation and represent a major issue. About 40,000 people are killed on the roads every year in the European Union, with around 1.7 million people incurring several injuries. Generally, these accidents are caused by faulty driver behavior, bad weather conditions, or mechanical problems. VANET could help to anticipate the road incidents and to improve the road visibility [11]. In addition to the road safety applications, driver services are another important issue for VANET. In some situations, they are more efficient than cellular networks for contextual dissemination of information related on traffic forecast, weather forecast, road works, and the like. Also, new services for the passengers have been proposed.

All these applications belong to the so-called *intelligent transportation systems* (ITS) [2]. The ITS are intended to improve the transportation in terms of safety, mobility productivity, impact on the environment, and the like. The underlying technologies encompass a broad range of communications and electronics technologies. They are integrated in the transportation system's infrastructure as well as in vehicles themselves.

ITS applications can be sorted as [17] (i) *infrastructure-oriented applications* for optimizing their management (such as transit management, freeway management, intermodal freight, and emergency organization), (ii) *vehicle-oriented applications* for increasing road safety (such as incident management, crash prevention, collision avoidance, and driver assistance), (iii) *driver-oriented services* for improving the road usage (such as traffic jam and road work information, traveler payment, and ride duration estimate) and (iv) *passenger-oriented applications* for offering new services on board (such as Internet access, distributed games, chats, tourist information, city leisure information, movies, and downloads announcements [37]).

The ITS motivations are multiple. With a better resource management (such as infrastructure, car fleets, and intermodal freight), transport productivity will increase. Regarding road safety, the U.S. Department of Transportation (DoT) launched a large initiative to reduce the number of deaths on the road (around 43,000 per year) [2]. The European Commission (EC) targets to halve the number of road fatalities by 2010 [1], and it launched large ITS projects. Some of the ITS applications are studied by car manufacturers to propose more and more equipped vehicles. A new business related to on-board services may appear in a few years. Finally, a better road management either by the infrastructure or by the drivers will contribute to environmental preservation by avoiding traffic congestion, optimizing the car speed, easing public transportation (intermodality), or organizing car-sharing services for instance. Since consumers are more and more concerned about safety and environmental issues, all these services become marketing arguments for car manufacturers.

6.3.2 Requirements of the Applications Regarding Routing

Taxonomy. By observing the different applications relying on VANET, we can notice that few of them need to establish a unicast communication. The sender could often send messages without knowing the receiver amid highly dynamic nodes and rapidly changing network topology.

Indeed, in many cases the receivers could be defined, such as those who are behind the sender (they will, for instance, encounter a *designated spot*), those who are in front of the sender (they could, for instance, send information related to the road), those who are in a given geographic area (close to a *designated spot* such as smog area), those who can offer a specific service (service discovery), those who have the identity x (a single receiver designated by its identity). These examples can be assimilated to one-to-many communications.

Moreover, some applications rely on local broadcast (in the neighborhood), which are also a kind of one-to-many application: traffic forecast, multiplayer games, and tourist information.

Finally, one-to-one communications are mainly used for car-to-infrastructure applications and very specific car-to-car applications (e.g., car following). Hence, the communication in VANET can then be sorted in three categories: one-to-one, one-to-many, and one-to-all (broadcast) as summarized in Table 6.1.

One-to-one communication. The one-to-one communication protocols rely mainly either on the topology (such as OLSR and AODV [13,15]) or on the geographical position (such as GPSR and DREAM [46]).

In a topology-based routing protocol (such as OLSR and AODV), the messages are routed with the help of the network topology. As explained in

Table 6.1 Communication Requirements of VANET Applications

One-to-One	One-to-Many	One-to-All
Internet connection	Alert messages	Location services
Communication between 2 vehicles	Weather forecast	...
Car following	Traffic information	
...	Cooperative driving	
	Multiplayer games	
	Talks	
	Tourist information	
	...	

the preceding section, the construction and maintenance overhead increase when the network dynamics increase. In a geographical position–based protocol, the forwarding decision is based on the knowledge of neighbors' positions, which are available to neighbors thanks to some location services that generally rely on one-to-all communication. Many solutions have been proposed for the VANET context (such as GPSR [46], DREAM [46], GSR [41], and MORA [25]). The routing algorithms using location information appear more appropriate for routing in VANET comparatively to topological-based routing [22].

Note however that one-to-one communication can be used when the connectivity between two vehicles has a long duration, which is not so frequent in a highly dynamic network. Moreover, few VANET applications require one-to-one communication.

One-to-all communication. One-to-all communication is the simplest way to disseminate information within a VANET. Each vehicle that receives a message retransmits it to the neighboring vehicles. This makes sure that as many vehicles as possible receive the message. This kind of communication is useful for location services (e.g., DREAM and GLS [46]).

In the one-to-all mechanism, each individual vehicle periodically broadcasts information about itself. Every time a vehicle receives a broadcast message, it stores it and immediately forwards it by rebroadcasting the message. This mechanism is clearly not scalable due to the large number of messages flooded over the network, especially in high traffic density scenarios.

We studied diffusions on a convoy of vehicles using the IEEE 802.11 standard [31]. The results showed that the flooding causes important collisions and contentions due to a high amount of redundant broadcasts. Specific strategies are useful to reduce the bandwidth waste. Note however that, by reducing the area in which vehicles are involved, the optimizing one-to-all communication becomes a one-to-many communication.

One-to-many communication. One-to-many communications can be divided in two classes: geocast approach and mobility-based approach.

In the geocast approach [38,44,54], the delivery of a message to nodes is done within a geographical region. Two area are defined: the target area and the forwarding area, which include at least the target area and a path between the sender and the target area. An intermediate node forwards a packet only if it belongs to the forwarding area. This first part is quite similar to the unicast-directed flooding approach; the node that is closer to the destination retransmits the message. When the target area is reached, the second part begins flooding the whole target area. A node broadcasts received packet to all neighbors provided that this packet was not already received before and that the node belongs to the target area.

Table 6.2   Taxonomy of Routing Protocols Based on Applications' Requirements

One-to-One		One-to-Many		One-to-All
Topology	Position	Geocast	Mobility	
AODV [13]	CAR [51]	DRG [54]	Epidemic [59]	DREAM [46]
CBRP [27]	DREAM [46]	GAMER [44]	LBF [53]	GLS [46]
DSR [21]	GPCR [42]	GeoGRID [44]	MDDV [63]	
Fast OLSR [9]	GPSR [46]	IVG [5]	OABS [4]	
HSR [26]	GSR [41]	LBM [44]	ODAM [8]	
OLSR [15]	Epidemic [59]	MGF [38]		
	LAR [46]		RBM [10]	
	LORA-CBF [56]		SB [20]	
	MDDV [63]		SOTIS [60]	
	MGF [38]		VTRADE [58]	
	MORA [25]		UMB [35]	
	MURU [49]		VADD [65]	
	VADD [65]			

In the mobility-based approach [8,20,35,58], the destination message can be defined by using the mobility information, such as the trajectory and the speed. The messages' destination is defined according to the nodes mobility, digital maps, or messages' exchanges.

We can notice that the one-to-many communication scheme is the most frequent one in VANET (Table 6.1). It can also be used for location services. Table 6.2 sorts the proposed protocols according to the kind of communication scheme offered. By comparing this table with Table 6.1, we can see a mismatch between applications' requirements and protocols' studies.

6.3.3   Toward a Use-Case Taxonomy

The traffic density is an important factor in VANET, and it seems important to investigate it in order to obtain an accurate taxonomy. The density of traffic can be divided in two categories: dense and sparse networks.

Dense network.   A dense network means high density of vehicles (e.g., rush hours in urban environment), which may cause contention and collisions. The communication can be optimized by using simple techniques such as hierarchical, probability-based, counter-based, and position-based approaches.

■ In the *hierarchical approach*, a hierarchy is created with cluster or road segmentation [7,47,56]. Some metrics can be used in order to define the clusters such as nodes ID, relative speed, or relative distance. Also, the vehicles can use digital road maps to segment the road.

However, the overhead needed to build such clusters increases with the network dynamics, while the clusters are less stable. In the segmentation approach, all vehicles must be equipped with digital road maps.

■ In the *probability-based approach*, each node receiving a message retransmits it to the neighboring nodes with probability p, and it uses a random delay to avoid collision [4]. Generally, the retransmission probability is computed on the basis of neighborhood knowledge.

However, the accuracy of the neighborhood knowledge decreases when the dynamics increases. The overhead needed to gather accurate data could significantly increase with the network dynamics. Note also that each node has to wait, a delay, before retransmitting the messages, whatever is the situation.

■ In the *counter-based approach*, the network dynamics is taken into account to decide whether or not to forward the message [53]. Nodes have a timer for each nonduplicated message they have received. The delay for each timer is randomly set when a node receives a nonduplicate message and is decremented afterward. The counter increases when the node overhears duplicate messages forwarded by its neighbor nodes. If the counter exceeds a given threshold (max-count) when the timer expires, then the node avoids forwarding the message by discarding the packet because a sufficient number of forwarding has been done by its neighbors.

In this approach each node must wait, a delay, before retransmitting the messages, whatever is the situation.

■ In the *position-based approach*, unnecessary forwarding nodes are eliminated by choosing the closest nodes to the destination. Each node uses a delay timer, which is inversely proportional to the distance between the sender and the receiver [4,8,53].

This approach is used without any neighborhood knowledge, but a delay is required before any retransmission.

Sparse network. In a sparse network, nodes of the vehicular network are often disconnected. This situation may cause a loss of messages when they cannot be retransmitted. In order to avoid loosing messages, two kinds of solutions have been proposed: store and forward (movement-based approach) and widening the retransmission region.

■ In the *store and forward* approach, the nodes use their movement in order to move on the messages toward the destination. A vehicle carries some messages until they meet their destinations or

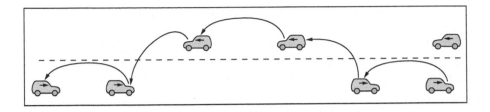

Figure 6.1   Widening of the retransmission region.

until the expiration of the waiting time, delay [3,36]. The message propagation can rely on epidemic schemes [59] or can rely on some subgraphs of the entire network. Optimizations can be performed when the context is known. We used the vehicles trajectory [16]. We exploited the predictable vehicles mobility [63,65] .

■ In the *widening region* approach, messages can be relayed by vehicles that are not concerned by their content [18]. For example, the retransmission of the alert messages in a convoy of vehicles (see Figure 6.1) can involve vehicles in the opposite convoy in case of necessity. The direction of the vehicles and the messages is defined by using the trajectory.

A use-case taxonomy of the routing protocols.   As a summary, we can improve the application-based taxonomy of Table 6.2, by introducing the previously described traffic characteristics. In Table 6.3 the routing protocols are sorted in three families: adapted to sparse networks (e.g., rural area), general (no traffic density specificities), and adapted to dense networks (e.g., rush hours in urban environment).

Such a taxonomy could still be refined by taking into account other parameters. For instance, in a multihop communication, the next relay can be decided either before the transmission (so-called *sender-oriented*) or after the transmission (*receiver-oriented*). When the decision is sender-oriented, the current relay indicates in the message header which neighbor will be the next relay, and the others will ignore the message (e.g., GPSR and GPCR). When the decision is receiver-oriented, each neighbor decides whether they retransmit the message or not (e.g., LAR and LBM).

In the sender-oriented approach, the current relay needs some information on its neighbors, and control messages are required to gather the information. But when the network is highly dynamic, the neighborhood is always changing and the frequency of the control messages should increase, leading to an unstable situation as for the proactive topological routing protocols [32].

Hence, in a highly dynamic network, the retransmission decision should not require information on the neighborhood (or with low accuracy). In the

Table 6.3   Application-Based Taxonomy for Routing Protocols According to Traffic Density in VANET

Traffic kind	Communication Kind				
	One-to-One		One-to-Many		One-to-All
	Topology	Position	Geocast	Mobility	
Adapted to sparse networks		Epidemic [59] MDDV [63] VADD [65]		Epidemic [59] MDDV [63] VADD [65]	Epidemic [59] MDDV [63] VADD [65]
General	AODV [13] DSR [21] Fast OLSR [9] OLSR [15]	DREAM [46] GSR [41] MGF [38] MORA [25] MURU [49]	DRG [54] GAMER [44] IVG [5] LBM [44] MGF [38]	RBM [10] VTRADE [58]	DREAM [46]
Adapted to dense networks	CBRP [27] HSR [26]	CAR [51] LORA-CBF [56] GPCR [42] GPSR [46]	GeoGRID [44]	LBF [53] OABS [4] ODAM [8] SB [20] SOTIS [60] UMB [35]	

receiver-oriented approach, the decision can be taken without any knowledge of the neighbors, avoiding the control messages. The drawback of this approach is that several nodes may decide to retransmit the message, leading to collisions and/or bandwidth wasting. Note, however, that simple techniques can be used to avoid these problems (e.g., lazy retransmissions).

Conclusion.   As we noticed, VANET applications can be sorted in three families depending on their communication needs: one-to-one, one-to-many, and one-to-all. Each of these three families contain several protocols designed for specific needs. Moreover, not all of them are able to support the variation of traffic density.

In order to design a general routing VANET solution, a strategy would be to propose several protocols and to choose the best on-the-fly. Indeed, suppose that a one-to-many routing protocol is chosen because most of the applications rely on this kind of communication. Then other applications relying on other communication scheme (e.g., following cars) may experience bad performances. Thus, to fulfill all application requirements, several protocols could be available. Some embedded software platforms allow this kind of cohabitation between protocols [17]. However, this could lead to more overhead and control messages.

Another approach is to design a versatile routing protocol capable of encompassing the three kinds of communication schemes we pointed out, which would be able to take into account road characteristic (e.g., density). In the rest of this chapter, we present a new strategy for routing in VANET and a new protocol named *conditional transmissions*. This is a particular case of content-based routing where receivers and transmitters are designated by means of conditions in the messages. Depending on the conditions, the number of involved vehicles varies (either for relaying or for receiving the message). By adapting the conditions to the traffic characteristic, it is possible to achieve good performances whatever is the traffic characteristics (such as providing that such characteristics have been detected by spy programs measuring density and uniformity).

6.4   Conditional Transmissions

In this section, we present a new approach for routing in VANET, which is well adapted to the network characteristics and to the applications requirements.

We first describe this approach. Then we introduce the conditional transmissions technique, and we give some examples of useful conditions. Then we discuss performances and implementations.

6.4.1   A New Approach for Routing in VANET

Avoiding the addresses.   Our team is involved in the development of an embedded platform allowing to carry out vehicle-to-vehicle communication tests on the road [12]. Beside the software communication core [17], we developed distributed applications such as alert diffusion, road foreseen (e.g., visibility and obstacles), and distributed entertainments (e.g., talks and games). This bottom-up approach shows that very few VANET applications need to establish a unicast communication. As explained above, most of the applications require, indeed, one-to-many communications, where the receiver is indicated by some kind of conditions (such as vehicles behind the sender for instance).

In fact, in a highly dynamic network, nodes can be designated by means of *identities* or *conditions*, but to our opinion it seems hazardous to denote them by means of *network addresses*. Each identity is unique, which can be given by the MAC address (already included in the 802.11 packet header). It can also be randomly and periodically chosen to preserve the privacy of the drivers (so-called *pseudonymity*). It is used only to distinguish two cars. On the contrary, a unicast address gives two kinds of information: unique identity of the node and some information related to the node's position in the network. However, managing the node's positions in the network seems too costly in a highly dynamic network. Indeed, both local (related to a neighborhood) and global (related to the entire network or landmark), either logical (e.g., hierarchical address) or geographical (e.g., GPS coordinates) positioning are difficult to maintain when the network dynamics increase. For instance, when sending a message toward a given position, it is likely that the receiver has moved and that the information related to its position is out-of-date. Managing network addressing often requires control messages that increase when the dynamics increase.

In order to deal with the highly dynamic network, we consider *conditional addressing* instead of *network addressing*, *path maintaining* instead of traditional unicast, and *conditional transmissions* instead of broadcast. Conditional transmissions are a kind of one-to-many communication, that can also be used for unicast communications [32].

Path maintaining. The path maintaining problem is maintaining a communication that began when the receiver was in the neighborhood of the sender. A communication between two cars is usually initiated in such a situation and not when they are far from each other. Indeed, the identity—and even the presence in the network—of a distant vehicle is not known from the potential sender, as it would be too costly to provide such information.

Sometime after a communication being initiated in the neighborhood, further to unforeseen events on the road, several hops may be needed to achieve it. This means that the path was initially of length 1, and has to be maintained to allow the communication within a reasonable length. Note that the receiver is known by the sender because it was initially in its neighborhood. No address is needed, and the communication can be maintained from neighborhood to neighborhood simply by using the unique identities. Indeed, if the path length increments regularly, then the last relay can deduce the new next hop in the path without a geographical non-network address (given that a unique identifier per vehicle is available).

Conditional transmissions. The conditional transmissions select the relay nodes by including the conditions in the messages. The receivers are also selected by including the conditions in the messages: only the nodes that fulfill the conditions will pass the message to their application layers.

Such a technique encompasses the VANET broadcast algorithms. Any conditions can be used, as well as combinations of them. But here receiver and transmitter conditions can be different, and in many cases no control messages are required in the neighborhood as explained below. The conditional transmission also encompasses the geocast algorithms when the condition defines a geographic area (being in a given area). Note that in some way, it also encompasses the unicast communication when the condition defines a unique receiver (having the identity x) and could be used for maintaining the path.

However, its main interest is the dynamic evaluation of the conditions carried out on message reception. The conditional transmissions can be seen as an intermediate solution between static addressing in the header of messages and mobile agents that may lead to dynamic behaviors. The former solution is not well adapted to dynamic networks. The latter may not be accepted due to security issues.[1] Here, the condition evaluation allows to deal with the dynamics in an efficient way, without compromising any security requirement. For instance, instead of designating the receivers as the cars which are in area x, the conditional transmissions designate those behind the sender and not farther than y meters from it. The durability of such a condition-based addressing is larger than the classical addressing approach, even when it relies on geographical position (such as being in area x).

Another important characteristic of the conditional transmissions is that the retransmission decision is *receiver-oriented*. As explained in the preceding section, in a dynamic network it is advantageous to delay the decision of retransmission, and to limit the use of information related to the neighborhood. Conditional transmissions allow such an optimization and avoid any control messages.

Note that the conditional transmissions principle could be used for optimizing other routing strategies. However, we present in this chapter a stand-alone version of the conditional transmissions. It is described in the next subsections. Performances studies are given in the Section 6.5.

6.4.2 Conditional Transmissions

Principle. A message is sent with two conditions, namely the *upward condition* (CUP) and the *forward condition* (CFW). When a message is received by a node, it is passed to the application layer if CUP is true and it is forwarded to neighbors if CFW is true.

The conditions are provided by the applications. Instead of asking the routing layer to send a message to a given address, the applications ask for sending a message to the nodes that satisfy a given CUP condition, which will be relayed by the nodes that fulfill a given CFW condition. Hence, the conditions are application-dependent and a distributed game would not use the same conditions as a safety alert application, for instance. A short logical language with specific keywords allows the sender to express conditions.

The conditions may require information related to the last relay. In this case, part of the control data in the messages could be modified by the

---

[1] Validating the code of an agent is not easy in a VANET, as the keys management is problematic.

relay nodes themselves. An example of such a *relay-dependent condition* would be "being 100 m behind the last relay." However, while we implemented them in our road test-bed, we do not consider such relay dependent conditions in this chapter, and specifically for the performances study.

As explained in Section 6.4.1, conditions requiring exchange of control messages to be evaluated should be avoided in highly dynamic networks. Indeed, the neighborhood becomes more and more unstable when the dynamic increases, leading to additional control messages and bandwidth consumption.

It is important to note that the conditional transmission has no impact on the security of the routing layer. By instance, for authentication purpose, the identities may be used in replacement of addresses. Both condition- and address-based routing can benefit from already proposed security solutions [55].

Conditions. Interesting conditions concern geographic information (e.g., distance from the sender, geographic position, or area) or time-related information (e.g., delay from the emission, date, or message duration). They can also concern trajectory-related information to determine whether a mobile is on the same trajectory as the sender or not. Of course, the conditions may also be related to the identity of nodes (e.g., sender, receiver, or relay) or to the kind of messages for instance (while this seems less useful in practice). Moreover, any combination of conditions can be used.

The trajectory-related conditions are often useful in VANET applications. For instance, they allow addressing a message only to the vehicles that will encounter a stationary car and not to those that are on the same road but in the opposite direction. In order to have an accurate evaluation of such conditions, we developed a specific trajectory-matching algorithm, well adapted to the road's linear characteristics.

Actually, using just the last GPS position of a vehicle to tag the messages is not efficient, as even vehicles following the same road would never have exactly the same GPS footprints. Furthermore, the relative error in the GPS position would not be necessarily identical to all the vehicles in close vicinity, even if they are covered by the same satellite constellation. So our trajectory-matching algorithm logs the last GPS positions of a vehicle in a circular buffer (ten positions are sufficient in practice). This GPS *positions history* is added to each exchanged message as the position tag of the emitter. When a node receives a message, it computes its geographical relevance by the following procedure: for each GPS position from the receiver GPS position's history, a perpendicular projection is issued to the corresponding position in the emitter GPS position's history (Figure 6.2). The lengths of these projections are, respectively, multiplied by weights, fetched from a predefined table, in order to attribute more weight to the recent positions if needed. Then, we obtain the sum of these products and

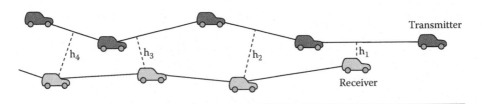

Figure 6.2 Trajectory matching based on successive GPS positions.

normalize it, yielding the *association factor*. Finally, the correlation between the received trajectory and the receiver's own trajectory is evaluated
by comparing the obtained *association factor* to a predefined empirical
threshold (in practice, a threshold of 10 m is sufficient, close to the GPS
mean relative positioning error). This approach is quite similar to the GPS
filtering solutions used in the map-matching phase of a geolocalization
procedure. See also [58] for a similar approach.

Control overhead. Besides the CUP and CFW conditions, other data are
added to the messages in order to evaluate these conditions. For instance,
when one of the conditions is related to the distance from the sender, the
geographical position of the sender is sent with the message. Hence, the
header of a message contains two conditions and some data necessary
for their evaluation. This leads to variable length headers, which are often
larger than those of classical routing algorithms. For instance, the condition used in Section 6.5 relies on distance and on trajectory matching. The
header is about 200 bytes (20 char for CUP and 20 char for CFW, 10 doubles for the latitudes and 10 doubles for the longitudes). For comparison,
the header of an OLSR message is about 16 bytes, and the header of the
geocast algorithms such as GAMER or LBM is about 80 bytes.

However HOP does not require control messages whereas OLSR needs
control messages to build the tables. For instance, "hello" messages contain
a neighborhood description with a minimum of 16 bytes and 4 bytes per
neighbor. Moreover AODV requires control messages to build the routes.
The RREQ messages are broadcasted; they contain 24 bytes.

The performance study presented in Section 6.5 takes into account the
control overhead of each protocol and shows very interesting results for
HOP.

Processing time. The processing time depends on the complexity of
the conditions. In [31], the impact of the *interpacket gap* (IPG) in intervehicle wireless communications is studied: a too short IPG gives a poor
throughput because it leads to many collisions [31]. Hence, the processing
time necessary to evaluate the conditions has generally no impact on the
performances in VANET.

Collisions. Designing the relay nodes by means of receiver-oriented conditions (see Section 6.4.1) implies that many mobiles in the same neighborhood may fulfill the CFW conditions. In this case, they could all resend the message, leading to bandwidth waste. To solve this problem, some already known solutions can be used. For instance, the conditions can take into account local parameters and some random techniques. A condition such as "rand() < 1/n" where n denotes the number of known neighbors could be added to the CFW condition for instance. A simple traffic analysis allows to approximate n. If some clusters are available for the purpose of another application, the conditions may forbid any retransmission in the vicinity of a cluster head.

However, better performances are reached if the transmission decision does not require an accurate knowledge of the neighborhood. For instance each node may wait for a random period, delay before sending the message, except if a neighbor has already resent it (lazy retransmission) [57]. This additional delay may not affect the performances because a reasonable interpacket gap (IPG) is needed.

In order not to distort the performance study presented in Section 6.5, such optimization has not been considered in this chapter.

6.4.3 Implementation

An implementation of a conditional transmissions service (called HOP) has been developed as part of our embedded distributed framework [12]. This framework is used for the purpose of tests on the road with some vehicles. We also implemented HOP under Network Simulator [52] for larger performance studies.

Embedded implementation. The core program of our platform (called *Airplug* [17]) manages all interapplication communications, either local to a vehicle or wireless between vehicles. It runs on Linux and is written in C for portability and efficiency purpose. On each vehicle, all the applications are launched by *Airplug*, and communicate through it, simply by writing to their standard output and by reading on their standard input. These local applications can be written in any language, providing it allows to perform asynchronous reading of the standard input. A distributed application is composed of a set of such local applications, distributed over the vehicles, and communicating via *Airplug* by means of a wireless protocol (IEEE 802.11 in our tests). HOP is implemented as a distributed application and is used by some other applications.

The conditional transmissions prototype HOP has been written in tcl/tk (for compatibility with ns-2, see hereafter). Consider a distributed application A composed of local applications A_i on each vehicles i.

1. If A requires multihop communication, each A_i needs to be registered by their HOP local instance denoted by HOP_i. This is done by means of a special message sent locally from A_i to HOP_i via *Airplug*.
2. To send a message m to some applications A_j on other vehicles, A_i gives to HOP_i (via *Airplug*) the message m, a forward condition CFW, and an upward condition CUP. By this way, the application determines the kind of communications, such as selective broadcast behind the car (alert) or 2-hops communication around the vehicle (e.g., passenger's entertainment).
3. Upon reception of such a request, HOP_i sends to the neighbor vehicles (via *Airplug*, wireless communication) the message m, the CUP, and the CFW conditions and some additional data needed to evaluate the conditions if needed.
4. The wireless message is received by the local HOP instance on neighbor vehicles via *Airplug*.
5. On a neighbor vehicle j, HOP_j evaluates the CUP condition (using the additional data if needed). If CUP is true, the message m is sent to the local application A_j via *Airplug* (local communication).
6. Next the CFW condition is evaluated. If it is true, the message m, the CFW, and the CUP conditions and the additional data are sent to the neighbors via *Airplug*. The additional data may be updated in case of relay-dependent conditions.

Some sequence numbers prevent the same message to be processed several times. We developed some additional programs to forge context-aware conditions that take into account traffic road characteristics and convoys when they exist. However, the description of these applications is out of the scope of this chapter; they are not considered in the rest of this chapter. Noncontext-aware conditions already give very interesting performance, as seen in Section 6.5. Several experiments on roads have been done with success.

On-the-road experiments. Among various experiments, let's detail an example of such experiments. A vehicle is stopped, the others are in a mobile convoy. All the k milliseconds, an application on the leader vehicle of the convoy generates a sequence of p messages. Depending on the inter-vehicles distance (IVD), the third vehicle in the convoy receives the messages either directly from the leader vehicle (one hop) or from the second vehicle in the convoy (two hops). Depending on the upward conditions (e.g., distance related conditions), the second vehicle in the convoy exploits or not the messages while it retransmits them. When the conditions are related to the trajectories, the stopped vehicle does not consider the messages because it does not share the same trajectory as the convoy.

Our implementation of the conditional transmissions also gave very interesting delays in joining of the last vehicle in a mobile convoy of six vehicles. We noticed that the delay was linear with the number of HOPs; the delay per HOP is around 240 ms [33].

Network simulator implementation. In order to study the performances of the conditional transmissions on a large number of vehicles, we also implemented them in ns-2 [52]. This has been done with a network layer structure: HOP is implemented as a routing agent (similarly to DSDV, AODV, and DSR), and can send packets toward the other layers. The sending and receiving code is written in C++, while the condition checking is written in tcl scripts, allowing to share some codes between the embedded platform and ns-2. The ns-2 implementation allows to take into account several scenarios with many cars and to perform comparisons with other routing techniques. This is presented in the following section.

6.5   Performances Studies

We present in this section a comparison study of the conditional transmission performances. Comparisons have been done with five well-known routing protocols in four road traffic scenarios. Before detailing the results, we begin by describing the simulation methodology.

6.5.1   Simulation Methodology

Road traffic scenarios. For this performances study, four road traffic scenarios have been considered. To facilitate the understanding, we focused on typical scenarios relying on convoys. Convoys appear frequently when road traffic is not sparse (they can be detected by piggybacking some information in the messages). Note that many routing optimizations become less useful in sparse traffic: a greedy broadcast algorithm possibly with messages storing may be sufficient (for either unicast or broadcast communication) without important drawback on the bandwidth. We then designed convoy scenarios with and without crossing and with and without taking into account stopped cars.

The four scenarios are illustrated in Figure 6.3. The first one corresponds to a simple convoy with 20 vehicles. In the second scenario, stopped vehicles are added each 300 m on both sides of the road. They represent vehicles stopped at red lights or at stop signs in crossing roads. The third scenario consists of two convoys with opposite directions that cross each other on the road. The relative speed of the second convoy varies. The communications are performed in the first convoy, but are affected by the vehicles of the second convoy. Finally, the last scenario consists of two convoys on two perpendicular roads, that cross each other (thanks to a bridge,

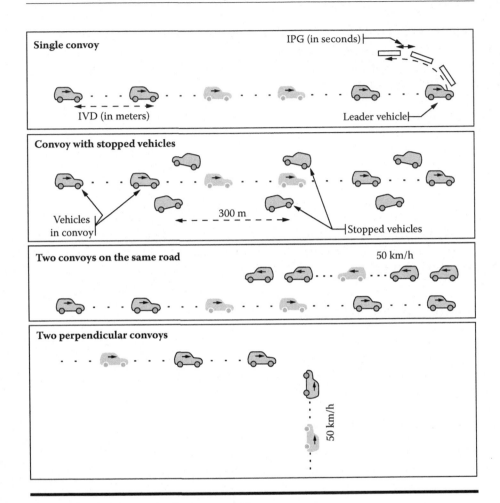

Figure 6.3 Different simulation scenarios.

for instance). The speed of the first convoy varies while the speed of the second is equal to 50 km/h; this low speed allows to maximize the disruption duration of the first convoy. To simplify, only rectilinear roads are considered. However the conditions used by HOP integrates our trajectory-matching algorithm that supports curved roads.

Each convoy is composed of 20 vehicles. The intervehicle distance takes successive values of 27, 50, 61, and 72 m, corresponding to the expected security distance (equivalent to 2 seconds) related to the legal speed limitations 50, 90, 110, and 130 km/h, respectively.

Simulator configuration and network traffic. The simulations have been done with ns version 2.28 [52]. The propagation model is the two-ray ground. This model simulates a direct communication until a given intervehicle distance and a communication with a single reflection on the ground if the distance is larger. Even though it can be improved [19], the ns-2 two-ray ground propagation model gives results close enough to our tests on the road for this comparison study. The communication range is uniform for each vehicle, and equal to 250 m while the interference range is equal to 500 m.

The first vehicle of the convoy regularly sends some packets to the others, with a constant interpacket gap (IPG). The IPG is a convenient way to specify the sending rate in a convoy [31]. It takes the successive values of 5.5, 11, 14, 20, 30, and 50 ms, corresponding to the sending rates of 2048, 1116, 804, 562, 375, and 225 kbit/s respectively. The size of all the packets is equal to 1440 bytes (maximal size of the LLC layer).

All the wireless communications, either broadcast or unicast, are performed with an emission rate of 2 Mbit/s. This prevents the *grey zone* phenomena [43]. Indeed, the IEEE 802.11 broadcast packets are always transmitted at a basic bit rate while the data packets can be sent at higher rates. Hence the broadcasted packets (such as the RREQ messages in AODV) can reach more remote nodes than data packets. As a consequence, a node may create a route or a routing table with some nodes that will be reached by data packets only if they are sent at the basic bit rate. This yields simulation results as close as possible to real world communications.

The transport protocol is UDP, while the routing protocol varies. We selected representative well-known routing protocols. One of the requirements for a protocol to be included in our comparison study was to be available under ns-2. The conditional transmissions technique (HOP) is compared with two proactive protocols (OLSR, Fast OLSR[2]), one reactive protocol (AODV) and two geocasts (LBM, GAMER). Note that with our scenarios, GPSR[3] would generally select the same next hop as OLSR. For all the simulations, the forward condition CFW used by HOP is a simple combination of matching the trajectory (see Section 6.4.2) and the distance: a vehicle retransmits a message only if it is behind the leader vehicle of the convoy and separated by at least 200 m. This is not a relay-dependant condition.

Note that, in the first scenario, the vehicles are not mobile and it can be easily checked that the mobility has no influence on the results of AODV, OLSR, Fast OLSR, and HOP. Moreover, a fixed convoy allows performing comparisons with the geocast protocols GAMER and LBM. These protocols define the receivers as those belonging to a fixed geographical area. In order to make comparisons with the other protocols, the last vehicle of the convoy should receive the messages. Due to the convoy mobility, a

---

[2] This is our own implementation of this specific protocol.
[3] We did not succeed to run GPSR under ns-2.

large area should then be defined (in order that the last vehicle remains inside the area of reception). However, with such a large area, the messages will not reach the end of the convoy because many vehicles will enter in the area and will not retransmit them. Hence, the comparisons with the geocast protocols require a fixed convoy. Such comparisons will only be done in the first scenario because the mobility factor is too high for the other scenarios.

Measures.   In our scenarios, packets sent by the first vehicle of the convoy are relayed by others until the last one. This may be seen as a kind of one-to-many communication (not a broadcast toward all the vehicles). However, we focus on the communication from the first vehicle to the twentieth and we can then compare the results with unicast protocols.

Two kinds of performance criteria have been considered in order to represent requirements of different applications. The first one is the end-to-end delay of the first packet sent by the first vehicle of the convoy and received by the last vehicle. This criterion is important for applications such as alert propagation. The second one is the ratio of the amount of data received by the last vehicle to the amount of data sent by the first one, expressed in percentage and noted *percentage of received data.* This criteria gives information about the end-to-end loss rate which is important for intervehicles applications. Moreover, since the amount of data sent by the first vehicle is the same for all protocols, this gives an indication on the end-to-end throughput as well.

As explained in Section 6.4.2, all the protocols do not have the same control overhead. While some protocols need many control messages (e.g., OLSR), others need few (e.g., AODV), or none (e.g., HOP). On the other hand, some protocols have a large packet header (e.g., HOP) while others have a small one (e.g., OLSR). A large header reduces the payload part of the message, but a large number of control messages impact the bandwidth. Hence, to obtain a fair comparison of the control overhead of different protocols, our approach was to compare both the delay of the first message and the amount of data received during a fixed period of time. Only the payload part of the messages have been taken into account when measuring the amount of sent data (and then the throughput).

Each simulation corresponds to a transmission of 60 seconds, a suffi-cient period to stabilize the results. Twenty simulations have been done for each case, and the displayed results represent the averages. We did not observe any significant dispersion of the results (except for Fast OLSR in a few cases). With 20 simulations per case, the evaluation of the four/six routing protocols on the four scenarios with different interpacket gaps, relative speed, and intervehicle distances leads to 2880 overall simulation runs.

6.5.2   Simulation Results and Performances Analysis

Single convoy. The objective of this first scenario is to compare HOP with five routing protocols (OLSR, Fast OLSR, AODV, LBM, and GAMER) on a single convoy of vehicles.

The percentage of received data is shown in Figure 6.4 for small and large intervehicle distances (IVD). We observe very good performances of the conditional transmissions. The performances decrease for all the vehicles when the sending rates increase as the interpacket gaps decrease and more collisions appear on the convoy [31].

The percentage of received data for AODV is influenced by the intervehicles' distance: performances are better with IVD = 27 m than with IVD = 72 m (Figure 6.4), as a large intervehicle distance implies more hops to reach the end of the convoy.

It seems that OLSR and Fast OLSR are penalized when the vehicle density increases, that is, when the intervehicle distance decreases (short IVD). Actually a high density increases the number of message collisions and then the delay for accessing the channel.

The proactive protocols need to regularly send control messages and seem more affected by the density than others. The performance of OLSR and Fast OLSR increases with the convoy speed because intervehicle distances also increase with the speed.

This is also the case of geocast protocols such as LBM. Since all nodes in the rectangle between the sender and the receiver retransmit the message, there are more collisions when the vehicle density is high. Hence, the performances decrease with the intervehicle distance.

The end-to-end delay of the first packet is shown in Figure 6.5. We also observe very good performances for the conditional transmissions. For instance, with an interpacket gap of 0.014 seconds (800 kbit/s), the delay is equal to 0.046 seconds with HOP and 0.103 seconds with AODV. The end-to-end delay is very important with OLSR (approximately 10 seconds with IVD = 72 m). This is due to the loss of TC messages, which are sent periodically by the nodes in order to define the MPRs. However, the delay for OLSR can be improved if the measures are done after the starting phase. For instance, after 20 seconds, the MPRs and routing tables become stable in this scenario, and we obtained a delay of 0.267 seconds. But this remains large compared to AODV and very large compared to HOP. The same phenomenon has been observed with Fast OLSR. Hence, the end-to-end delay of the first packet is not displayed in the figure for OLSR and Fast OLSR in order to keep a linear scale.

Convoy with stopped vehicles. This scenario allows us to study the impact of stopped vehicles along the road on the communications inside a

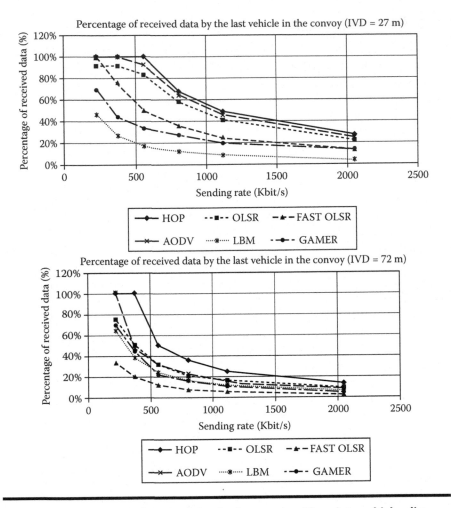

Figure 6.4 Percentage of received data in the convoy with an intervehicles distance (IVD) equal to 27 m and 72 m.

mobile convoy. As explained in Section 6.5.1, the geobroadcast protocols cannot be compared here.

We noticed the same phenomenon related to the end-to-end delay for OLSR and Fast OLSR. The end-to-end delay for AODV and HOP is not affected by the stopped vehicles. Hence, we only present the percentage of data received by the last vehicle in the convoy (Figure 6.6).

The performances of HOP are not affected by the stopped vehicles because the forward condition (CFW) is true only for the vehicles that belong to the convoy.

On the contrary, as the intervehicle distance increases, AODV uses the stopped vehicles to route the messages. Indeed these vehicles become,

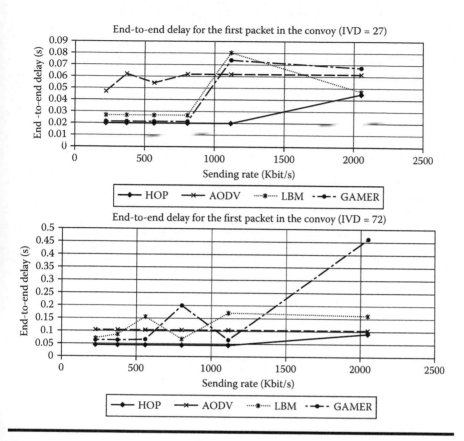

Figure 6.5 **End-to-end delay for the first packet with an intervehicles distance equal to 27 m and 72 m.**

more often, attractive to build a route when the intervehicle distance is large. As a consequence, the percentage of routes breaking is larger when IVD increases. This impacts the percentage of received data but not the delay of the first packet.

Table 6.4 gives the percentage of received data in the first and second scenarios, for an intervehicle distance of 72 m and an interpacket gap of 0.014 seconds. These results confirm that the performances of AODV, OLSR, and Fast OLSR are affected by the stopped vehicles along the road, contrary to HOP.

Crossing convoys. The objective of this scenario is to compare the performances of HOP, AODV, OLSR, and Fast OLSR in a mobile convoy that crosses another one on the same road.

The same comments related to the end-to-end delay still apply in this scenario. Figure 6.7 illustrates the percentage of received data by the last vehicle in the convoy with 27 m and 72 m of intervehicle distance.

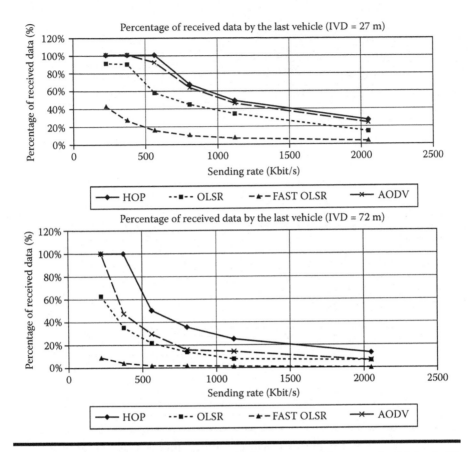

Figure 6.6 Percentage of received data in the convoy with stopped vehicles (with an intervehicles distance equal to 27 m and 72 m).

The performances of OLSR are bad compared to other protocols. This can be explained by the crossing convoy that increases the density and the collisions of control messages.

The performances of AODV is affected by the second convoy. Indeed some (unstable) routes are built with the vehicles of the second convoy, increasing the number of routes breaking.

Table 6.4 Percentage of Received Data in a Single Convoy and in a Convoy with Stopped Vehicles (with an Intervehicle Distance Equal to 72 m)

	OLSR	Fast OLSR	AODV	HOP
Single convoy	21 %	8 %	23 %	36 %
With stopped vehicles	14 %	2 %	15.7 %	36 %

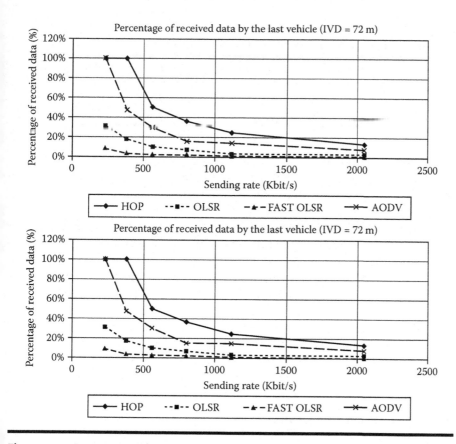

Figure 6.7 **Percentage of received data in the convoy with a crossing convoy and with an intervehicles distance equal to 27 m and 72 m.**

The performances of HOP are not disturbed by the crossing vehicles, because only the vehicles of the first convoy are involved, thanks to the CFW condition.

Table 6.5 gives the percentage of received data of the first and third scenarios, for an intervehicle distance of 72 m and an interpacket gap of 0.014 seconds. These results confirm that the performances of AODV, OLSR, and Fast OLSR are affected by the vehicles in the crossing convoy. This is particularly true for OLSR (21 % compared to 7 %) and Fast OLSR (8 % compared to 2 %). These results show that only HOP is not affected by the crossing convoy.

Perpendicular crossing convoys. The objective of this scenario is to compare the performances of HOP, AODV, OLSR, and Fast OLSR in a principal convoy of vehicles with a perpendicular crossing convoy (see Section 6.5.1).

Table 6.5 Percentage of Received Data in a Single Convoy and in a Convoy with a Crossing Convoy (with an Intervehicle Distance of 72 m)

	OLSR	Fast OLSR	AODV	HOP
Single convoy	21%	8%	23%	36%
With crossing convoy	7%	2%	15.7%	36%

The same comments related to the end-to-end delay still apply in this scenario, and we only present the percentage of data received by the last vehicle in the convoy with an intervehicle distance of 27 m and 72 m (Figure 6.8).

The performances of OLSR in this scenario are better than in the previous one. This may be explained by the lower vehicles' density in case of perpendicular convoys compared to parallel convoys.

HOP is not affected by the crossing convoy thanks to the trajectory-related condition, which is not fulfilled by the vehicles of the second convoy.

Table 6.6 gives the percentage of received data of the first and fourth scenarios, for an intervehicle distance of 72 m and an interpacket gap of 0.014 seconds. These results confirm that the performances of AODV, OLSR, and Fast OLSR are affected by the vehicles in the perpendicular convoy.

6.5.3 Conclusions on the Simulations

In the four scenarios, HOP obtains better results than other protocols. The conditional transmissions allow to perform communications in a convoy without being affected by the road traffic. While the ad hoc network topology is highly dynamic, HOP always offers very acceptable performances for the applications, as opposed to proactive, reactive, and geocast routing algorithms. This can be explained by the fact that HOP does not need any knowledge of the neighborhood nor control messages. The small messages overhead of other protocols (compared to HOP) is counterbalanced by their need of information, collected by means of control messages. The conditions encompasse address- or position-based routing and lead to a more stable routing scheme. While a network address, a position, a routing table, or a route are changing when the topology is dynamic, a road-adapted condition remains stable for a longer time.

By comparison, the other protocols suffer because of the high dynamics. Since the topology is very unstable, the routing tables are always inaccurate, and require many messages to be updated. Our simulations show poor performances for the proactive routing protocols (OLSR and Fast OLSR). More generally, one may say that protocols requiring a knowledge of the neighborhood would suffer because of the high dynamics.

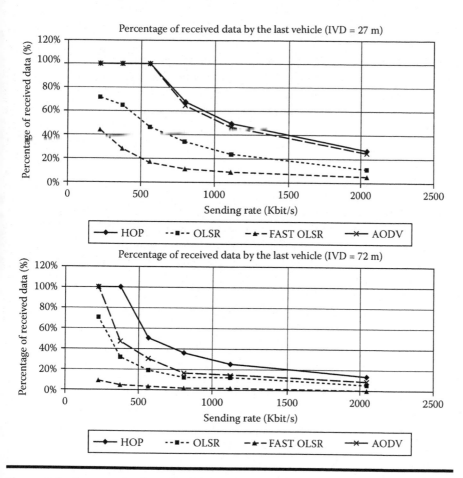

Figure 6.8 Percentage of received data in the convoy with a perpendicular crossing convoy and with an intervehicle distance equal to 27 m and 72 m.

On the other hand, reactive protocols (such as AODV) need to seek a route from the sender to the receiver. But the durability of the routes is short, and many control messages are needed to maintain the communication. The performances of the reactive protocols are better than the proactive

Table 6.6 Percentage of Received Data in a Single Convoy and in a Convoy with a Perpendicular Crossing Convoy (with an Intervehicle Distance Equal to 72 m)

	OLSR	Fast OLSR	AODV	HOP
Single convoy	21%	8%	23%	36%
With perpendicular convoy	12%	2%	15.7%	36%

ones, but suffer from disturbing vehicles (e.g., stopped vehicles and crossing convoy).

Finally, while the geocast routing protocols (such as LBM and GAMER) are adapted to send a message in a specific geographical area, they cannot efficiently be used to send a message to a highly mobile node because the destination area would be too large (and too many vehicles would be involved). As a consequence, they should be used to reach some fixed destinations such as infrastructure relays.

6.6   Conclusions

This chapter has presented an analysis of the routing problem in vehicular networks centered on the users' (or applications') needs and introduced a new routing strategy called *conditional transmissions*, which offer very good performances. It also presented a new taxonomy of the VANET routing protocols as well as a rigorous performance comparison of OLSR, Fast OLSR, AODV, LBM, and GAMER.

Several routing protocols have been proposed or adapted for the VANET. We summarized them in a brief survey organized according to their functionalities. For each family of protocols, the impact of the dynamic has been analyzed. Some of the drawbacks we pointed out were confirmed by the simulations. They showed poor performances for the proactive routing protocols (OLSR and Fast OLSR) because of routing table updates. The performances of the reactive protocols (such as AODV) are better than the proactive ones, but are affected by disturbing vehicles. The geocast routing protocols (such as LBM and GAMER) are not efficient to send a message toward a highly mobile node.

We studied the application requirements in term of communication schemes. We noticed three of them: one-to-one, one-to-all, and one-to-many. We then sorted the routing protocols according to the communication scheme they address. We pointed out the fact that many studies deal with one-to-one communications while most of the applications require one-to-many communications. We also refined our taxonomy with traffic density considerations and obtained 12 cases.

Having the context in mind (applications' requirements, road traffic characteristics) it is important to design new routing algorithms or to efficiently use existing solutions. It is worth noting that a general routing solution would encompass about 12 cases. A strategy would consist of using several protocols in the network. While some embedded software platforms allow such a cohabitation, it may lead to bandwidth wastage by increasing control in the network. Another approach would be to design a versatile routing scheme.

We then proposed a novel approach for routing in VANET, relying on conditional transmissions. Instead of transporting addresses or positions, a message is sent with some conditions used for retransmission or reception. Thanks to the dynamic receiver-oriented evaluation of the conditions, this solution can efficiently support the high dynamics of the networks. The conditions can rely on the time or the message duration, the position or the distance, the speed or the trajectory, and any combination of such conditions. Moreover, the use of conditions allows adapting the routing to the applications' requirements and to the traffic characteristics without increasing the overhead in the network.

We discussed the implementation of such a routing strategy, and we reported road experiments with six vehicles. Beside the prototype dedicated to the road experiments, an implementation for network simulator (called HOP) has been developed in order to scrutinize the performances in convoys of 20 vehicles in four different traffic scenarios and with several parameter values. A rigorous simulation methodology has been developed.

The simulation results show that the conditional transmissions offer performances that are significantly better than those of (i) the proactive algorithms OLSR and Fast OLSR, (ii) the reactive algorithm AODV, and (iii) the geocast algorithms LBM and GAMER. Obviously, the performances of conditional transmissions depends on the conditions used. However, with simple conditions—that did not change for all the simulations while the scenarios and the parameters were varying—we observed that the end-to-end delay of the first packet is very short with HOP and is not affected by the road traffic scenario. Moreover the end-to-end ratio of received data to sent data is also not affected by the dynamics. It logically decreases when the interpacket gaps of the source decreases as with other protocols (see [31]), but it remains very interesting.

References

[1] The eSafety initiative, http://www.esafetysupport.org.
[2] The Intelligent Transportation System, http://www.its.dot.gov/its overview. htm.
[3] G. Allard, P. Jacquet, and B. Mans, "Routing in extremely mobile networks," In *Proc. of 4th Annual Mediterranean Ad Hoc Networking Workshop*, Ile de Porquerolles, France, 2005.
[4] H. Alshaer and E. Horlait, "An optimized adaptive broadcast scheme for inter-vehicle communication," In *Proc. of IEEE Vehicular Technology Conference*, Stockholm, Sweden, 2005.
[5] A. Bachir and A. Benslimane, "A multicast protocol in ad-hoc networks: Inter-vehicles geocast," In *Proc. of IEEE Semiannual Vehicular Technology Conference*, Jeju, Korea, 2003.
[6] I. Basagni, S. Chlamtac, and V. Syrotiuk, "A distance routing effect algorithm for mobility (DREAM)," In *Proc. of Fourth Annual ACM/IEEE International Conference on Mobile Computing and Networking*, Dallas, TX, 1998.

[7] P. Basu, N. Khan, and T. Little, "A mobility based metric for clustering in mobile ad hoc networks." In *Proc. of International Conference on Distributed Computing Systems Workshop*, Mesa, AZ, 2001.

[8] A. Benslimane, "Optimized dissemination of alarm messages in vehicular ad-hoc networks (VANET)," In *Proc. of 7th IEEE International Conference*, Toulouse, France, 2004.

[9] M. Benzaid, P. Minet, and K. Agha, "Integrating fast mobility in the OLSR routing protocol," In *Proc. of IEEE Conference on Mobile and Wireless Communications Networks*, Stockholm, 2002.

[10] L. Briesemeister and G. Hommel, "Role-based multicast in highly mobile but sparsely connected ad hoc networks," In *Proc. of 1st ACM International Symposium on Mobile Ad Hoc Networking & Computing*, Piscataway, NJ, 2000.

[11] Car-to-car communication consortium, http://www.car-to-car.org.

[12] Caremba platform, http://www.hds.utc.fr/caremba.

[13] I. Chakeres and M. Belding-Royer, "AODV routing protocol implementation design," In *Proc. of the International Workshop on Wireless Ad Hoc Networking (WWAN)*, Tokyo, Japan, 2004.

[14] I. Chatzigiannakis, E. Nikoletseas, and P. Spirakis, "An efficient communication strategy for ad-hoc mobile networks," In *Proc. of 15th International Conference on Distributed Computing (DISC)*, London, U.K., 2001.

[15] T. Clausen, P. Jacquet, A. Laouiti, P. Muhlethaler, A. Qayyum, and L. Viennot, "Optimized link state routing protocol," In *Proc. of IEEE International Multitopic Conference INMIC*, Pakistan, 28–30 December, 2001.

[16] J. Davis, A. Fagg, and B. Levine, "Wearable computers as packet transport mechanisms in highly-partitioned ad-hoc networks," In *Proc. of 5th IEEE International Symposium on Wearable Computers (ISWC)*, Washington, DC, U.S., 2001.

[17] B. Ducourthial, "About efficiency in wireless communication frameworks on vehicular networks," In *Proc. of Workshop ACM WIN-ITS with IEEE ACM QShine*, Vancouver, Canada, August, 2007.

[18] B. Ducourthial, Y. Khaled, and M. Shawky, "Conditional transmissions, a strategy for highly dynamic vehicular ad hoc networks," In *Proc. of 8th IEEE International Symposium on a World of Wireless, Mobile and Multimedia Networks*, Helsinki, Finland, 2007.

[19] N. Eude, B. Ducourthial, and M. Shawky, "Enhancing ns-2 simulator for high mobility ad hoc networks in car-to-car communication context," In *Proc. of the 7th IFIP International Conference on Mobile and Wireless Communications Networks (MWCN 2005)*, Marrakech, Morocco, 2005.

[20] E. Fasolo, A. Zanella, and M. Zorzi, "An effective broadcast scheme for alert message propagation in vehicular ad hoc networks," In *Proc. of IEEE International Conference on Communication*, Istanbul, Turkey, 2006.

[21] L. Feeney, *A Taxonomy for Routing Protocols in Mobile Ad Hoc Networks*. Technical Report, Swedish Institute of Computer Science, October 1999.

[22] H. FuBler, M. Mauve, H. Hartenstein, M. Kasemann, and D. Vollmer, "A comparison of routing strategies for vehicular ad hoc networks," In *Poster at ACM MobiCom: ACM MC2R*, Atlanta, U.S., 2002.

[23] T. Fukuhara, T. Warabino, T. Ohseki, K. Saito, K. Sugiyama, T. Nishida, and K. Eguchi, "Broadcast methods for inter-vehicle communications system," In *Proc. of IEEE Wirelesss Communication and Networking Conference*, New Orleans, LA, 2005.

[24] M. Gerharz, C. de Waal, M. Frank, and P. Martini, "Link stability in mobile wireless ad hoc networks," In *Proc. 27th Annual IEEE Conference LCN*, Tampa, FL, 2002.

[25] F. Granelli, G. Boato, and D. Kliazovich, "MORA: A movement-based routing algorithm for vehicle ad hoc networks," In *Proc. of 1st IEEE Workshop on Automotive Networking and Applications 2006*, San Francisco, CA, 2006.

[26] A. Iwata, C. Chiang, G. Pei, M. Gerla, and T. Chen, "Scalable routing strategies for ad hoc wireless networks," *IEEE Journal on Selected Areas in Communications, Special Issue on Ad-Hoc Networks*, 7(8) (August 1999): 1369–1379.

[27] M. Jiang, J. Li, and Y. Tay, *Cluster Based Routing Protocol (CBRP)*. Technical report, IETF, 2001. Internet draft.

[28] S. Jiang, "An enhanced prediction-based link availability estimation for manets," *IEEE Transactions on Communications*, 52(2) (2004):183–186.

[29] M. Joa-Ng and I.-T. Lu, "A peer-to-peer zone-based two-level link state routing for mobile ad hoc net-works," *IEEE Journal on Selected Areas in Communications* 17(8) (1999):1415–1425.

[30] B. Karp and H. Kung, "GPSR: Gready perimeter stateless routing for wireless networks," In *Proc. of MobiCom'00*, Boston, MA, 2000:43–54.

[31] Y. Khaled, B. Ducourthial, and M. Shawky, "IEEE 802.11 performances for inter-vehicle communication networks," In *Proc. of the 61st IEEE Semiannual Vehicular Technology Conference*, VTC, Stockholm, Sweden, Spring 2005.

[32] Y. Khaled, B. Ducourthial, and M. Shawky, "A usage oriented taxonomy of routing protocols in vanet," In *Proc. of 1st UBIROADS workshop with IEEE GIIS*, Marrakech, Morroco, July 2007.

[33] S. Khalfallah, M. Jerbi, M.-O. Cherif, S.-M. Senouci, and B. Ducourthial, "Experimentations des communications inter-vehicules," In *Proc. of CFIP 2008*, Les Arcs, France, March 2008.

[34] Y. Ko and H. Kung, "Location-aided routing (LAR) in mobile ad hoc networks," In *Proc. of Fourth Annual ACM/IEEE International Conference on Mobile Computing and Networking*, Dallas, TX, 1998:66–75.

[35] G. Korkmaz, E. Ekici, F. Ozguner, and U. Ozguner, "Urban multi-hop broadcast protocol for inter-vehicle communication systems," In *Proc. of the 1st ACM International Workshop on Vehicular Ad Hoc Networks*, Philadelphia, PA, 2004.

[36] T. Kosch, *Technical Concept and Prerequisites of Car-to-Car Communication*. Technical report, BMW Group Research and Technology, 2002.

[37] K. Lee, S.-H. Lee, R. Cheung, U. Lee, and M. Gerla, "First experience with cartorrent in a real vehicular ad hoc network testbed," In *Proc. of ACM VANET MOVE'07*, Anchorage, AK, May 2007.

[38] M. Legner, "Map-based geographic forwarding in vehicular networks." Master's thesis, Stuttgart University, 2002.

[39] Q. Li and D. Rus, "Sending messages to mobile users in disconnected ad-hoc wireless networks," In *Proc. of 6th Annual International Conference on Mobile Computing and Networking (MOBICOM)*, 2000.

[40] T. Little and A. Agarwal, "An information propagation scheme for VANETs," In *Proc. of 8th International IEEE Conference on Intelligent Transportation Systems*, Vienna, Austria, 2005.

[41] C. Lochert, H. Hartenstein, J. Tian, H. Fler, D. Herrmann, and M. Mauve, "A routing strategy for vehicular ad hoc networks in city environments," In *Proc. of IEEE Intelligent Vehicles Symposium*, Columbus, OH, 2003.

[42] C. Lochert, M. Mauve, H. FuBler, and H. Hartenstein, "Geographic routing in city scenarios," *In Proc. of ACM MOBICOM*, Philadelphia, PA, 2004.

[43] H. Lundgren, E. Nordstrom, and C. Tschudin, "The gray zone problem in IEEE 802.11b based ad hoc networks," *ACM SIGMOBILE Mobile Computing and Communications Review* 6 (July 2002):104–105.

[44] C. Maihfer, "A survey of geocast routing protocols," *IEEE Communications Surveys and Tutorials* 6 (2nd quarter), 2004.

[45] C. Maihoer, W. Franz, and R. Eberhardt, "Stored geocast," In *Proc. of 13th Fachtagung Kommunikation in Verteilten Systemen (KiVS)*, InformatikAktuell, Leipzig, Germany, 2003.

[46] M. Mauve, J. Widmer, and H. Hartenstein, "A survey on position-based routing in mobile ad hoc networks," *IEEE Network Magazine*, November/December 2001.

[47] B. McDonald and T. Znati, "Design and performance of a distributed dynamic clustering algorithm for ad-hoc networks," In *Proc. of Annual Simulation Symposium*, Seattle, WA, 2001.

[48] H. Menouar, M. Lenardi, and F. Filali, "A movement prediction-based routing protocol for vehicle-to-vehicle communications," In *Proc. of 1st International Vehicle-to-Vehicle Communications Workshop*, colocated with MobiQuitous, San Diego, CA, July 2005.

[49] Z. Mo, H. Zhu, K. Makki, and N. Pissinou, "MURU: A multi-hop routing protocol for urban vehicular ad hoc networks," In *Proc. of 3rd Annual International Conference on Mobile and Ubiquitous Systems: Networks and Services*, California, 2006.

[50] V. Namboodiri and L. Gao, "Prediction based routing for vehicular ad hoc networks," *IEEE Transactions on Vehicular Technology* 56(4) (November 2007):2332–2345.

[51] V. Naumov and T. Gross, "Connectivity-aware routing (car) in vehicular ad hoc networks," In *Proc. of IEEE INFOCOM*, Anchorage, Alaska, U.S., 2007.

[52] Network simulator-2. http://www.isi.edu/NSNAM/NS/.

[53] S. Oh, J. Kang, and M. Gruteser, "Location-based flooding techniques for vehicular emergency messaging," In *Proc. of 2nd International Workshop on Vehicle-to-Vehicle Communications*, San Jose, CA, 2006.

[54] Harshvardhan P.J., "Distributed robust geocast: A multicast protocol for inter-vehicle communication," Master's thesis, Dept. of Electrical and Computer Engineering, NCSU, 2006.

[55] M. Raya, P. Papadimitratos, and J. Hubaux, "Securing vehicular communications," *IEEE Wireless Communications Magazine*, Special Issue on Inter-Vehicular Communications 13. October 2006.

[56] R. Santos and A. Edwards, "Performance evaluation of routing protocols in vehicular ad hoc networks," *International Journal of Ad Hoc and Ubiquitous Computing* 1(2005).

[57] I. Stojmenovic, M. Seddigh, and J. Zunic. "Dominating sets and neighbor elimination-based broadcasting algorithms in wireless networks," *IEEE Transactions on Parallel and Distributed Systems* 13(1) (January 2002):14–25.

[58] M. Sun, W. Feng, T. Lai, K. Yamada, H. Okada, and K. Fujimura, "GPS-based message broadcasting for inter-vehicle communication," In *Proc. of International Conference on Parallel Processing*, Washington, DC, 2000.

[59] A. Vahdat and D. Becker, *Epidemic Routing for Partially Connected Ad Hoc Networks*. Technical Report CS-200006, Computer Science Dept, Duke University, 2000.

[60] L. Wischhof, A. Ebner, and H. Rohling, "Information dissemination in self-organizing intervehicle networks," *IEEE Transactions on Intelligent Transportation Systems* 6(1) (2005).

[61] L. Wischhof and H. Rohling, "Congestion control in vehicular ad hoc networks," In *Proc. of IEEE International Conference on Vehicular Electronics and Safety*, Xi'an, Shaanxi, China, 2005.

[62] L. Wischhof and H. Rohling, "On utility-fair broadcast in vehicular ad hoc networks," In *Proc. of 2nd International Workshop on Intelligent Transportation*, Hamburg, Germany, 2005.

[63] H. Wu, R. Fujimoto, R. Guensler, and M. Hunter, "MDDV: A mobility-centric data dissemination algorithm for vehicular networks." In *Proc. of the 1st ACM Vehicular Ad Hoc Networks*, Philadelphia, PA, 2004.

[64] J. Yu and P. Chong, "3HBAC (3-hop between adjacent clusterheads): A novel non-overlapping clustering algorithm for mobile ad hoc networks," In *Proc. of the IEEE Pacific Rim Conference on Communications, Computers and Signal Processing*, Victoria, B.C., Canada, 2003.

[65] J. Zhao and G. Cao. "VADD: Vehicle-assisted data delivery in vehicular ad hoc networks," In *Proc. of 25th Conference on Computer Communications (INFOCOM)*, Barcelona, Spain, 2006.

Chapter 7

Data Dissemination in Vehicular Networks

Markus Strassberger, Christoph Schroth,
and Robert Lasowski

Contents

Exchanging information among vehicles is one of the key technological enablers to support foresighted driving, thus making traffic, in total, safer, more comfortable, and more efficient. In order to make the vision of a widespread intervehicle network come true, where arbitrary data can be transfered among nodes and reliably disseminated within certain geographical regions with low latency, a number of technological challenges have to be solved. This chapter gives an overview on those challenges and presents various approaches with respect to data dissemination in vehicular networks. In particular, an integrated approach is presented that explicitly aims to maximize the overall network utility in all traffic situations and thereby seamlessly scale between high and low traffic densities.

7.1 Introduction

Making cars talk to one another offers new opportunities to the world of cars. Intervehicle communication bears great potential to improve both road traffic safety and comfort. Possible use-cases reach from safety-related warning systems to amended navigation mechanisms and information- and entertainment applications.

7.1.1 Scope and Motivation

In particular, active safety systems will greatly benefit from car-2-car communication. Being aware of a hazard situation, a car may, for example, notify other cars in advance. Cars having access to remote information from other cars in the vicinity are able to foresee a hazardous situation and warn the driver accordingly in time. The main idea is to share the individual knowledge about a car's driving condition with the vehicles in the proximity, or, to be more precise, with those vehicles that benefit from this information.

In order to accomplish this task, the following three steps are necessary:

- ◼ Detection of a local hazardous situation (i.e., at the current position of the car) without interaction of the driver, utilizing only widely deployed on-board sensor systems.
- ◼ Exchanging corresponding information with other vehicles nearby.
- ◼ Predicting remote hazardous situations along the route by means of that received information.

This chapter focuses on the second step, namely, the controlled dissemination of information within the vehicular ad hoc network. However, as we will see, in this particular domain of automobile active safety, it is sensible not to fully decouple the process of information dissemination from the overall application domain and its dedicated requirements.

7.1.2 Terminology

In contrast to routing, which usually refers to the unicast transport of data packets from a source node to a predefined destination via multiple intermediate nodes, information dissemination strategies are concerned to deliver a certain piece of information to all nodes of a certain group—most often all nodes within a specific geographic region. In the context of vehicular ad hoc networks, this typically affects the dissemination of safety-related information within a certain area. Whereas routing algorithms are mostly concerned with route discovery, maintenance, and flow optimization, key focus of information dissemination is typically on minimization of network load while still ensuring that the respective information reaches all necessary recipients in time. The concept of information dissemination is often also referred to as message or information diffusion, pointing out that the information artifacts spread through the network. Information dissemination can therefore also be seen as a way of limited and controlled flooding of information in the network. However, it should be noted that—in contrast to simple flooding—the shape and size of the affected network region can be specifically controlled.

7.2 Specific Requirements and Prerequisites

7.2.1 The Scalability Problem—Two Contrasting Scenarios

In large and especially distributed systems or networks scalability is a very crucial characteristic. Neuman established the term [1] and defined scalability as the ability to handle the addition of nodes, objects, or network size, without suffering a noticeable loss in performance or increase in administrative complexity. Usually some kind of bottleneck can be identified causing scalability problems for a system or network. In the scenario of vehicular ad hoc networks scalability issues arise in several different contexts. The number of active nodes (vehicles) has an impact on network connectivity as well as the blocking probability on the wireless channel. In addition, protocol design has a great impact on scalability related to the wireless transmission. Typically, the most crucial bottleneck is the limited capacity of the wireless channel. Li et al. [2] presented the capacity limitations for ad hoc networks (we will further discuss those capacity limitations in the context of network utility maximization in Section 7.4.

Due to the shared wireless channel and the multihop communication between distant nodes information dissemination further faces the problem of low node density. Scalability is especially becoming an issue in very large network scenarios. In networks with a low node density, connectivity is the biggest challenge. In a network with few nodes, it becomes very hard to distribute information using multihop connections due to the highly fragmented network. Therefore, the protocol and system design have to deal with both the connectivity and the scalability challenge:

- Connectivity. In scenarios where the intervehicle network is not very dense, for example, when detecting a single icy spot on a rural road, it must be ensured that the information about the road condition is not lost once it has been sent by the detecting vehicle. Vehicles may even leave the dangerous area without having met another network node to communicate with. In this case, intelligent store and forward algorithms must handle the transport of information back into the desired area. Low connectivity will be very typical during the first years of deployment and in areas of low traffic density.

- Network Load. In city scenarios or in highway traffic, when the intervehicle network is quite populated, information about a traffic jam may be detected and sent out simultaneously by multiple vehicles. When trying to forward this information to other vehicles, the channel can easily be jammed by the big number of forwarders. The communication system has to apply intelligent repeating mechanisms in order to prevent the network from being overloaded. The number of messages that have to be sent over the shared medium is predominantly influenced by the number of vehicles and the number of applications deployed in these vehicles. However, network load is additionally influenced by the fact that active safety messages have to be rebroadcasted to ensure reliable packet delivery in fragmented networks. This ensures the availability of the message for new vehicles entering the area after the initial broadcast. Further, the utilization of channel capacity will not be uniformly distributed over the whole network. In particular, traffic originated by safety-related warning applications usually has a bursty traffic profile. Hence, a great number of information messages will be generated and sent within a certain area in a short period of time.

Future vehicular ad hoc systems therefore have to perform well in both network characteristics. Hence, information dissemination in vehicular ad hoc networks must seamlessly scale between high network load and fragmented networks. This requires an integrated approach (as will be presented in Section 7.4) that does not solely focus on one aspect.

7.2.2 Requirements and System Constraints

In addition to the problem of scalability and connectivity, the specific characteristics of automobile active safety impose several further important requirements on information dissemination in vehicular ad hoc networks.

Altruisms and joint fairness. Common sense of fairness usually guarantees all network participants a certain amount of bandwidth (either the same for all nodes or staged based on priority considerations), that is, no node is precluded from accessing the communication channel. However, vehicular ad hoc networks differ from mobile ad hoc networks not only in the dynamics of node movements. It is important to note that the network is designed with particular emphasis to increase road safety. One main consequence is an inherent altruism in the network system, that is, a single node does not have any advantage when offering an information artifact to others. Each node only gains benefit from participating in the network, if all participants are willing to cooperate in an altruistic way. The existence of the explicit network objective in vehicular ad hoc networks rationales a different interpretation of fairness. Especially if the network operates at its limit, it must be ensured that

- Messages comprising safety critical information can access the channel with very low latency, if necessary (i.e., the message carries, so far, unknown and critical information and the intended receivers are close to the critical incident)
- Messages supporting the drivers' tasks can (but need not necessarily to) be prioritized over unspecified messages (deployment applications), if the carried information has significant utility to others
- If the channel capacity is insufficient to transmit all desired messages, explicitly those messages that provide the least utility may starve. Because of the network objective, this will typically concern (but is not limited to) messages of deployment applications.

Therefore in certain critical situations certain nodes should be temporarily suppressible for the sake of others, if they cannot contribute to the altruistic network objective. This principle was termed joint fairness in order to emphasize the difference to common sense fairness.

Traffic Differentiation and Prioritization. A priority mechanism for active safety messages must be given in order to keep delay times for warning messages low when accessing the channel. A study by Matheus et al. [3] in the context of the German research project Network-on-Wheels [4] (NoW) has shown that a solution depending on active safety applications that aim at reducing the severity and number of accidents alone will not be able to penetrate the market successfully. Therefore, an intervehicle communication system must also be able to support so-called deployment applications.

Applications of this type cover passenger comfort, entertainment, or improvement of the driver's range of information. Usually, active safety applications should have higher priority over deployment applications when the network is congested. Even among active safety applications, an order must be established as different emergencies might have different levels of urgency. In addition to that, even identical messages may have different priorities in different situations. For that reason, priority is not only dependent on the application types, but also on the situation a message is sent or received. Therefore, prioritization must also be context-adaptive.

Optimal bandwidth usage. On the one hand, the available channel capacity is limited and has to be shared with many other applications. It is obvious that it should be used in a way as optimal as possible. On the other hand, the decision on when to warn a driver about an upcoming danger typically has to deal with probabilities. Applications like a local danger warning system, as rationalized in Section 7.1, can reduce their error rate when more data is available for consideration and thus reduce uncertainties. For this reason, the network should allow for an adaptable mechanism that can inject more data into the network as long as capacity limits are not reached. That means it should be possible that available network capacity can consequently be used by various applications until a certain threshold is reached. In contrast to existing applications in wired networks, it is not always reasonable to reduce the network usage as far as possible. However, when reaching a reasonable upper bound, it must be ensured that messages of lower relevance are dropped before messages that could have contributed with a higher utility.

Dissemination area and message lifetime. First, there is a need to keep the information about dangerous situations "alive" for certain amounts of time in the vicinity of the event. Since the lifetime and the area of relevance of such an event cannot be easily foreseen, the system must offer means to dynamically adapt the lifetime and dissemination area for warning messages.

Situation adaptivity. Furthermore, weather conditions and dangerous situations may change or disappear over time. The system must provide means to react to changing context. Additionally, it must be possible to inject newer information into the network, updating old events or marking them as past events.

Network dynamics. Due to the nature of vehicular ad hoc networks, the communication system also has to deal with the high mobility of nodes. In scenarios as volatile as highway traffic the communication windows are limited to less than 10 seconds. For that reason, it is sensible to exchange messages with high relevance first.

7.2.3 Classes and Levers of Information Dissemination

Typically, most sophisticated dissemination strategies address only one of those fundamentally different problems, that is, focus only on minimizing the overall network load or ensuring packet delivery at low penetration rates, respectively. For the first, three different classes of levers can be distinguished, namely,

- Adaptive selection of forwarding nodes (intelligent flooding). Here again, two competing paradigms can be distinguished: sender- versus receiver-centric dissemination. In addition, in this context recently the classification of packet- versus information-centric forwarding was introduced, where information-centric forwarding takes the characteristics of the specific information artifact that is disseminated into account for the forwarding decisions.
- Adjustments of transmit power. The transmit power is usually directly adapted according to the current node density. The basic idea is to keep the number of neighbors that are within direct communication range small and therefore reduce interferences and packet collisions on the wireless channel.
- Constraints on message lifetime and geographical validity (geobroadcast), limiting both the area and the time a message is distributed and thus reducing the number of circulating messages and in turn the network load.

For the latter, most often physical store and forward strategies are applied, that is, information artifacts are stored in the vehicles and rebroadcasted whenever sensible. In addition, two classes of information have to be distinguished, namely, event data (e.g., warning messages) and periodic messages comprising, for example, the vehicles' speed, heading, and acceleration. Periodic messages are typically only valuable for vehicles in the immediate vicinity. Therefore, in contrast to event data, those messages usually will only be broadcasted once and not distributed or disseminated within the network. However, it is important that both types of data can be commonly handled by one approach, which in particular affects the process of packet prioritization during medium access.

7.3 State-of-the Art Algorithms and Strategies for Information Dissemination in Vehicular Ad Hoc Networks

The basic dissemination scheme used to realize broadcast-based vehicular ad hoc networks is the so-called flooding, where each node rebroadcasts every single message it receives. The main problem inherent to the flooding mechanism is the huge amount of superfluous transmissions leading to network congestion. This effect aggravates with an increasing node density and network size, leading to the scalability problem. Hence, the limited network resources are partially absorbed by redundant traffic, while highly relevant and time-critical messages are prevented from accessing the medium.

7.3.1 Intelligent Flooding Mechanisms

In order to transmit packets from one source node to a huge number of other nodes in both reliable and efficient ways, rather sophisticated methodologies have been developed during the last couple of years. Williams and Camp provide a comprehensive overview on numerous fundamental algorithms that all aim at minimizing the number of such redundant rebroadcasts and to make sure packets are reliably propagated to all desired destination nodes [5]. Basically, these so-called intelligent flooding algorithms can be divided into three categories: "probability based," "area based," and "neighbor knowledge based."

> **Probability-based algorithms.** In the context of probabilistic algorithms, nodes only rebroadcast data packets with an adjustable probability, thereby limiting the number of retransmissions. Ni et al. [6], for example, prove that, especially in dense networks, node and network resources can be saved by limiting the number of retransmissions even on a random basis. One specific example works as follows. Upon the reception of a packet, nodes initiate a counter with a value of one and then wait for a randomly determined period of time (random assessment delay [RAD]). During that time, they increment the counter by one for each redundant packet received. If the counter is less than an adjustable threshold when RAD has expired, the packet is rebroadcasted. The introduction of a probability-based discrimination of nodes bears the major disadvantage of inflexibility. In rather dense networks, the overall dissemination efficiency can certainly be improved by restricting the number of forwarders. In sparse network environments, however, where the density of wireless enabled vehicles is very low, this approach would even deteriorate the system performance since transmissions are omitted that would have been necessary for a fast dissemination throughout the network.
>
> **Area-based algorithms.** Nodes following an area-based approach only relay messages in case their transmission coverage area overlaps the coverage area of the message's sender to not more than a predetermined extent. In this way, nodes reaching a high number of new receivers are more likely to rebroadcast the information, which considerably helps saving network resources without

reducing transmission reliability. In this context, two basic mechanisms are proposed [6]: In the frame of distance-based schemes, nodes compare the mere distance between themselves and the senders of received messages. Upon the reception of previously unknown packets, a timer is initiated. During countdown, redundant packets are cached. After expiration, the algorithm checks whether any of the redundant packets' senders were closer than some adjustable threshold. If so, the data packet is not broadcasted. The location-based scheme, in contrast, uses a more precise technique for rebroadcast limitation: When nodes receive a certain packet, they extract the sender's geographic location and compute the additionally covered area in case they forward the packet. If lower than a certain threshold, the packet is not broadcasted. If higher, again a timer is initiated. During this timer run, all redundant packets are evaluated with respect to the new, additionally covered area. This area calculation and threshold comparison are conducted with all redundant broadcasts received until the packet reaches either its scheduled send time (at the expiration of the timer) or is dropped.

A further methodology that can be assigned to the field of area-based methods is proposed by Fuessler et al. [7]: especially in highly mobile ad hoc networks, the scheme helps delivering messages efficiently to a certain geographic region while using broadcast communication. The proposed algorithm is called "contention-based forwarding" and works without the help of any beaconing or the maintenance of information about direct neighbors. In the frame of this approach, not only the sender of a packet determines the next forwarder, but also all neighbors of a source node are included in the decision-making process of finding the most suitable next sender. The algorithm consists of two major parts: the forwarder selection process and the suppression of the other forwarding candidates. First, the source node transmits the packet as a single-hop broadcast to all of its neighbors within communication range. The receivers read out the geographical location of the actual packet destination and compare it with their own position (which is assumed to be available via GPS positioning devices). Only in case they are closer to the destination than the last forwarder, they start the contention process with the other potential next forwarders. Timers are started, whose values depend on the extent of geographical progress the respective node provides toward the destination. The lower the progress a node provides, the higher the initial timer value. In this way, it can be guaranteed that the timer of the node that is closest to the packet destination expires first. Upon expiration, the node initiates the packet rebroadcast. Figure 7.1 visualizes how well a node is suited for being the next forwarder of a packet to its

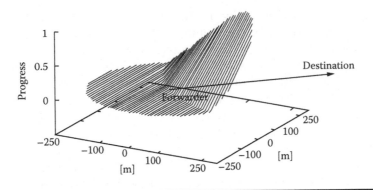

Figure 7.1 The "progress"-measure used in the contention-based forwarding algorithm [7].

destination. A progress value of zero means that a node is not adequate for rebroadcasting a packet (since the packet would not approach its destination), where a value of one represents the optimal case.

After selecting the best-suited packet forwarder, it must be ensured that all other nodes that have also received the packet pass on broadcasting it (which would lead to a superfluous absorption of network resources and congest the wireless channel). The simplest strategy proposed works as follows: As soon as a timer expires, the respective node assumes that it is the most appropriate one to act as the next forwarder and broadcasts the packet. All neighboring nodes that still have a timer running and receive this rebroadcast, cancel their timers and do not forward the packet. One of the considerable drawbacks of area-based dissemination approaches again is the inherent inflexibility. Retransmissions should not be discriminated according to static, predetermined thresholds. In fact, to optimally leverage network resources, each node receiving a message should try to retransmit it, supposed there is no data packet waiting to be sent which provides a higher utility to the neighbors. An inflexible transmission limitation only helps in overloaded network situations to a certain extent, but does not work as a comprehensive and adaptive approach for efficient message dissemination.

Chiasserini et al. [8] also deal with possibilities to improve the efficiency of broadcast-based communication in multihop vehicular networks. By modifying channel access mechanisms, the broadcast delay can be reduced and safety-critical messages can be delivered on time. The basic approach works as follows. Through the introduction of spatial traffic differentiation within the MAC layer, vehicles that are about to rebroadcast a certain message access the medium with different priorities, depending on their distance from the last vehicle that transmitted. With the help of simulative studies, the implications of adapting the magnitude of the so-called contention windows to this distance between sender and receiver of a certain message are evaluated.

Neighbor-knowledge–based approaches. The third category of restricted flooding methodologies postulates that all nodes know about the identity of all their neighbors (network nodes within communication range). One exemplary realization of a neighbor-knowledge–based method is the so-called "flooding with self-pruning": each node includes the list of all its neighbors in every packet broadcasted. Upon receiving such a message, nodes compare this list with their respective neighborhood and only initiate a rebroadcast if new network participants would be reached.

Kim et al. [9], for example, presents one specific, neighbor-knowledge–based approach to make flooding of messages in vehicular ad hoc networks more reliable, efficient, and fast. To reduce redundant retransmissions to a minimum, nodes perform two techniques called dynamic delay and priority checking. Nodes thereby have to know the distance between them and their one-hop neighbors to decentrally manage the decision whether to forward a packet or not.

The distributed power control [10] represents a second example for restricted flooding mechanisms based on neighbor knowledge. In this context, nodes have to adapt their transmission power to the density of other network nodes in their vicinity. This is a valuable approach both in sparse and in dense networks. In scenarios with only a low penetration of wireless-enabled vehicles driving around, an increased transmission power helps improve the overall connectivity and thus the reachability of nodes. On the other hand, a transmission power decrease in very dense networks supports the reduction of connection redundancy and amends medium access efficiency due to a reduced number of collisions. It is suggested that vehicles periodically broadcast beacons to notify potential neighbor nodes about their location. Every beacon features an entry indicating the transmission power with which it has been sent. Cars receiving those beacons can then, in turn, compare the received power with the original send power and thus deduce information about the actual path loss. Every car is supposed to maintain a table containing the average path loss to each of its neighbors. In case vehicles have a large number of neighbors available that have found to be reachable via low-loss paths, the power for the next packet transmission can be reduced.

This mechanism for a decentrally organized power control helps significantly reduce message redundancy in dense networks and improving node reachability in sparse networks. However, the specific content of the messages disseminated is not taken into account at all when deciding upon forwarding a message or not. For example, a local danger warning generated miles away from the receiver is treated exactly the same way as one created by the direct network neighbor (which intuitively is of a way higher relevance for the recipient).

Further algorithms in the field of neighbor-knowledge–based methods are scalable broadcast algorithms (SBA), dominant pruning, multipoint relaying, ad hoc broadcast protocol, CDS-based broadcast algorithm, and LENWB, which are all well documented [6].

All three categories presented above offer excellent and well-evaluated solutions for realizing reliable and efficient broadcast communication of messages in mobile ad hoc networks. Having available such algorithms for ensuring the delivery of one message to a huge number of destination nodes, one has to think about the next step toward a scalable, situation-adaptive, and utility-based information propagation strategy in consideration of the limited availability of network resources.

In Section 7.4, we present a comprehensive approach that allows for a provision of vehicles with the information they are currently most interested in, while optimally leveraging the available channel capacity both in sparse and in dense networks.

7.3.2 Traffic Differentiation, Quality of Service, and Utility Optimization

As pointed out, prioritizing data traffic is crucial for ensuring that car drivers are provided with safety critical information with low latency. Numerous different strategies have been proposed allowing for prioritizing specific data traffic over less relevant data. In the following discussion, we will focus on traffic differentiation in IEEE 802.11 networks, since this will most likely be the basis of the prevailing communication standard in the domain of ad hoc inter vehicle communication. Various approaches to differentiating traffic in IEEE 802.11 can be found in different texts [11–13], where time slots are reserved to enable quality of service (QoS) as well as to achieve a fair access of data to the shared wireless medium especially in case the network is heavily loaded.

Provision of QoS. Lindgren et al. [14] provide an overview of different further mechanisms for realizing quality of service in 802.11-based wireless networks. By means of simulative studies, they are all compared with respect to medium utilization, access delay of packets, and scalability.

- The method proposed by Deng and Chang [15] mainly applies two mechanisms: arbitration interframe spaces (AIFS) and backoff algorithms are used for traffic differentiation.
- The distributed fair scheduling (DFS) uses the size of the currently treated packets to determine backoff timers. Traffic flows with smaller packets are allowed to transmit more often, thereby introducing a certain level of global fairness.
- The black burst algorithm mainly aims at minimizing the delay for real-time traffic. After sensing the medium to be idle for at least PIFS, high-priority stations jam the medium for a certain period of time (black burst), where the length of this burst depends on the time the station has waited to access the medium. After transmitting the black burst, the station listens to the medium for a short period of time to see if any other station is sending a longer black burst and thus must be granted the privilege to transmit a frame. If the medium is idle, the respective frame can be transmitted. In contrast, low-priority stations use the ordinary CSMA/CA access method of the legacy IEEE 802.11 standard.

In 1999, Deng and Chang [15] were the first ones to elaborate on a possibility to realize data traffic prioritization in wireless networks that are based on distributed coordination function (DCF) as implemented in the 802.11 standard. According to Torrent-Moreno et al. [16], the scheme proposed was the basis for the subsequent work on the enhanced distributed coordination function (EDCA), which is part of the IEEE 802.11e [17] specification. Since then, numerous studies have been conducted with respect to performance and possible further improvements of 802.11e. For example, some texts [18–20] provide an overview of 802.11e performance with respect to unicast data flows in wireless networks. Some other texts [21–23] give a good understanding of the main levers underlying the EDCF mechanism and their individual effects on the overall network performance.

Due to the necessity to disseminate safety-critical messages throughout an ad hoc network as fast as possible, where other data is not very susceptible to packet delays, Torrent-Moreno, Jiang, and Hartenstein propose and evaluate a possibility to realize a mechanism that favors the transmission of particularly relevant data [16].

By deploying four different traffic categories (TCs) of which each applies different AIFS and CW_{min}, the prioritized transmission of different kinds of traffic can be simulated. The simulation results show that highly prioritized nodes (using short values for AIFS and CW_{min}) may access the channel faster then nonprioritized ones, resulting in shorter channel access times. The effects of adjusting CW_{min} and AIFS are investigated in terms of the probability that packets are received. In a highly loaded network such as the one considered in the frame of this work, a huge percentage of packets are normally lost due to collisions. By applying shorter AIFS and CW_{min}, the probability of reception of packets sent by a prioritized node can be significantly increased.

It could also be shown that the channel access time (defined as the period of time from packet creation to sending it to the channel) can be reduced, where the probability of reception is increased for data packets broadcasted from a prioritized station.

Adaptive service differentiation. Romdhani, Ni, and Turletti describe an adaptive service differentiation scheme for the realization of QoS in wireless ad hoc networks [24]. The focus thereby lies on a modified version of the existing IEEE 802.11e EDCA mechanisms for the differentiation of traffic stemming from different applications. The main goal is to leverage network resources as efficiently as possible by adapting the nodes' contention windows not only to the class of the currently treated traffic, but also to the actual network load. The motivation for this extension to the static data packet differentiation is as follows. Especially in high-load scenarios, the time used for channel access negotiation (both the contention for the virtual internal medium and the real shared medium) significantly increases. Thus, the latency of time-critical packets may rise and the overall network performance deteriorates. Therefore, the following, twofold scheme is proposed to efficiently support time-bounded traffic. First, the resetting of contention windows to CW_{min} after each successful transmission attempt is not conducted at once any more, but in several, adaptive steps. Motivated by the assumption that when a collision occurs, a new one is likely to occur in the near future, the authors propose to update the contention window slowly, for example, by a static factor of 0.5. Since such a static factor is not an optimal solution in all network conditions, however, an adaptive factor considering the estimated collision rate at the respective node is used. This adaptive, slow contention windows decrease is a trade-off between wasting some backoff time and risking a collision followed by a whole packet retransmission. The value of the expected collision rate at the stations is calculated using the number of collisions and the total number of packets sent during a constant period of time. Second, the adaptation of the persistence factor (PF) is proposed as a further means to reduce the probability of collisions and to decrease packet delays.

Utility-fair broadcast rate assignment. Wischhof and Rohling propose another scheme for allocating data rates to network nodes in a broadcast-based vehicular ad hoc network, which leverages the utility the respective nodes' packets provide to its neighbor nodes [25]. Each node is supposed to continuously monitor its environment and estimate the utility of transmitting its own packets based on the information obtained from other packets that have been sent by other nodes beforehand. First of all, each node is assumed to have packets p_k available that are of size s_p. If K packets are stored within the packet queue of node i, the maximum utility per byte (η_i) that this node has available is governed by

$$\eta_i = \max_{k=1...K} \left[\frac{u_{tx}(p_k)}{s_{p_k}} \right] \qquad (7.1)$$

Each node is assumed to be able to calculate a value $u_{tx}(p_k) \in [0, 1]$ corresponding to the utility of transmitting the packet p_k at the current point in time. The overall objective of the algorithm is to assign each network node a data rate ρ_i that is proportional to the maximum utility it is currently able to provide (η_i) and thus realize the utility-fair broadcast rate assignment:

$$\frac{\rho_i}{\eta_i} \approx \frac{\rho_j}{\eta_j}, \forall i, j \in [1, N_{2hop}] \qquad (7.2)$$

The equation may be motivated as follows. The higher the maximum utility a node can currently provide compared to the utility its neighbors can provide, the higher the bandwidth the node is granted. To determine this rate ρ_i, the parameter β_{local} has to be calculated [see Equation (7.3)], which represents the locally common ratio of effective data rate to utility. β_{local} can be computed as follows.

$$\beta_{local} = \min_{i=1...N} (\beta_{1hop,i}) \qquad (7.3)$$

Based on information attached to all previously transmitted packets in the neighborhood, a local utility average μ can be found. For each node i, $\beta_i = \frac{B_{eff}}{n\mu}$ then denotes the available effective data rate per node per utility. n is the total number of network nodes within node i's transmission range, where B_{eff} is the monitored locally valid effective data rate. After determining β_{local} as the minimum of all the β_i in the neighborhood, all necessary information has been gathered to finally define ρ_i [see Equation (7.4)]. The lower the utility of the neighbor nodes, the higher is β_{local} and thus also the assigned data rate ρ_i. In other words, the lower the relative utility of the other network nodes' information, the more packets is node i allowed to send. Θ is a system-specific factor characterizing the network's target load.

$$\rho_i = \eta_i \Theta \beta_{local} \qquad (7.4)$$

This approach aims at maximizing the overall aggregate utility (also referred to as utility) in a broadcast-based wireless network. The methodology proposed by Wischhof and Rohling [25] takes into account utility values of previously transmitted data packets, but does not guarantee an up-to-date packet schedule. Also, the high dynamics inherent to vehicular ad hoc networks are not taken into account. In real-world vehicular ad hoc network deployments, there is no static neighborhood exiting, where parameters such as the number of network nodes n remain stable for a longer time span. Instead, the number of adjacent (within radio range) nodes, their available information, and thus β_{local} will strongly vary over time.

Content-based forwarding. Carzaniga and Wolf propose an algorithm for propagating information within networks according to the individual nodes' needs [26]. In the frame of the content-based forwarding (CBF) algorithm, each network node determines if and where to forward data packets. This decentrally organized information routing is designed for wired and stable networks with fixed links and can hardly be operated in a vehicular ad hoc network environment. The CBF algorithm passes on giving messages explicit destination addresses. As opposed to a conventional unicast routing protocol, nodes decide autonomously upon forwarding a message to certain neighbor nodes or not. This decision is made based on message content and the knowledge about the interest other nodes have in certain kinds of information. Receivers declare their interest by means of selection predicates, where information sources simply broadcast messages. CBF is intended to provide any receiver with messages that match the selection predicates declared by them before. Messages distributed in a content-based network must be structured as a set of attribute-value pairs as depicted in Figure 7.2. Ideal applications for the CBF communication scheme are publish/subscribe event notification services, services for system monitoring and management, network intrusion and detection, service discovery, data sharing, and many more. The routing functionality of a content-based network follows an approach based on two different

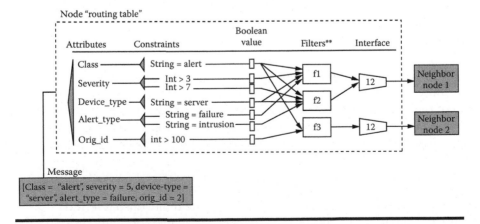

Figure 7.2 Content-based forwarding [26].

forwarding tables. The first comprises constraints imposed by the physical topology, whereas the second contains selection predicates, which are continuously being updated by the applications. Selection predicates have been defined as logical disjunctions of elementary constraints over the values of individual attributes. Carzaniga and Wolf [26] define the forwarding function of each nodes as follows. The next-hop destinations of a message are determined by applying the constraints of the physical topology first, and by matching message content with the selection predicates specified for each of the interfaces (connections to adjacent nodes). CBF can be separated into the routing information update algorithm and the predicate-matching algorithm.

As mentioned above, nodes maintain detailed information about their neighbor nodes' information interest. Two different mechanisms are defined to keep routing information up-to-date in all the nodes of the network. The first one is called the "push" mechanism, where potential receivers periodically broadcast their individual interest in information (encoded as receiver advertisements [RAs]) throughout the whole network. These RAs are evaluated at each receiving node and predicates are updated accordingly. In more detail, at each of the nodes, the predicate announced by an RA is combined with the already existing predicate, which is attached to the interface that is on the path back to the issuing node. For combining the predicates, a logical disjunction is used. Only if this combination generates a new predicate for this interface, the node rebroadcasts the RA. If not, the RA is considered as redundant and is not going to be forwarded any more. In this way, the dissemination of RAs is stopped as soon as nodes do not gain new knowledge from it any more. The second possibility to keep routing information up-to-date in content-based forwarding network works as follows. Nodes are able to issue sender requests (SRs) to gather content-based routing information from other nodes: SRs are broadcasted throughout the whole network; recipients of an SR-generate update replies (URs) and send them back to the issuer on the reverse path. Leaf nodes of the network immediately return an UR to the issuer, whereas intermediate nodes combine their proprietary interest reported by the nodes downstream of the issuer. Finally, the original issuer of an SR receives one UR for each of its interfaces. CBF basically follows a conventional broadcast communication scheme. However, forwarding is limited to those nodes having announced predicates matching the actual message content. The predicate-matching mechanism is defined as a two-part, left-to-right structure. The left part of the graph in Figure 7.2 shows the constraints—featured in any of the predicates attached to any of the interfaces of a node. The following works as an example for an individual constraint. In case the attribute "severity" of a message has been determined to be greater than 3, the second of the constraints depicted in Figure 7.2 can be considered as fulfilled. Boolean values then indicate the result of applying a constraint on a message (1: fulfilled, 0: not fulfilled). The right-hand side comprises logical connections representing conjunctions of constraints into filters and their disjunction into predicates of interfaces.

7.4 An Integrative Approach: Network Utility Maximization in Vehicular Ad Hoc Networks

As previously pointed out, information dissemination in vehicular ad hoc networks must seamlessly scale between high network load and fragmented networks. Hence, it is necessary that all requirements as stated in Section 7.2 are equally taken into account within an integrated approach. In addition, due to the long product life cycles of cars it is an important issue to use the strictly limited channel capacity in all situations as optimal as possible. Most often, network protocols aim at optimizing the average network performance, taking into account typical use-cases or situations. Usually this leads to certain trade-offs between different requirements imposed by different use-cases. The specific requirements of vehicular ad hoc networks, in particular imposed by active safety applications, typically lead to a nonoptimal system performance under usual conditions. This is because significant network capacity is reserved in order to ensure low-latency dissemination of safety critical messages.

Exploiting a standardized utility metric on a per packet basis that takes into account the specific situation the vehicles are currently in, thereby enables a much better overall network throughput under usual conditions and thereby in particular supports the need of fine-grained prioritization of safety critical information.

7.4.1 Network Utility Maximization in Common Networks

In this context, the maximization of the global, aggregate utility (network utility maximization [NUM]) as described by Kelly et al. [27] has also become an important principle underlying several rate allocation schemes and congestion control protocols for diverse communication networks. The basic approach to realize utility-based data rate control and thereby maximize the aggregate utility of all network nodes can be studied in detail in La and Anantharam and GAO and Bharghavan [28, 29]. Consider a system of J links (also referred to as resources) and a set of I users, where C_j is the restricted capacity of link $j \in J$. Each user i utilizes a fixed set of J_i links in its path, which is a nonempty subset of J. The authors then define a matrix A consisting of ones and zeros, where $A_{i,j} = 1$ in case link j is part of user i's actual route, J_i and $A_{i,j} = 0$ otherwise. The data throughput of user i is denoted as x_i, such that the user receives utility $U_i(x_i)$. $U_i(x_i)$ is an utility function that has to be defined according to the mission-specific goals of the respective applications. For example, the mere data rate may work as an adequate means for measuring the utility delivered by a file transfer application. In their initial mathematical model, increasing, strictly concave and continuously differentiable functions of x_i, where $x_i \geq 0$ are assumed [27]. For elastic traffic as introduced above, concave functions may adequately reflect the utility provided to end users. The higher the data rate, the more useful is the application for the user. In case of inelastic traffic with minimal performance requirements, however, convex or sigmoid functions are more suitable.

Getting back to the utility-based rate control model, utilities are assumed to be additive such that the aggregate utility of rate allocation $x = (x_i, i \in I)$ is $\sum_{i \in I} U_i(x_i)$. The resulting optimization problem can then be described as follows:

$$\text{Maximize} \sum_{i \in I} U_i(x_i) \tag{7.5}$$

$$\text{with } A_x^T \leq C \tag{7.6}$$

$$\text{where } x \geq 0 \tag{7.7}$$

Thereby $C = (C_j, j \in J)$ represents the capacities of the respective links. The first constraint [Equation (7.6)] says that the total data rate of a link may not be larger than its capacity. Assuming that utility functions are known, this problem can be solved mathematically. However, in practical scenarios, a central system coordinator knowing all the individual utility functions is unlikely to exist. For that reason, Kelly [30] proposed methods to allow for a fully decentral, utility-oriented rate allocation based on a pricing scheme, where the users are to solve the problem of maximizing their individual utility, which is the utility received by applications less the amount they pay. Kelly, Maulloo, and Tan [27], proposed a rate-based algorithm, which also solves the system of equations above. The underlying methodology relies on a shadow price charged by links depending on the total traffic load going through it. User's willingness to pay per unit time is denoted by P_i, where $x_i(t)$ is the current data rate at time t. Each link $j \in J$ is supposed to charge a price per unit data flow μ_j, which is calculated according to Equation (7.8), where b_j increases with the total data rate going through it. The function relating the provided data rate to the price the node has to pay is b_j. System constant is represented by κ. In other words,

$$\mu_j(t) = b_j \left[\sum_{i: j \in J_i} x_i(t) \right] \tag{7.8}$$

$$\frac{d}{dt} x_i(t) = \kappa \left[p_i - x_i(t) \sum_{j \in J_i} \mu_i(t) \right] \tag{7.9}$$

Differential Equation (7.9) can be motivated as follows. The links (re-sources) in the network are assumed to continuously give feedback to the users and thereby notify them about the actual price per unit data flow $\mu_j(t)$. If the product of a user's current data rate $[x_i(t)]$ and the price per unit data flow is lower than the price the user is willing to pay $[p_i(t)]$, the user may increase its data rate. By means of a mathematical proof, it has been shown that each user reaches an equilibrium state, where the price he is willing to pay equals its aggregate cost. Also, the algorithm proposed by Kelly et al. solves a relaxation of the problem defined by Equation (7.5), that is, the aggregate, global utility of the whole network can be maximized.

While this approach assumes that the network is able to provide nec-essary feedback to the users, and users then adjust their rates based on the feedback info to maximize the global, aggregate utility, La and Anantharam [28] introduced an extension of this utility optimization scheme that com-pletely passes on any kind of feedback from the network. Assuming that the data rate through one of a network's links is smaller than its capacity, there is no contention since each user receives its desired rate. As soon as the rate approaches the capacity, however, the increase of the data rate of any of the users leads to a backlog at the link (packet queues emerge). The resulting increased queuing delay at that link can then be regarded as an increase of the implicit cost the users pay due to the larger delay. The network has to account for this change by recovering the increased system cost through a new pricing scheme. q_j, $j \in J$ is denoted as the current backlog at link j. As soon as the link is congested (the total rate through it equals its capacity), the link has to charge the users a price, where the price per unit flow per unit time g_j is determined as the queuing delay at the link, $g_j = q_j/C_j$. J is governed by Equation (7.10), where the total price per unit time a user i pays is determined by Equation (7.11). Equation (7.12) defines the net utility received by a user i. To make explicit feedback from the network superfluous, the actual size of the users' packet queues is the basis for the computation of link-backlogs q_i.

$$\text{Total price per unit flow per unit time} = \sum_{j \in J_i} g_i \qquad (7.10)$$

$$\text{Total price per unit time} = x_i \sum_{j \in J_i} g_j \qquad (7.11)$$

$$\text{Net utility received by user i} = U_i(x_i) - x_i \sum_{j \in J_i} \frac{q_j}{C_j} \qquad (7.12)$$

7.4.2 Network Utility Maximization in Vehicular Ad Hoc Networks

The algorithms presented in the numerous publications regarding network utility maximization provide an excellent background for the considerations of network utility maximization in the context of vehicular ad hoc networks, but also feature major weaknesses regarding the specific requirements in the automotive domain. Most of the schemes assume the existence of certain utility functions, but do not elaborate on a comprehensive model of user needs and their perception of utility provided by different applications. However, especially with regard to cooperative vehicle safety systems, utility provided to users cannot be quantified with the help of simple convex or concave functions, which only depend on measures such as packet delay, throughput, and jitter.

In particular, there are the following significant differences when transferring the NUM problem as described above to vehicular ad hoc networks:

No fixed links. Connectivity in vehicular ad hoc networks is highly dynamic. Connections may only last for a few seconds, and there is no per link capacity. Instead, all nodes within mutual radio range have to share the same medium and have to share that capacity.

No rate-based utility calculation. Since connectivity is only short lived, occasions to emit a whole flow of data are rare. The calculation of utility must cannot be based on the rate of transmission but rather on the information that is transmitted.

Utility functions are manifold. Nevertheless, in general the types of utility functions are not known beforehand. Utility calculation must be possible using arbitrary functions. In addition, utility functions may change and are not static. This is particularly true for applications that depend strongly on context information, like active safety applications.

Hence, the information contained in data packets must be evaluated against the background of the vehicles' respective contexts as will be explained later. Apart from that, NUM schemes mostly head for a globally fair allocation of resources to the different network participants. However, in order to meet the requirements of critical road safety, stations should not be granted access to the medium on an equal basis, but depending on the specific utility of their available information (according the principle of joint fairness in altruistic systems as introduced above). Thereby, the constraint of the optimization problem as argued in Equation (7.6) is not valid in vehicular ad hoc network environments. Since individual data flows, which can be assigned specific data rates, do not exist in this case, the utility or utility functions leveraged cannot be a function of data rate, jitter, or delay, but of a single data packet and the utility provided to its potential recipients.

The constraint to the maximization problem is, as briefly described before, the per node available channel capacity. In some texts [31,32], the capacity of wireless ad hoc networks is treated in detail. The main goal is to estimate the "useful" bandwidth that each node can expect for its own traffic.

The dominant factors influencing the network's capacity are network size, traffic patterns applied, and detailed local radio interactions. The total one-hop capacity of an ad hoc network is governed by the amount of spatial reuse possible in the network. Assuming a constant radio range, this spatial reuse is proportional to the network's physical area. Further assuming that the node density δ is uniformly distributed, the physical area of the network, A, is related to the total number of nodes by $A = n/\delta$. Therefore, the total one-hop capacity of the network, C, is also proportional to the area, as $C = kA = kn/\delta$ for some constant k. If each node originates packets at a rate λ, and the expected physical length of a communication path between source and destination nodes is L, the minimum number of hops required to deliver data packets is L/r, where r is the fixed radio transmission range. Thus, the total one-hop capacity in the network required to send and forward packets can be described by Equation (7.13):

$$C \succ n\lambda\frac{L}{r} \tag{7.13}$$

Combining this with $C = k\frac{n}{\delta}$, we wind up with a formula describing the capacity available at each node, λ.

$$\lambda \prec \frac{kr}{\delta}\frac{1}{L} = \frac{C/n}{L/r} \tag{7.14}$$

Due to the decentrally organized contention for the shared, wireless medium, nodes "waste" a lot of time while backing off, deferring their medium access, and resolving channel collisions. In the case of unicast communication data flows, where one node communicates with exactly one other node over one or several hops, the per node capacity is even more decreased due to the requirement that nodes forward each others' packets. For a unicast-based ad hoc network, the expected path length is governed by Equation (7.15)

$$L = \frac{2\sqrt{A}}{3} \tag{7.15}$$

This is also true for information dissemination that is typically based on intelligent flooding. In particular event-based messages are rebroadcasted by different vehicles in order to ensure sufficient dissemination of the information within a certain area and over a sufficient period of time. Taking into account the scalability and connectivity problem as described in Section 7.2.1, the number of necessary rebroadcasts in particular depends on the current network density. Thus, the parameter L is dynamic and again must be considered situation-dependent.

In order to optimize the overall utility of the vehicular ad hoc network, two fundamental methodologies form the basis of network utility maximization in vehicular networks. First, the utility data packets provided to potential recipients in the local neighborhood must be quantified. Second, message transmission must be prioritized according to the resulting utility values to maximize the utility received by all the vehicles participating in the network.

To describe the basic characteristics of optimizing the overall network utility in vehicular ad hoc networks, the following idealized and simplified scenario can be studied. Suppose there are a set NB of wireless stations (e.g., with $|NB| = 6$) that are in a neigborhood and therefore within mutual communication range. At one specific point of time t, each of them has at least one packet ready for broadcast. Since all network nodes have to share the wireless medium, only one of them is able to transmit at the same time. Each of the nodes ($n_i \in NB$) has a dedicated interest in certain information (in the form of packets $p_{n_i, j}$) that can be quantified with the help of a utility function $ur_i(p, t)$ that varies with time. To maximize the utility provided to the whole network, a global scheduler that has available detailed knowledge about all the nodes' packets and their respective utility functions would conduct a two-step process. First, a node-internal packet utility ranking is established. For each of the data packets $p_{n_i, j} \in P_{n_i}$ that are enqueued at a specific node n_i, the utility provided to all the adjacent nodes is computed. This normalized network utility ($\bar{U}$) as shown in Equation (7.16) is computed as the sum of all the $|NB|$ nodes' individual utility values provided by packet $p_{n_i, j}$. It should be noted that the utility for the broadcasting node n_i is obviously zero, since the information is already known.

$$\bar{U}(n_i, p_j, t) = \frac{1}{1 - |NB|} \left(\sum_{k=1}^{|NB|} ur_k(p_{n_i, j}, t) \right) \qquad (7.16)$$

The packet $p_{n_i}^{max}(t)$ of node n_i with

$$p_{n_i}^{max}(t) = \left\{ p \in P_{n_i} \,\middle|\, \bar{U}(n_i, p_j, t) = \max_{p \in P_{n_i}} [\bar{U}(n_i, p_j, t)] \right\} \qquad (7.17)$$

providing the maximum utility to all of the reachable neighbors at time t is considered the packet the node n_i should broadcast as the next one.

However, the internal packet ranking is not sufficient to maximize the global, aggregate utility provided to the whole network. In fact, the 6 packets in the example, which have won the node-internal ranking process, probably provide different utility to the rest of the network. Thus, as a second step, an external ranking is to be established that ensures that the node n_k whose highest ranked packet $p_{n_k}^{max}(t)$ provides the highest utility to the network (the maximum normalized network utility, $\hat{U}$) finally is granted access to the shared, wireless medium, transmitting packet p_{NB} with

$$p_{NB}(t) = \left\{ p \in \bigcup_{n_k \in NB} p_{n_k}^{max}(t) \, \middle| \, ur(p) = \max_{n_i \in NB} \left(ur\left(p_{n_i}^{max}(t) \right) \right) \right\} \quad (7.18)$$

Thereby, the overall network utility $\hat{U}(n_i, p_j, t)$ provided by transmitting packet p_{NB} is

$$\hat{U}(i, j, t) = \max_{n_i \in NB} \left\{ \max_{p_j \in P_i} \left[\frac{1}{1 - |NB|} \left(\sum_{k=1}^{|NB|} ur_k(p_{n_i, j}, t) \right) \right] \right\} \quad (7.19)$$

As mentioned, for the application in a vehicular ad hoc environment, however, a global, central packet scheduler that is able to see the network from a bird's eye view does not exist. Since ad hoc networks are operated decentrally, packet scheduling must be realized without the help of any central entity. In fact, each node will have to approximate the utility each of its packets provides to its adjacent nodes and thus establish a node-internal packet broadcast sequence. The necessarily limited knowledge of the nodes also prevents the realization of a perfect schedule. None of the network nodes exactly knows about its neighbors' individual interest in information (which is represented by the respective *ur*-functions).

However, when thinking of an efficient, utility-based message dissemination algorithm, the utility provided by a certain message must be computed prior the transmission. It must therefore be emphasized that there is a difference between actual receiver utility *ur* and the utility estimated by a transmitter (ut). *ut* is a value between zero (no utility) and 1 (maximum utility) and must be computed by a node prior to transmitting a data packet. It represents the normalized utility the sender assumes one of its packets currently would provide to its neighborhood. The sender will then leverage all the traffic differentiation mechanisms as described in the latter according to this value. Highly beneficial packets, for example, will be favored with respect to access to the medium. *ur*, on the other hand, is the actual utility received by one specific vehicle; on the basis of all existing contextual information the receiver has available at the time of reception. While the sender of packets can only estimate the utility provided to potential receivers due to their necessarily limited knowledge, the receiver is able to comprehensively evaluate how beneficial a specific piece of information is. In this context, we refer to the sum of all individual utility accounts of all vehicles at a certain point in time as the global benefit. It represents the benefit provided to all wireless-enabled cars within a scenario and is the measure to be maximized. The computation of the global network benefit is shown by Equation (7.20). By integrating over all the receiver benefit values ur_{n_i} computed by node n_i, one winds up with the aggregate benefit provided to node n_i. The global benefit [$GB(t)$] is the sum of all the respective values for all $|N|$ nodes participating in the network.

$$GB(t) = \sum_{i=1}^{|N|} \int_0^t ur_{n_i}(p, t)dt \tag{7.20}$$

A second, important assumption underlying the differentiating traffic according to its respective utility for other nodes is that adjacent nodes evaluate certain messages in a similar way, that is, the utility values estimated by a sender and provided to a receiver of the same message must be similar. This rationalizes that the expected utility can be properly estimated prior to the packet transmission. Thereby, two major assumptions must be fulfilled to allow for a sound utility approximation. First, functions for determining ut and ur must be similar in all the different vehicles produced by different car manufacturers. If cars estimate information utility completely different, the intrinsic paradigm of altruism is obviously violated. Second, we assume that the majority of vehicles in a neighborhood (within mutual communication range) are within a similar context as well. Interest in certain information can thus be regarded as similar, too. Provided that these postulates hold, an explicit prioritization of data according to ut helps meet the requirements of a scalable and efficient information dissemination vehicular ad hoc network.

7.4.3   Quantification of Traffic Information Utility

Utility provided by traffic information could be quantified with the help of measures such as reduced number of accidents, reduced environmental pollution, or shorter travel times [33]. It should be noted that the specific benefit of receiving certain traffic data is also affected by emotional impacts and is therefore often subjective. However, especially when regarding cooperative safety systems, the utility of a specific piece of information is rather objective. Given the current vehicle context, driver alerts or autonomous actions typically have to be performed in a well-defined way in order to prevent critical driving conditions.

Thereby, the specific utility a specific piece of traffic-related information provides to a receiving vehicle or the driver, respectively, depends on how much this information influences the further decisions [34]. To illustrate this main principle, imagine the following two example scenarios:

> **The Well, I don't care anyway scenario.** Vehicle A has encountered a critical road condition, for example, an icy spot on a curvy road. The information is disseminated through the ad hoc network. Vehicle B receives the information. Obviously, the utility for vehicle B is zero, if the B will not pass the critical location. This is, in particular, true if B has already passed the location, or it is not on its further route. The utility is also zero, if B is just approaching the critical location and there is no time left to intervene accordingly. Similarly, the utility of information about a traffic jam a couple of hundred kilometers away is also close to zero, even if it is on the planned route. This is obviously because the travel time to the reported area is comparably long, and therefore it is very likely that the traffic situation has already changed again until arrival.
>
> **The Yes, I already know scenario.** Again, vehicle A has encountered a critical road condition, for example, an icy spot on a curvy road. The existence of the event has been confirmed by various other vehicles. Vehicle B receives all of those confirmations. Obviously, even if there is still some additional utility in sharing those confirmations, the overall utility approximates again zero, because vehicle B is already sufficiently informed and confident.

Those two short examples make it obvious that even the same piece of information may have different utility for different vehicles, depending on the individual previous knowledge and the current driving task, the latter in particular comprising the planned route. The utility of an information artifact can therefore be quantified as weighted influence upon the individual local view on the vehicle's driving context, where the weight correlates to the impact on the further decisions on the system. For a formal mathematical description of the quantification of the information utility please refer to Strassberger's thesis [34].

The utility thereby depends on a variety of different environmental parameters, leading to a noticeable computational overhead. However, it can be realized that in most cases the utility depends on a similar parameter in a similar way. For example, typically the utility decreases with increasing distance to the origination of the information and its age. It is therefore reasonable to estimate the expected utility based on a variety of different heuristics taking into account specific sets of context parameter, with individual specific configuration. The configuration thereby deals with the usual dynamics of a specific context parameter changing over time. A spot of aquaplaning, for example, usually does not change within minutes whereas the position of the end of the traffic jam does. It is important to note that beside the estimation of the utility for the immediate receiving nodes, also the further dissemination progress must be considered.

Based on those considerations, the expected utility of traffic-related messages *ut*, can heuristically be quantified taking into account various parameters that can be associated with the informational, the environmental, or message context. The informational context ensures that the specific data contained in the messages is adequately evaluated with regard to its utility, where parameters from the environmental context take into account the vehicular and network-specific situation, which may influence message utility as well. Last, message-specific characteristics such as message age play an important role for utility determination as well. All of these parameters must be converted into preliminary utility values with the help of subfunctions. A subfunction is devoted to quantifying the utility a message would provide to neighbor nodes only with regard to one specific parameter. By summing up the weighted results of the subfunctions, a single utility value per message can be computed. The exact functionality of the utility computation will be explained later.

Message context. T first parameter defining the message context is a message's age, which can be easily computed as the difference between actual system time and the timestamp contained by a message. Kosch presented an exemplary function for quantifying the utility messages of different age provided to recipients [35]. Intuitively, a piece of information that was generated half an hour ago may be, depending on the respective application, not very useful any more. For demonstrating the application dependency of the exact utility subfunction, consider the following example. The utility provided by a message warning against a recent rear-end collision ahead certainly drops rather fast, since the danger conventionally is eliminated after a relatively short period of time. A notification about an oil flick, however, is certainly valid for a longer time span.

As explained above, vehicles are assumed to try to broadcast all messages contained in their local memory in case a new neighbor has been detected. For determining the utility provided by each of the stored messages for the potential recipients in the neighborhood, the latest time of broadcast is an important parameter as well. The longer a message has not been transmitted, the higher the likelihood that neighboring cars do not know it so far. The reception of a new message will provide high utility to the recipients. As a result, the subfunction incorporating this parameter increases from zero to one with an increasing period of time that has passed since the latest broadcast. The exact distribution of the function again strongly depends on the concerned application.

Also, if a specific message has been received only a short while ago a new broadcast will only add low additional utility to the neighbors since it can be assumed that the message is well known to most of them. The subfunction that takes into account the latest time of reception of a message thus also increases from zero to one with an increasing time span that has passed since the latest reception.

Environmental context. Similar to approaches like the contention-based forwarding [7], different receivers of the same message are differently appropriate for relaying the message to other nodes. The higher the distance between sender and receiver of a message, for example, the greater the chance to reach as many as possible new neighbor nodes. The respective subfunction must rise from zero to one with an increasing forwarder distance. By applying this subfunction, the selection of the most suitable forwarders is supported, thereby ensuring an as fast as possible dissemination of a message. With the help of the two parameters "forwarder distance" and "latest reception," static approaches [7,8] can adaptively be incorporated. As soon as the node that is the farthest away from the last sender has finished its broadcast, the other candidates for rebroadcasting the message will overhear this transmission and automatically adjust the estimated utility of their respective packets to rather low due to the application of the parameter "latest reception." They assume that most of the vehicles in their neighborhood know the packet now, making a further transmission superfluous. The subfunction for evaluating this parameter must increase from zero to one with an increasing distance between sender and potential forwarder.

The current number of reachable neighbor nodes is one further parameter that is important for determining the utility a message may provide to the rest of the network. Traffic prioritization must be realized both among packets that wait for transmission within one node and among those waiting at different nodes. Suppose there are two vehicles close by a road crossing and both intend to broadcast the same message. One of them is located exactly in the middle of the crossing, reaching all vehicles driving on each of the four adjacent roads. The other one stands in a street canyon, where high building walls prevent it from reaching as many other vehicles as its "competitor" can. To maximize the utility provided to the overall network, the packet waiting at the car standing right in the middle of the crossing should be assigned a higher utility value than the other one. Methodologies allowing for a realization of such an intervehicular data contention will be presented and analyzed later. A proper subfunction accounting for the evaluation of this parameter increases from zero to one with a growing number of reachable neighbor nodes.

For some of the applications, the distance to adjacent nodes and their respective driving directions is crucial for the utility evaluation of broadcasted packets. In the context of a crossing collision warning application, for example, the utility provided by periodically broadcasted beacons depends on whether there are close-by vehicles that drive into different directions or not. If so, emitted beacons warn the adjacent nodes of an imminent collision and thus provide significant utility for them.

Driving flexibility is a further important criterion for quantifying the utility provided by a message for a vehicle. A congestion warning, for example, is of limited use, if no alternative route exists which could be chosen instead of the road the recipient is currently driving on. Also, the individual properties of the respective vehicles must be considered when evaluating whether there is an alternative route available or not. A huge truck, for example, may not be able to drive off a congested highway, if it is not authorized to use the existing by-passes. Such information regarding the availability of alternative routes, depending on vehicle properties can be obtained with the help of the navigation system. In its simplest form, a proper subfunction for assessing this parameter is a step-function (0, if no alternative route exists; 1, if an alternative route is available).

Last, the scheduled travel route plays a role for the utility determination. An accident that happened besides the route is not necessarily relevant for a car. An appropriate subfunction can thereby leverage information from the standard navigation interface that has both road maps and its car's driving direction available. The last two parameters (driving flexibility and travel route relation) can hardly be correctly estimated by cars for their neighbor nodes. These parameters can only be used in the frame of a context translation scheme, where cars exchange detailed information about their individual contexts to notify others about their specific information interest prior to a data transmission. An appropriate subfunction accounting for the travel route relation parameter again could be a step-function. Only if an event is on or at least close to the scheduled travel route, the message is assumed to provide utility.

Informational context. The distance to information source is one of the major parameters determining the utility of messages. It can be assumed that a driver's interest in a certain piece of information decreases with an increasing distance between himself and the information source. The notification about an accident that has occurred several miles away can be assumed to be rather irrelevant for a driver. Kosch proposed an exemplary distribution of the utility depending on the distance between original sender and recipient Figure 7.3 [35]. Again, the exact subfunction heavily depends on the concerned application type. One could argue that applying the two parameters "forwarder distance" and "distance to information source" could be conflictive since they relate geographical distance and message utility differently. This is not completely right; while the forwarder distance parameter only considers the distance between the last forwarder of the packet and the receiver, the distance to information source parameter processes the distance between the original issuer and the actual receiver.

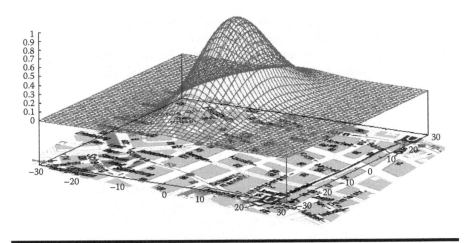

Figure 7.3 Exemplary utility sub-function for computing the parameter "distance to information source" as presented in Kosch [35].

Thus, the first parameter only locally influences the estimated utility of a message, where the latter determines information utility from a more global perspective.

To only leverage the distance of information source would not be sufficient, however. The utility of pieces of information may vary with the time of day. During the rush hour, a notification about a free parking lot that is located 500 m away from the driving target provides certainly more utility than during the night where there are many parking lots available close-by the target.

The purpose of traveling is an extremely important criterion as well. When driving for shopping, for example, information regarding parking lots is regarded as very useful. Depending on what drivers are planning to do at their destinations, messages differ with respect to utility. For vehicles following the altruistic approach introduced above, however, it is very difficult to determine the utility of messages provided to neighbor vehicles based on their respective purposes of traveling. Again, this can only be realized with the help of a context transfer prior to the broadcast of messages. To be able to comprehensively evaluate the utility of all its messages, a vehicle would need to have detailed knowledge about all of the adjacent nodes available, thereby generating lots of overhead traffic.

Next, the quality and credibility of the provided information plays an important role for the utility determination process. The more comprehensive and credible the data contained in a message, the more utility is provided to the recipient. Credibility can be quantified by comparing previously received information or own sensor values with the newly arriving message, where different degrees of quality must be defined specifically for each of the envisioned applications.

The rate of change of information can influence the message utility as well. Receiving a message notifying about an item lying on the road may be outdated after a shorter period of time (since the item may have moved) than a message warning against an oil flick (whose location is rather stable). Continuous or step functions representing the degree of correlation between received information and data obtained by the own sensors will work as proper subfunctions.

To account for a general difference of utility values between various applications, the information category is introduced as a further factor for quantifying message utility. Data packets stemming from a real-time, safety-of-life application, for example, should always be considered to provide more utility to the recipients than messages informing about free parking lots. A step-function comprising different utility values for each of the envisioned applications is to be set up.

Last, the news value of a certain piece of information must be evaluated. For example, a congestion warning's utility strongly depends on whether the receiver knows about the congestion already. In case a newly received message is exactly equal to a previously received notification, there is no additional utility provided at all. If the information contained slightly differs from already known data, some utility is provided, whereas a message containing fully unknown information delivers maximum utility. Exact distributions of the subfunctions again depend on the respective application. In the context of a traffic congestion warning service, for example, the deviation of the congestion's starting and end points compared to previously known data can serve as an appropriate measure. Nodes will not be able to exactly quantify the news value a message will provide to its neighbor nodes. However, they may assume a similar level of information available in their neighborhood and take the news value provided to them as an indicator for the utility provided to adjacent nodes.

As argued above, all these parameters have to be translated into preliminary utility values with the help of the subfunctions. The results are then weighed and summed up to generate one single utility value (ut) per packet p at a specific point in time t. Equation (7.21) highlights the process to quantify the utility a specific piece of information is expected to provide to other nodes (ut). Suppose there are Π parameters such as distance to information source and message age. They all have to be computed with the help of the message-(m), vehicle-(v), and information context (i). The data sources for computing all parameters from the three contexts are the actual message, previously received and stored messages, and the vehicle-specific situation. All Π parameters are evaluated with application-dependent heuristic subfunctions $b_i (b_i \in [0, 1])$. The results of these functions finally have to be weighed with application-dependent factors a_i. In

Benefit based message dissemination

General principle

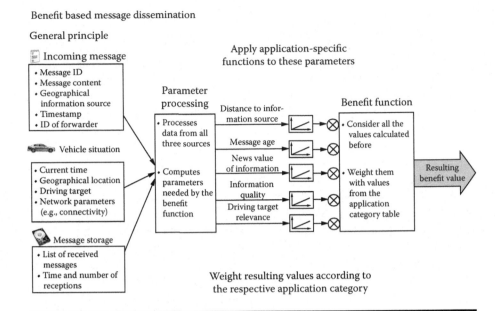

Figure 7.4 Computation of the packet utility value.

the context of a crossing collision warning, for example, the message age parameter can be assumed to be more important for the message utility determination than the parameter evaluating the information quality. Last, all weighed parameters have to be summed up and divided by the sum of the a_i. Figure 7.4 shows the general principle of determining one single utility approximation for a certain piece of information.

$$ut(p, t) = \frac{1}{\sum_{i=1}^{\Pi} a_i} \sum_{i=1}^{\Pi} a_i b_i(m, v, c, t) \qquad (7.21)$$

All the heuristic subfunctions b_i and also the weights a_i differ with regard to the respective application. Certain parameters play a major role in determining the utility of messages from one application, where the same parameters are irrelevant in the context of another service. The utility provided by crossing collision warning messages, for example, does not vary with the purpose of traveling of individual vehicles.

Another important insight is that there is not one generally valid and optimal function for determining *ut* or *ur*. In fact, car manufacturers will have to agree on a common understanding of utility quantification and thereby set up a comprehensive model featuring all required subfunctions and weights covering all envisioned applications. In the context of traffic-related data dissemination, car manufacturers will also have to set up similar field tests with a significant amount of test drivers to find out about the experienced utility of various information artifacts/pieces of information, e.g., break down warnings, weather information, traffic jams, etc.

7.4.4 Intervehicle Contention for the Shared Medium

As mentioned, the intervehicle contention for the shared medium must be adapted such that the most relevant packet all nodes within mutual communication range have available is granted access to the medium. For the purpose of traffic differentiation, the dynamic adaptation of the contention window (CW) within the MAC layer is considered as a major lever. Each station uses its individual CW to select both backoff and defer timers, which represent a key feature of the CSMA mechanism. The CW size is initially set to CW_{min}, and is increased each time a transmission attempt fails with an upper limit of CW_{max}. In this case, another backoff is performed using the new CW value. After each successful transmission, the CW value is reset to CW_{min}, and the station that successfully completed the transmission defers its activities for another DIFS (distributed coordination function interframe space) and conducts a post-backoff.

By selecting short contention windows in case a highly relevant packet requests access to the medium, the node's likelihood to win the contention process is increased. The smaller the CW, the shorter the timers defining the period of time after a node may try or retry to access the shared medium. Equation (7.22) shows the modified, utility-based computation of a defer timer, whose starting integer value is randomly determined from a uniform distribution over the interval [0,CW].

$$\text{defer timer} = (\text{rand mod} CW_{bb} + 1) \cdot \text{slottime} \tag{7.22}$$

Rand R is a random number and CW_{bb} represents the node's current, utility based CW. Then

$$CW_{bb} = f_A\{ut[p_{n_i}^{max}(t_i)]\} \tag{7.23}$$

with t_i is the point in time the contention is initiated, and f_A is a dedicated mapping function. The most simple example would be linear mapping of the expected utility ut and the CW, as, for example,

$$CW_{bb} = [1 - ut(p_{n_i}^{max})] \cdot (CW_{max} - CW_{min}) + CW_{min}$$

Thus, the CW a node applies can always be adapted to the utility ($ut \in$ [0, 1]) of the currently handled message. For a more detailed elaboration of different mapping function, please refer to Strassberger's thesis [34].

Besides the mere adaptation of the nodes' actual CW sizes according to estimated packet utility values, one could think about taking one step further and passing on the transfer of remaining timer values into the next contention periods. Instead, timers are newly computed for each contention period, reflecting the expected utility (ut) of the currently treated data packets. This helps avoiding the following situation. A node trying to transmit a rather redundant packet has initiated a relatively long timer some contention periods ago. After waiting for a sufficient amount of periods, its remaining timer is short and even may be shorter than the one started by another node that tries to transmit a highly beneficial packet. This ensures a medium access strategy that is fully utility-oriented and lacks any utility-agnostic, fair resource allocation tendency. Not the time a packet has already been waiting, but its up-to-date utility value decides upon its chance to get medium access. Note that timers are still continuously counting down also in the frame of this scheme as long as the channel is idle. In fact, they are only newly set each time a transmission occurred on the shared medium. As a consequence, a total starvation of packets cannot occur as long as the medium is idle for a sufficient period of time.

Instead of adapting the contention windows, the defer timers could also be adapted with respect to the estimated packet utility. This approach eliminates any random effects and therefore optimally maps the principle of joint fairness as introduced above. In order to keep packet collisions on the wireless channel low, however, it must be ensured that the resulting utility values distribute enough, so that it is unlikely that two or more vehicles within a neigborhood end up with the same utility value.

7.5   Conclusions

Information dissemination is one crucial task to accomplish for intervehicle communication, supporting active safety through foresighted driving. A variety of sophisticated approaches to either solve the scalability or the connectivity problem have been proposed during the last years. Due to the dedicated characteristics of intervehicle communication, in particular, the long product life cycles of cars, the principle of network utility maximization, which is well known from common IP-based wired networking, provides a valuable concept to share the available channel capacity. However, the process of network utility maximization in vehicular networks inherently differs from common approaches; in particular the utility has to be quantified per packet in contrast to usually deployed rate allocation schemes. In addition, the packet utility must be estimated prior to the packet transmission. An explicit situation-dependent utility consideration thereby provides a well-argued metric to delay or discard specific packets in case of network congestion.

An integrative utility maximization framework that enables to exploit arbitrary optimizing proposals for both intelligent flooding and data traffic differentiation thereby is key enabler to satisfy all dedicated requirements of vehicular ad hoc networks; in particular, the potential conflicting goals of minimizing the network load (that is, related to the scalability problem in dense networks) and maximizing the reliability of information dissemination (that is, related to the connectivity problem in sparse intermittent connected networks).

It could be shown that the overall utility of the network can be significantly increased by an explicit utility maximization if the medium access strategy is consequently mapped to the expected utility measure [34,36,38–40].

An explicit utility maximization as described in this chapter features the following qualitative characteristics:

Altruism. Altruism is the major overall characteristic of cooperative vehicles. Nodes do not primarily aim at maximizing their own utility, but head for transmitting information such that their neighbors are provided with the data they are most interested in. Before transmitting a data packet, each node evaluates the utility of its currently available messages and broadcasts the one currently providing the most utility to its neighbors. Suppose the neighbors' interest in information is reasonably met, the utility provided to these nodes is higher than in the case of an undifferentiated packet scheduling.

Application-oriented information differentiation. Most approaches to improve the performance of message dissemination in mobile ad hoc networks as introduced in Section 7.3 mostly rely on packet-specific data such as the geographical location of the last forwarder or the time at which the packet was generated. However, taking into account application-level data enables an in-depth utility-evaluation of single messages. Two different packets notifying about a rain field, for example, do not necessarily provide the same utility to a receiver. Depending on the news values of the two messages with respect to the current knowledge, the utility values may significantly vary. If a receiver has not heard of the rain field at all so far, the data packet is assumed to be of great value. In case the vehicle has received ten similar messages already, only slightly varying regarding the exact position of the rain field, the newly arriving information is of low utility. This information evaluation is not foreseen by most of the existing approaches so far.

Decentrally controlled message scheduling. The concept of utility maximization as presented does not comprise one single, central schedule for packet transmission. In fact, each vehicle autonomously evaluates the utility certain data may provide to potential recipients only based on its local and necessarily limited knowledge. One of the great advantages of this approach is that maintenance of scheduling tables or any connection to a central controller is superfluous, thereby significantly increasing the scalability of such framework. Due to the limited cognition of the individual nodes, the proposed system cannot ensure ideal message dissemination, but a significant performance improvement in comparison to information-agnostic message dissemination strategies.

Democratically inspired utility evaluation scheme. Every single vehicle has to decide which packets to broadcast with what priority. Depending on the nodes' interest in a certain packet, its dissemination can be accelerated or discriminated. Suppose a network intruder broadcasts useless packets just to overload the ad hoc network, the propagation of its packets is automatically limited to a minimum.

Sustainability through flexibility. The presented framework for utility maximization provides full flexibility to changing requirements and will work both in dense and sparse networks. Due to nonstatic transmission restrictions such as the in case of many existing approaches like "Geocast" [37], network resources can be optimally leveraged in different network environments. "Hard" limitations in both the local and time domain are avoided. Messages are relayed anyway, where traffic prioritization ensures that low-priority messages do not absorb the wireless medium and prevent high-priority messages from being transmitted. Thus, an explicit utility maximization will provide an excellent basis for the initial deployment of vehicular ad hoc networks, but is equally appropriate for future scenarios where the density of wireless-enabled vehicles is high.

Integrating approach. To make broadcast communication reliable and efficient and to guarantee optimal information dissemination, there are many specific approaches existing that all tackle part of the challenges inherent to vehicle ad hoc networks. The described utility-oriented approach thereby works as a comprehensive substitute for many of these approaches. Only to deploy a static methodology based on probabilistic, area- or neighbor-knowledge–based techniques does not fully meet the varying requirements of highly dynamic vehicular ad hoc networks. As mentioned, in sparse networks, a static retransmission limitation prevents from an as fast as possible message propagation. Although the channel is idle, a node might pass on broadcasting a data packet because it currently is beyond a geographical transmission threshold. In the frame of an explicit utility-oriented approach as presented, in contrast, the number of message transmission underlies flexible limits. If no other, more relevant piece of information requests access to the medium, a message is broadcasted even if the expected utility for potential receivers is extremely low. The proposed framework for utility maximization integrates ideas such as those incorporated in the contention-based forwarding approach, for example. As explained before, the distance between last forwarder and the receiver influences message utility values. The higher this distance, the more utility a data packet is assumed to have for the network. The suppression of other forwarders can then be realized with the help of the parameter "latest time of reception." As soon as the node that is the farthest from the last forwarder has finished broadcasting a packet, the adjacent nodes overhearing this transmission assume that most other nodes have received this piece of information as well. The expected utility of a further transmission is thereby significantly reduced.

Next, neighbor-knowledge–based approaches as introduced above are leveraged as well. The number of neighbors, for example, is one criterion for determining the utility of messages different vehicles have available. The larger the number of new nodes reached by one car, the higher its priority with respect to medium access. Numerous ideas for realizing traffic differentiation presented above can be incorporated into the presented utility maximization framework as well. We also establish some kind of global schedule and ensure that nodes with particularly relevant data are favored regarding medium access. We, however, pass on the complex establishment and maintenance of scheduling tables (such as proposed for DPS) and rely on a completely autonomous scheme. The basic principle of the IEEE 802.11e protocol, the realization of internal and external contention for the shared medium, is comprised by the presented approach as well, where a rough-granular classification can be passed on into a limited amount of traffic categories.

Finally it should be noted that an explicit utility metric on a per packet basis as described in the chapter can also be used for a fine-grained data differentiation on different channels or even different networks like cellular systems. The underlying metric can be uniformly applied and arbitrarily supplemented with further context aspects and heuristics, for example, taking into account dynamic adaptations of transmit power, specific favorable road locations like intersections or potential transmissions costs as in cellular networks.

References

[1] B. Clifford Neuman, "Scale in distributed systems," *Readings in Distributed Computing Systems*, Los Alamitos, CA: IEEE Computer Society, 1994: 463–489.

[2] J. Li, C. Blake, D. S. J. De Couto, H. I. Lee, and R. Morris, "Capacity of ad hoc wireless networks," In *Proc. of the 7th ACM International Conference on Mobile Computing and Networking (MobiCom 01)*, Rome, Italy, July 2001.

[3] K. Matheus, R. Morich, and A. Luebke, "Economic background of car-to-car communication," In *Proc. of 2. Braunschweiger Symposium Informationssysteme fuer mobile Anwendungen (IMA 2004)*, Braunschweig, Germany, 2004.

[4] NoW: Network on Wheels Projekthomepage, http://www.network-on-wheels.de (accessed July 7, 2008).

[5] B. Williams and T. Camp, "Comparison of broadcasting techniques for mobile ad Hoc networks," In *Proc. of the ACM International Symposium on Mobile Ad Hoc Networking and Computing (MOBIHOC)*, Lausanne, Switzerland, June 9–11, 2002.

[6] S. Ni, Y. Tseng, Y. Chen, and J. Sheu, "The broadcast storm problem on a mobile ad hoc network," In *Proc. of the ACM/IEEE International Conference on Mobile Computing and Networking (MOBICOM)*, Seattle, WA, (1999):151–162.

[7] H. Fuessler, J. Widmer, M. Kaesemann, and M. Mauve, "Contention-based forwarding for Mobile Ad-Hoc Networks," *Ad Hoc Networks* 1 (2003):351–369.

[8] C.-F. Chiasserini, R. Gaeta, M. Garetto, M. Gribaudo, and M. Sereno, "Efficient broadcasting of safety messages in multlhop vehicular networks," In *Proc. of 5th International Workshop on Performance Modeling, Evaluation, and Optimization of Parallel and Distributed Systems (PMEO-PDS 2006)*, Rhodes Island, Greece, April 2006.

[9] K. Kim, Y. Cai, and W. Tacanapong, "A priority forwarding technique for efficient and fast flooding in wireless ad hoc networks," In *Proc. of the ICCCN '05*, San Diego, CA, October 2005.

[10] M. Ruffini and H.-J. Reumermann, "Power-rate adaptation in high-mobility distributed ad-hoc wireless networks," *In Proc. of 2nd International Workshop on Intelligent Transportation*, Hamburg, Germany, March 15–16, 2005.

[11] V. Kanodia, C. Li, A. Sabharwal, B. Sadeghi, and E. Knightly, "Ordered packet scheduling in wireless ad hoc networks: Mechanisms and performance analysis," *In Proc. of the ACM International Symposium on Mobile Ad Hoc Networking and Computing (MOBIHOC)*, Lausanne, Switzerland, June 2002.

[12] V. Kanodia, C. Li, A. Sabharwal, B. Sadeghi, and E. Knightly, "Distributed priority scheduling and medium access in ad hoc networks," *Wireless Networks* 8 (September 2002): 455–466.

[13] J. Sheu, C. Liu, S. Wu, and Y. Tseng, "A priority MAC protocol to support real-time traffic in ad hoc networks," *Wireless Networks* 10 (January 2004): 61–69.

[14] A. Lindgren, A. Almquist, and O. Schelen, "Quality of service schemes for IEEE 802.11—A simulation study," In *Proc. of the Ninth International Workshop on Quality of Service (IWQoS 2001)*, Karlsruhe, Germany, June 6–8, 2001.

[15] J. Deng and R. Chang, "A priority scheme for IEEE 802.11 DCF access model," *IEICE Transactions on Communications* E82-B, (1) (January 1999).

[16] M. Torrent-Moreno, D. Jiang, and H. Hartenstein, "Broadcast reception rates and effects of priority access in 802.11–based vehicular ad-hoc networks," *In Proc. of the First ACM Workshop on Vehicular Ad hoc Networks (VANET)*, Philadelphia, PA, October 2004: 10–18.

[17] IEEE, IEEE 802.11 Wireless Local Area Networks, The Working Group for WLAN Standards. http://grouper.ieee.org/groups/802/11/ (accessed December, 2008).

[18] S. Choi, J. Prado, S. Shankar, and S. Mangold, "IEEE 802.11e contention-based channel access (EDCF) performance evaluation," *In Proc. of the IEEE International Conference of Communications (ICC)*, Anchorage, AK, May 2003.

[19] P. Garg, R. Doshi, R. Greene, M. Baker, M. Malek, and X. Cheng, "Using IEEE 802.11e MAC for QoS over Wireless," Technical Report, Computer Science Dept., Stanford University, 2002.

[20] A. Lindgren, A. Almquist, and O. Schelen, "Quality of service schemes for 802.11 wireless LANs, an evaluation," *Mobile Networks and Applications (MONET)* 8 (2003): 223–235.

[21] I. Aad and C. Castelluccia, "Differentiation mechanisms for IEEE 802.11," *In Proc. of the IEEE INFOCOM*, Anchorage, AK, April 2001.

[22] B. Li and R. Battiti, "Performance analysis of an enhanced IEEE 802.11e distributed coordination function supporting service differentiation," *In Proc. of the International Workshop on Quality of Future Internet Services (QoFIS)*, Barcelona, Spain, October 2003.

[23] H. Zhu and I. Chlamtac, "An analytical model for IEEE 802.11e EDCF differential services," In *Proc. of the International Conference on Computer Communications and Networks (ICCCN)*, Dallas, TX, October 2003.

[24] L. Romdhani, Qiang Ni, and T. Turletti, "Adaptive EDCF: Enhanced service differentiation for IEEE 802.11 wireless ad hoc networks," In *Proc. of IEEE Wireless Communications and Networking Conference*, New Orleans, LA, March 16–20, 2003.

[25] G. Wischhof and H. Rohling, "On utility-fair broadcast in vehicular ad hoc networks," In *Proc. of the 2nd International Workshop on Intelligent Transportation (WIT 2005)*, Hamburg, Germany, March 2005: 47–51.

[26] A. Carzaniga and A. L. Wolf, "Forwarding in a content-based network," In *Proc. of ACM SIGCOMM 2003*, Karlsruhe, Germany, August 2003: 163–174.

[27] F. P. Kelly, A. Maulloo, and D. Tan, "Rate Control for communication networks: Shadow prices, proportional fairness and stability," *Journal of Operations Research Society* 49(3) (March 1998):237–252.

[28] R. J. La and V. Anantharam, "Utility-based rate control in the Internet for elastic traffic," In *Proc. of IEEE/ACM Transactions on Networking*, (TON) 10(21) (April 2002): 272–286.

[29] X. Gao, T. Nandagopal, and V. Bharghavan, "Achieving application level fairness through utility-based wireless fair scheduling," In *Proc. of the IEEE Global Telecommunications Conference*, San Antonio, TX, 2001.

[30] F. Kelly, "Charging and rate control for elastic traffic," *European Transactions on Telecommunications* 8(1) (January 1997):33–37.

[31] J. Li, C. Blake, D. S. J. De Couto, H. I. Lee, and R. Morris, "Capacity of ad hoc wireless networks," In *Proc. of the 7th ACM International Conference on Mobile Computing and Networking*, Rome, Italy, July 2001.

[32] P. Gupta and P. R. Kumar, "The capacity of wireless networks," *IEEE Transactions on Information Theory* IT-46(2) (March 2000):388–404.

[33] R. Schwarz, W. Schaufelberger, L. Raymann, H. Merz, F. Zaugg, T. Kloth, and P. Farago, "Wirksamkeit und Nutzen von Verkehrsinformation," Forschungsauftrag SVI 2000/386 auf Antrag der Vereinigung Schweizerischer Verkehrsingenieure (SVI), August 2004.

[34] M. Strassberger, "Kontextbereitstellung in automobilen Ad-hoc Netzen," PhD thesis, University of Munich, Germany, 2007.

[35] T. Kosch, "Situationsadaptive Kommunikation in Automobilen Ad-hoc Netzen," PhD thesis, Munich University of Technology, Germany, 2005.

[36] C. Maih, W. Franz, and R. Eberhardt, "Stored geocast," *Kommunikation in Verteilten Systemen (KiVS)* (2003):257–268.

[37] C. Schroth, "An altruistic approach for message dissemination in vehicular ad hoc networks," Masters Thesis, Munich University of Technology, Germany, 2006.

[38] C. Adler, S. Eichler, T. Kosch, C. Schroth, and M. Strassberger, "Self-organized and context-adaptive information diffusion in vehicular ad hoc networks," *In Proc. of the International Symposium on Wireless Communication Systems (ISWCS)*, Valencia, Spain, September 2006.

[39] C. Adler, R. Eigner, C. Schroth, and M. Strassberger, "Context-adaptive information dissemination in VANETs—Maximizing the global benefit," In *Proc. of the Fifth IASTED International Conference on Communication Systems and Networks (CSN 2006)*, Palma de Hallorca, Spain, August 2006.

[40] S. Eichler, C. Schroth, T. Kosch, and M. Strassberger, "Strategies for context-adaptive message dissemination in vehicular ad hoc networks," In *Proc. of the Second International Workshop on Vehicle-to-Vehicle Communications (V2VCOM)*, San Jose, CA, July 2006.

Chapter 8

Message Scheduling within Vehicular Networks

Mohamed Shawky and Mohamed Salah Bouassida

Contents

The basic objective of dynamic message scheduling is to avoid and control congestion within network, while taking into account the quality of service required for each type of transmitted data. Congestion control or avoidance aims to best exploit the available network resources while preventing sustained overloads of network nodes and links [3]. Appropriate congestion control mechanisms are essential to maintain the efficient operation of a network. The congestion indicators are the size of the neighbor table, the number of retransmissions "heard," the message transmission delay, the size of the messages stack, and so on. Ensuring congestion control within vehicular ad hoc networks VANETs addresses special challenges, due to the characteristic and specificities of such environment (high dynamic and mobility of nodes, high rate of topology changes, high variability in nodes density and neighborhood, broadcast/geocast communication nature, and the like). In this chapter we present a congestion control approach, based on the concept of dynamic priorities-based scheduling to ensure a reliable and safe communication architecture within VANET. Message priorities are dynamically evaluated according to their types, the network context, and the neighborhood.

8.1 Context and Motivations

Dynamic message scheduling is a challenging issue within vehicular ad hoc networks. Indeed, it should take into account the characteristics of VANET while ensuring the quality of service required by the applicative level (Figure 8.1). Vehicular ad hoc networks are a form of MANETs used for communication among vehicles and between vehicles and roadside equipment (Figure 8.2). In addition to the challenging characteristics of MANETs (such as lack of established infrastructure, wireless links, multi-hop broadcast communications), VANET bring new challenges to achieve safe communication architecture within such environment. Indeed, within VANET networks, nodes are characterized by high dynamics and mobility, in addition to the high rate of topology changes and density variability.

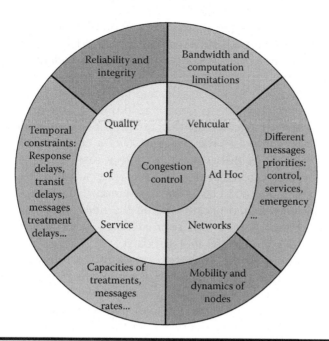

Figure 8.1   Congestion control context within VANET.

Stibor et al. [15] evaluate the neighborhood nature of vehicular networks within a four highway lane context (two lanes for each direction). They carried out simulations and analysis that show that the average number of potential communication neighbors is approximatively four. In addition, in 50% of all occurrences, the maximum potential communication duration is 1 sec; in 90% of the occurrences, the upper boundary for the communication time is 5 sec. Another important constraint in the multihop

Figure 8.2   Vehicular ad hoc network [17].

intervehicular communications is the limited bandwidth within such environment. Indeed, the wireless channel can be occupied by competitive nodes for many reasons (collisions, interferences, insufficient signal strength, duration of the transmission sequence, and the like) [7].

To deal with these environment constraints and to ensure safe and optimized communication architecture (to guarantee required services on a "best effort" network) the establishment of a quality of service policies becomes mandatory, which requires the conception of a congestion control approach within VANET. In this context, we propose in this chapter a new congestion control approach dedicated to operate within vehicular networks, integrated within the 802.11p underway standard, and based on dynamically scheduling transmitted packets within the network according to their priorities. This congestion control approach was proposed by the C2C consortium to be integrated to the ISO standard.

To present our contributions, this chapter is structured as follows. Section 8.2 presents a taxonomy of congestion control approaches within wireless and vehicular networks. In Section 8.3, we present our congestion control approach, based on dynamic scheduling of messages according to their priorities. We start by giving an overview on the virtual channeling concept of the 802.11p underway standard. We present then our dynamic priorities assignment and messages transmission scheduling approach. The real applicability of our approach is validated, via formal verification, implementation, and measurements, which we present in Section 8.4. Finally, Section 8.5 concludes this chapter.

8.2 Congestion Control Approaches

8.2.1 State-of-the-Art Congestion Control Schemes

During the last few years, several congestion control approaches have been presented, dedicated to operate within wireless networks. In this section, we cannot claim to present an exhaustive study of these approaches. However, we distinguish two congestion control schemes for wireless networks: end-to-end and hop-by-hop families. End-to-end protocols aim to ensure flow fluidity between senders and receivers, without worrying about the internal relay nodes, whereas hop-by-hop congestion control methods take into account the capacities of the internal links. We present in the following the principal protocols belonging to these two approaches.

8.2.1.1 End-to-End Approaches

- Raghunathan and Kumar [8] show that congestion control schemes dedicated to operate within wired networks are not suitable for wireless environment and that a cross-layer redesign is needed to take into account the nature of wireless communications. Indeed, with TCP, a flow can only obtain congestion feedback from every directed link along its path. However, nearby links not directly on the path can interfere with this flow because the fundamental difference between wireless and wired environment is that wireless communications are of a broadcast nature.

■ In the proposal of Rath and Sahoo [9], an end-to-end congestion control technique is presented, carried out by TCP and physical layers. The adaptive windows-based congestion control mechanism used by TCP for a wired network may not be appropriate for a wireless network. This is due to the time varying nature of a wireless channel and interference from other nodes causing packets loss, which is different from packet loss due to congestion. But, TCP's congestion control mechanism does not discriminate packet loss due to congestion and due to bad channel or interferences; it rather applies the same congestion control mechanism for both. For this reason, within the proposed cross-layer approach, the MAC layer changes transmission power as per the channel condition and interference received from the neighboring nodes, whereas the TCP layer controls congestion using Reno-2 windowing flow control.

■ The protocol presented by Chen, Abate, and Sastry [2] updates the flow control model for a TCP-like scheme and extends it to the context of a wireless network. It proposes two new congestion control schemes. The first scheme employs a static algorithm while the second applies a dynamic one. Both algorithms modify the number of connections that a single user has with the network and thus provide an appropriate number of connections, opened at the application layer by a sender.

8.2.1.2 Hop-by-Hop Approaches

■ Yi and Shakkottai argue that hop-by-hop schemes are feasible over a wireless network [10]. Such schemes provide feedback about the congestion state at a node to the hop preceding it. This preceding node then adapts its transmission rate based on this feedback. Feedback is typically provided based on the queue length at the congested node. If the queue length exceeds a threshold, congestion is indicated and the preceding node is notified in order to decrease its transmission rate.

Hop-by-hop schemes require per flow state management in intermediate nodes, which generates scalability problems. However, in a wireless network, the number of flows per node is of much smaller order than in the Internet. Further, wireless networks usually have per flow queueing for reasons of packet scheduling, and the fact that different users are at different locations, thus requiring different physical layer strategies (such as the channel coding and modulation scheme of the power level).

The congestion control proposed in Yi and Shakkottai [10] is thus based on a hop-by-hop scheme, which is shown to converge in the absence of delay and allocates bandwidth to various users in a proportionally fair manner. The proposed hop-by-hop algorithm is established according to the queue length and the feedback delays.

■ Zang et al. [11] investigate congestion control problems in multi-hop wireless networks, with time-varying link capacities. By modeling time variations of capacities as perturbations of a constant link, the authors propose a primal-dual congestion control algorithm and prove its stability without delay consideration. In the presence of delay, they define theoretical conditions for the technique to be locally stable. Both proposals presented by Yi and Shakkottai [10] and Zhang et al. [11] are based on the studies carried out by Kelly et al. [6].

■ Zang et al. [12] present a congestion control approach in wireless networks for vehicular safety applications. The basic idea of this approach is to detect congestions using event-driver detection and measurement and to manipulate MAC transmission queues IEEE 802.11p, in order to ensure the safety messages sent on the control channel (Figure 8.3). The event driven congestion detection is triggered whenever a high priority safety message is detected to guarantee the QoS of the safety applications, while the

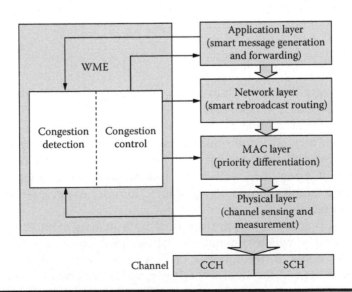

Figure 8.3   Cross layer congestion control architecture [12].

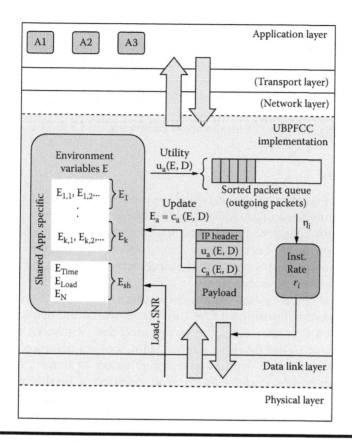

Figure 8.4   Implementation of a node [13].

measurement-based congestion detection consists of measuring the channel usage and comparing it with a defined threshold.

■ The congestion control approach proposed by Wischhof and Rohling [13], dedicated to operate within vehicular ad hoc networks, consists of adapting transmissions to the available bandwidth in a hop-by-hop manner. Thus, nodes transmitting information with a high utility for VANET will be allowed to consume a larger share of the available bandwidth (Figure 8.4). A priority is evaluated for each packet, depending on its utility and size. Then, an instantaneous data rate is determined, according to the computed priority. This approach requires context exchange between neighbor nodes, which generates a communication overhead.

■ Torent-Moreno et al. present a fair bandwidth sharing approach for VANET [16]. This approach consists of limiting the wireless load resulting from the periodic messages by requiring a strict fairness among the vehicles. The authors assume a constant packet generation rate and propose a centralized power control algorithm that provides the optimum transmission range of every node. This proposal was formally validated. However, simulations have been carried out under ideal conditions assuming that interferences between node transmissions follow a deterministic model. Moreover, the proposed algorithm requires a synchronization between vehicles, which generates a communication overhead.

8.2.2 Discussions

On the one hand end-to-end congestion control approaches are not suitable for vehicular ad hoc networks [10]. Indeed, within these approaches, relay nodes context are not considered, and thus, interferences, collisions, and transmission problems are not taken into account. However, the required quality of service of a transmission can be defined by the sender with end-to-end congestion control approaches.

On the other hand, hop-by-hop approaches suffer from lack of scalability when the number of transmitted flows increases within the network. However, it is known that the size of the transmitted data within VANET is not considerable due to the dynamic nature of this environment and to the node limitations in terms of storage and computation capacities. It is thus proved that hop-by-hop congestion control approaches are the most suitable for VANET, while taking into account the required quality of service of the transmitted data as for the end-to-end approaches. Hop-by-hop congestion control approaches, described above, present some drawbacks (generation of communication and computation overheads, reactive congestion control techniques, idealistic verification frameworks, and so on).

We propose in the next section a congestion control approach, considering these drawbacks, whose design should ensure the following objectives:

- Adaptive congestion control approach: An approach dynamically adaptable to the context of VANET, while taking into account the required QoS metrics (in terms of reliability and delays).
- Hop-by-hop approach: To take into account relay nodes, while considering the required quality of service as for the end-to-end congestion control approaches.
- Cross-layered approach: While remaining adapted to the IEEE 802.11p underway standard.
- Applicative layer congestion control approach: To define packet priorities according to their application and their utilities in the network.
- Definition of policies and directives: To fix a maximum duration of a transmission per VANET node and a maximum number of transmissions per time interval.

■ An efficient compromise: A compromise between proactive nature (to treat high-priority safety and emergency messages without delays) and low energy and computation consumption overheads. No additional equipment could be required or bandwidth consumption due to communication overhead should be generated.

8.3 Dynamic Message Scheduling Approach within VANET

In this section, we present a congestion control approach dedicated to operate within VANET. The basic idea of this applicative layer approach is to define congestion control policies, based on dynamically scheduled message transmission in the network. Message scheduling is carried out according to priorities, evaluated as a function of the utility of the concerned messages, the sender application, and the neighborhood context. To present our contributions, we describe first the 802.11p underway standard environment and its influence on the congestion control policies. We then present the principle steps of the approach: dynamic priority assignment, message scheduling, and message transmission. Finally, we summarize these steps in an algorithm.

8.3.1 802.11p Multichanneling

VANET architecture is dealing with a single media based on the European version of IEEE 802.11p (together with IEEE 1609.x). This media is currently under standardization. The Car-2-Car Communication Consortium (C2C-CC) working groups PHY/MAC and NET made the following assumption on the usage of the 20 MHz (multichannel operation):

> The CEC-CC compliant communication system shall be able to receive simultaneously on two neighbored channels with 10 MHz bandwidth. The received signals may originate from different transmitters. The two channels correspond to the 2*10 MHz channels that are claimed for critical road safety in ETSI TR 102 492-1 V1.1.1 Part1. The C2C-CC compliant devices shall have adjacent channel rejection Category 2 (IEEE 802.11p), i.e. they shall have an adjacent channel rejection of at least 37 dB for the 3 Mbits/s data rate. A C2C demonstrator (and vehicles in the phase of market introduction) may emulate the two channels by mapping two virtual channels on a single physical channel or by using two channels that are well separated in frequency. This emulation must be kept in mind in the design process for the dual-receiver system.

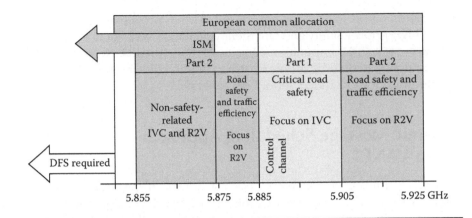

Figure 8.5   802.11p channeling.

The multichannel operation, as specified in the C2C-CC decision (Figure 8.5), implies that a simultaneous reception on two channels should be possible, in case the incoming signal levels do not differ by more than the adjacent channel rejection value (37 dB). Transmitting on both channels simultaneously will not be required. To ensure compatibility with 802.11a, the system might be emulated using IEEE 802.11a or single receiver IEEE 802.11p.

The two VANET wireless channels (control and service channels) are used for different traffic as such further congestion control mechanisms will differ slightly.

8.3.1.1   Control Channel

The control channel (CCh) is primarily used for the transmission of beacons and high first hop priority traffic. All messages that are necessary to maintain the VANET are transmitted on this channel, especially the network layer beacons. Furthermore high-priority messages (emergency notifications) are sent on this channel. Normally, such messages occur on an event basis. With multihop communications, only the first hop will require high priority. Jiang et al. define the basic link layer behavior of safety communications in the control channel as single-hop (direct communication among vehicles, within range of one another), uncoordinated (distributed communications without coordinator), broadcast messaging (self-contained short messages) in an unbounded system, consisting of all neighboring equipped vehicles in a dedicated channel [14].

8.3.1.2   Service Channel

Service Channel (SCHl) is available for safety applications with lower priority. Here periodic messages could be sent. This channel should also be used by forwarders of multihop and geocast messages. A second service channel (SCH2) is dedicated to short distance peer-to-peer VANET communications with reduced power level. However, this service channel is currently unused.

8.3.2 Priority Assignment

Packets will be assigned a priority by the message generators, when they are created. The relative time of transmission of each priority level will, however, vary as network density increases, with medium- and low-priority packets, being delayed more to allow high-priority packets to be sent without excessive delays. The priority of a packet is composed of two fields: the first is static, deduced from the application type, and the second is dynamic, obtained from the specific context of the VANET (vehicles density and variance of the mean speed of the neighborhood) and determined by the congestion control module. The dynamic and the static fields are combined to obtain the overall priority indicator.

Lets $Pri_{message}$ indicates the priority of a message. $Pri_{message}$ is evaluated as follows:

$$Pri_{message} = Dynamic\_factor \times Stat\_Pri_{message}/Msg\_size \qquad (8.1)$$

Where $Stat\_Pri_{message}$ denotes the static priority of the message, $Dynamic\_factor$ denotes the dynamic factor of the priority to take into account network context and *Msg_size* denotes the size of the message.

8.3.2.1 Static Factor from Application Class

The static priority factor is defined according to the sender application and the content of the message. We adopt five priority levels, defined hereafter:

■ $PRI_{Emergency}$ is the priority affected to single-hop emergency messages to notify an important event without delay. The safety of vehicles is dependent of this kind of message.
■ PRI_{VANET} is the priority affected to the network layer beacons.
■ PRI_{HIGH} is the priority affected to high-priority safety applications.
■ PRI_{MID} is the priority affected to normal safety applications.
■ PRI_{LOW} is the priority affected to low-priority safety applications.

Regarding multihop and geocast communication, it is assumed that only the first hop can have the priority $PRI_{Emergency}$. It is assumed that these priority levels are associated with the application classes, but when communicating in geocast or multihop modes, this describes only the communication priority on the first hop.

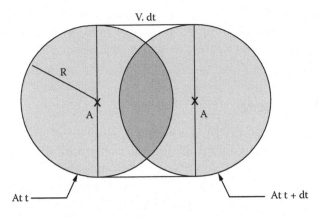

Figure 8.6 Node speed consideration.

8.3.2.2 Dynamic Factor from Network Context

We present the basic policies of our congestion control approach within VANET, to evaluate the dynamic priority factor of sent messages in the following list:[1]

- **Node speed consideration:** The dynamic factor takes into account the node speed, according to the covered zone at each dt, as illustrated in Figure 8.6. The priority of a message increases when the speed of the sender increases. Thus, at each dt the dynamic factor is reevaluated as follows:

$$\text{Dynamic\_factor} = \frac{\pi R^2 + 2RV.dt}{\pi.R^2} \qquad (8.2)$$

  where R is communication range and V is mean speed.

- **Message utility consideration:** The dynamic factor also takes into account of the utility of the sent message, according to the number of its retransmissions by the neighborhood, in case of periodic or geocast messages. Thus, when a node A has to send a periodic or geocast message M, and receives the same message M sent by another node B (Figure 8.7), it should calibrate the dynamic factor of the message M, in order to take into account the zone covered by the node B, compared to its communication range zone (= $\pi.R^2$ where R is the communication range). The smaller the covered zone, the higher the priority to send the message. The dynamic factor is thus

---

[1] A part of this work has been carried out jointly with the VANET research group in the HEUDIASYC laboratory

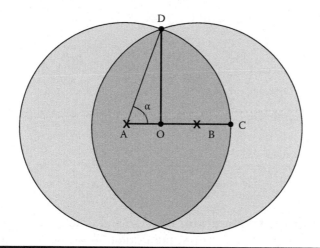

Figure 8.7   Message utility consideration.

equal to the ratio between the total zone covered by the receiver and the already covered zone (CZ):

$$\text{Dynamic\_factor} = \frac{\text{Total\_Zone}}{\text{Covered\_Zone}} = \frac{\Pi R^2}{CZ} \qquad (8.3)$$

To evaluate the covered zone CZ by the node B (doubly hatched in Figure 8.7) at the side of the node A, we carried out the following operations (we note d the distance between A and B, and R the range of A and B):

$$CZ = 4 \times \text{surface\_ODC}$$

$$= 4 \times (\text{surface\_ADC} - \text{surface\_AOD}) \qquad (8.4)$$

$$\text{surface\_ADC} = \frac{\alpha \Pi R^2}{2\Pi} = \frac{\alpha R^2}{2} \qquad (8.5)$$

$$d(O, D) = \sqrt{R^2 - \left(\frac{d}{2}\right)^2}$$

$$\Rightarrow \text{surface\_AOD} = \frac{d}{2} \times \sqrt{R^2 - \left(\frac{d}{2}\right)^2} \qquad (8.6)$$

$$\cos(\alpha) = \frac{d(A, O)}{d(A, D)} = \frac{d}{2R} \qquad (8.7)$$

Equations (8.4), (8.5), (8.6), and (8.7) $\Rightarrow$

$$CZ = 4 \times \left(\arccos\left(\frac{d}{2R}\right) \times \frac{R^2}{2} - \frac{d}{2} \times \sqrt{R^2 - \left(\frac{d}{2}\right)^2} \right) \qquad (8.8)$$

Equations (8.3) and (8.8) $\Rightarrow$

$$\text{Dynamic\_factor} = \frac{\Pi R^2}{4 \times \left(\arccos\left(\frac{d}{2R}\right) \times \frac{R^2}{2} - \frac{d}{2} \times \sqrt{R^2 - \left(\frac{d}{2}\right)^2} \right)} \tag{8.9}$$

■ **Message validity consideration:** The dynamic factor takes into account the message validity (maximum duration of the message). As for the EDF scheduling approach (earliest deadline first), the message whose deadline is earliest holds the highest priority. The dynamic factor is thus computed as follows:

if remaining time to deadline $\neq 0$
             Dynamic factor = 1/remaining time to deadline
     else
             Dynamic factor = 1

end if

8.3.3 Message Scheduling

Each node schedules its messages according to their priorities. We divide the scheduling process into two types: static and dynamic, presented in the following sections.

8.3.3.1 Static Scheduling

The static scheduling process consists of affecting messages according to their priorities into the suitable communication channel queues. $\text{Pri}_{\text{Emergency}}$, $\text{PRI}_{\text{VANET}}$, and PRI_{HIGH} priority messages are affected to the control communication channel queue, whereas PRI_{MID} and PRI_{LOW} priority messages are affected to the service queue one.

8.3.3.2 Dynamic Scheduling

Periodically, each node triggers a rescheduling process, consisting of scanning the message queues, and computing the overall priority indicator for each message (taking into account the dynamic factor of each priority, presented above). The rescheduling process then reorders the messages according to their new computed priorities.

Considering that the number of messages sent within the control channel is smaller than the number of messages sent within service one, we adopt the following policy. When the service channel is overloaded and the control channel is free, messages within the service queue are switched to control one and considered as high-priority messages. We consider that the service channel is overloaded if the number of messages in the queue exceeds a defined threshold called "service channel congestion threshold."

8.3.4 Message Transmission

The message transmission process sends the highest priority message within the corresponding channel, whenever it is free. However, sending high-priority packets via the control channel is preemptive, comparing to packets sent via service channel. Indeed, in order to send high-priority packets with the minimum delays, lower-priority packets sending is freezed, even if their corresponding channel is free. In addition, when a node receives messages of higher priority than messages that it will send (the first message in its queues), it freezes its sending.

Concerning the dynamic use of the bandwidth within VANET, the IEEE 802.11p underway standard supports three mandatory user data rates 3 Mbit/s, 6 Mbit/s, and 12 Mbit/s within a 10 MHz channel, and some optional data rates up to 27 Mbit/s. Obviously, the most robust data rate is the 3 Mbit/s. This rate must be shared among all applications and vehicles inside the interference range. Note that the interference range is a multiple of the communication range. Not to saturate the provided bandwidth and to allow a reliable transmission of the emergency messages, the bandwidth offered to VANET application per 10 MHz is equal to the half of the total bandwidth. We present hereafter how a vehicle i can compute the effective bandwidth it can use.

Let n denote the number of neighbors of the node i, a denote the average number of applications per node, defined at the bootstrap. The number of all the applications N_A within an interference range, reaching the bandwidth, is thus equal to

$$N_A = 2a.(n+1) \qquad (8.10)$$

The effective bandwidth that an application can use within the vehicular network is thus computed as

$$\text{Effective\_Bandwidth} = \frac{\text{selected\_bandwidth}}{2a.(n+1)} \qquad (8.11)$$

Note that the effective bandwidth computed at the side of a node is equal to the selected bandwidth multiplied by the percentage of the active sending time:

$$\text{effective\_Bandwidth} = \text{Selected\_Bandwidth} \times \text{Active\_Time} \qquad (8.12)$$

The percentage of the active sending time, in which a node can offer the bandwidth to an application is thus

$$\text{Active\_Time} = \frac{100}{2a.(n+1)}\% \qquad (8.13)$$

As an example, if the number of vehicles within a neighborhood is equal to 10, with 5 running applications in each vehicle, the active time is equal to 1 percent. The effective bandwidth for each application is equal to 15 Kbit/s.

The two parameters defining the active sending time are the maximum data transmission duration and the minimum interval between data transmissions, as follows:

$$\text{Active\_Time} = \frac{\text{Maximum\_Data\_Transmission\_Duration}}{\text{Minimum\_Interval\_Between\_Data\_Transmissions}} \qquad (8.14)$$

To continue with the same numerical example presented above (effective bandwidth = 15 Kbit/s), and if we define the minimum interval between data transmissions equal to 1 sec, the maximum data transmission duration is evaluated at 5 ms.

8.3.5 Priority-Scheduled Transmissions Algorithm

We summarize the different congestion control policies presented above in Algorithm 8.1. This algorithm contains four principle procedures. The first one "Initialization" is responsible for initializing the control and channel queues and triggering a unit timer, used by a second procedure "Expire_Timer," which reevaluated periodically and dynamically the priorities of the different messages (evaluation of the speed, utility, and validity dynamic factors). The third procedure "Receive_Message" ensures the reception of generated messages to be scheduled within the waiting queues. Finally, the procedure "Send_Message" is responsible for sending the highest-priority message to the transmission engine, to be sent effectively through the corresponding channel.

Algorithm 8.1 Dynamic Message Scheduling Procedures

Initialization( )
{
Control_Queue = ∅ ; Service_Queue = ∅ ; Trigger (Unit_Timer)
} // **End Initialization**

Receive_Message(Message M)
// Message M received from the input modules, to be scheduled
{
if (Message.Channel == Control) **then**
    Add(M, Control_Queue)
    //The addition of a message within a queue is carried out according to its priority level, in an ascending order
else
    Add(M, Service_Queue)
end if
} // **Receive_Message**

Send_Message(Message M)
// Sending Message M to the transmission module, to be sent effectively
{
if (Control_Channel is Free) **then**
    Send(Control_Queue(Last))
    Delete(Control_Queue(Last))
end if
if (Service_Channel is Free && Control_Queue == ∅) **then**
    Send(Service_Queue(Last)) // according to the defined data rate
    Delete(Service_Queue(Last))
end if
} // **Send_Message**

Expire_Timer (Unit_Timer)
// Reevaluate dynamically the messages priorities in the control and service queues
{
for M ∈ Control_Queue **do**
    M.Speed_Factor = M.Speed_Dynamic_Factor * $\frac{\pi R^2 + 2RV.dt}{\pi.R^2}$
    if (M.Remaining_time_to_deadline ≠ 0) **then**
        M.Validity_Factor = 1/M.Remaining_time_to_deadline
    end if
    if (M.Type == Geocast or M.Type == Periodic) **then**
        //At the reception of the same message, sent by another node at a distance d
        M.Utility_Factor = $\Pi R^2 / \left(4 \arccos\left(\frac{d}{2R}\right) * \frac{R^2}{2} - \frac{d}{2} * \sqrt{R^2 - (\frac{d}{2})^2} \right)$
    end if
    M.Dynamic_Priority = M.Speed_Factor*M.Vaidity_Factor*M.Utility_Factor
    M.Priority = M.Static_Priority * M.Dynamic_Priority / M.Size

end for
Sort (Control_Queue) // according to an ascending order
// Similar Dynamic reevaluation for the Service_Queue
Trigger (Unit_Timer)
} // **End Expire_Timer ()**

8.4 Analysis and Validation

8.4.1 Formal Verification

The quality of service of a system is composed of a constraints set (delays, response time, data rate, and so on). The system operation ensures the required QoS if it verifies the needed constraints. Automata allows to model dynamic control aspects of system operation. These models, in addition to temporal mechanisms (called temporized automata), are suitable to specify temporal QoS constraints [1,4].

To verify and validate our congestion control technique, we propose use of temporized automata, through the UPPAAL[2] integrated tool environment for modeling, validation, and verification of real-time systems. UPPAAL is developed in collaboration between the Department of Information Technology at Uppsala University, Sweden and the Department of Computer Science at Aalborg University, Denmark. To present our validation process, we first start by presenting an overview of temporized automata and the UPPAAL tool. Then, we present our objectives, simulations, and results.

8.4.1.1 Temporized Automata

In a temporized automata, transitions between states are conditioned by temporal constraints on clock variables. An elementary temporal constraint is a boolean property in which a clock variable is compared to a constant integer. Extended temporized automata enhances temporized automata by providing the possibility of manipulation of nontemporal variables. In the following description, we do not distinguish between "temporized automata" and "extended temporized automata."

A temporized automata is a sextuplet (Σ, S, S_0, C, V, E) where:

- ■ Σ is a finite set of actions.
- ■ S is a finite set of states.
- ■ $S_0 \in S$ is the initial state.
- ■ C is a finite set of clocks.
- ■ V is a finite set of variables.
- ■ E is the set of transitions. A transition is a tuple ($s, \mu, \gamma, \lambda, \lambda', s'$) indicating that starting by state s, the automaton executes the action

---

[2] http://www.uppaal.com.

μ, if the constraint γ is satisfied; clocks of λ are reset, variables of λ' are updated, and the new state is s'.

8.4.1.2 The Uppaal Modeling, Validation, and Verification Tool of Real-Time Systems

UPPAAL is a tool box for validation (via graphical simulation) and verification (via automatic model-checking) of real-time systems. It consists of two main parts: a graphical user interface and a model-checker engine. The idea is to model a system using temporized automata, simulate it, and then verify properties on it. A real-time system in UPPAAL is composed of concurrent processes, each of them modeled as an automaton. The automaton has a set of locations. Transitions are used to change location. To control when to fire a transition, it is possible to have a guard and a synchronization. A guard is a condition on the variables and the clocks saying when the transition is enabled. The synchronization mechanism in UPPAAL is a hand-shaking synchronization. Two processes take a transition at the same time, one will have a **!a**, and the other **a?**; **a** being the synchronization channel. The verification tool provided by UPPAAL checks for the following properties:

1. Reachability properties: These properties ask whether for a given state formula φ, there exists a path starting at the initial state, such that φ is eventually satisfied along that path. Reachability properties do not by themselves guarantee the correctness of the system, but they validate the basic behavior of the model.
2. Safety properties: These properties ask whether a bad result will never happen or an awaited result is invariantly true.
3. Liveness properties: These properties are of the form "an awaited result will eventually happen." With this type of properties, response conditions can be verified as "whenever φ is satisfied, eventually ψ will be satisfied."
4. No deadlock property: This property verifies if there is not any state where there are no outgoing action transitions, neither from the state itself or any of its delay successors.

8.4.1.3 Simulations and Results

To simulate, via UPPAAL, our priority-based scheduling mechanism within VANET, we divide our system into four independent subsystems, represented each by an automaton that we present hereafter:

■ The message manager automaton (Figure 8.8): This subsystem is responsible for generating messages and sending them to the congestion control module, which will process them according to their priorities. To simplify our simulation platform, and without loss of generality, we consider hereafter two levels of priorities

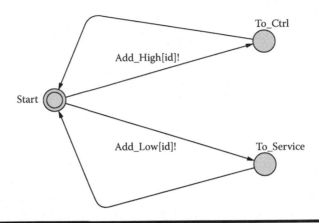

Figure 8.8   Message manager.

(high-priority packets sent within the control channel and low-priority packets within the service channel).

■ The congestion control message enqueueing automaton (Figure 8.9): This subsystem is responsible for the reception of messages from the message manager and for their addition to the appropriate queues.

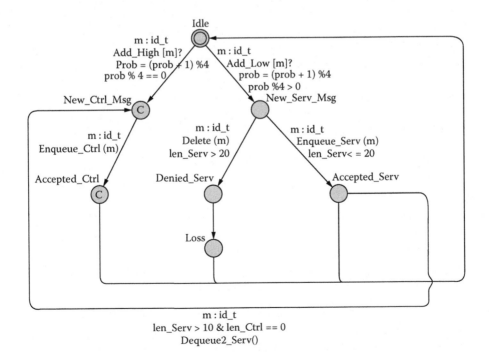

Figure 8.9   Congestion control message enqueueing.

A Add_High synchronization message is required to transit from the Idle state to the New_Ctrl_Msg one. The automaton switchs to the Accepted_Ctrl state after adding the high-priority message to the control channel queue. The addition of a low-priority message follows the same process, with the difference that the congestion control module can deny sending a message if the service channel is overloaded (number of low-priority messages > 20). Note that when the service channel is overloaded and there are no high-priority packets to send, a low-priority message can be sent via the control channel. This message is thus dequeued from the service queue list and enqueued in the control channel list.

■ The congestion control message dequeueing automaton (Figure 8.10): This subsystem is responsible for withdrawing messages from the control and service queues and transmitting them to the transmission engine. The synchronization messages between the congestion control message dequeueing automaton and the transmission engine one are Send_Ctrl and Send_Serv.

■ The transmission engine automaton (Figure 8.11): This subsystem is responsible for the messages effective transmission on the appropriate channels. A sending error rate is chosen equal to 20 percent. When an error occurs during message sending, the automaton switches to the *Error* state.

The first step of validation is the simulation via graphical interface. We thus randomly simulate all the possible transitions of the four automata. Figure 8.12 illustrates an example of the execution of a low-priority packet

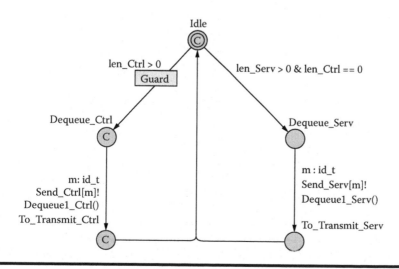

Figure 8.10   Congestion control message dequeueing.

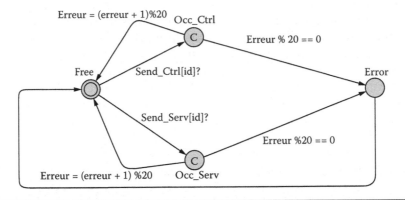

Figure 8.11 Transmission engine.

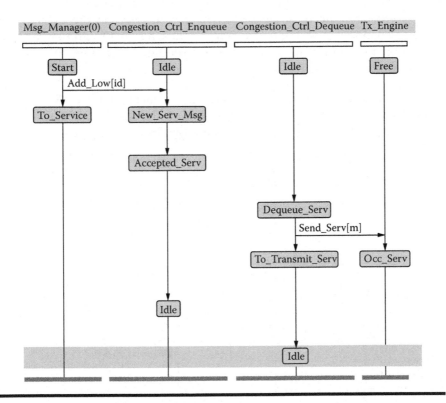

Figure 8.12 Service packet sending diagram.

sending (activity diagram generated by UPPAAL). This verification step validates the following results:

- No deadlock in the operation of our messages priority–based scheduling approach. All states of the modeled automata have successors.
- All states of the modeled automata are eventually reachable.
- All the high-priority messages are effectively sent on the control channel. However, some low-priority packets can be deleted, due to service channel congestion (bounded service messages queue).
- The sending of high-priority messages is preemptive compared to the emission of low-priority messages. Service messages are sent only when there is not any control message in the queue of the control channel.
- In addition, the emission of a high-priority message is carried out without delay. All the high-priority messages are considered as emergency, requiring thus to freeze the emission of lower-priority messages.

We tried then to carry out the second validation step provided by UPPAAL, the model checking verification. We start by specifying our verification objectives (queries) via a description language. Although this verification step did not succeed due to memory limitations,[3] we can affirm that our priority-based scheduling approach is correct, considering the results of the first verification step.

8.4.2 Implementation and Real Measurements

We have developed a congestion control module, setting up the priority-based scheduling and transmission techniques within VANET. Figure 8.13 shows our congestion control module and its interaction with the other modules of a VANET router. We present hereafter the principles of our module and its interactions:

- **Congestion control module:** Within the congestion control module, two queues are implemented: one for the control channel messages and one for the service channel messages. The dotted edge in Figure 8.13 represents the possibility of switching messages from the service channel queue to the service channel one, when it is needed. In addition, four threads are implemented, one main thread

---

[3] The number of states and variables in UPPAAL can make the model-checking verification process very constraining and complex [5]

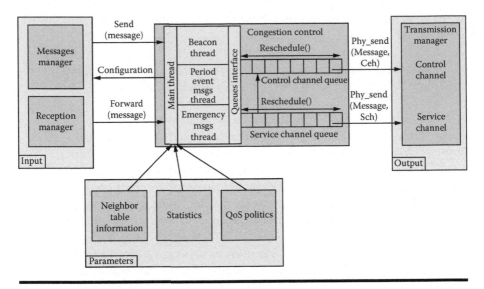

Figure 8.13 Congestion control module.

and others are dedicated for each type of messages received from the input modules.

■ **Input modules:** Input modules send messages to the congestion control module. The message manager module generates new messages to be sent within the control or service channel, and the reception engine sends received messages to be forwarded by the congestion control module.

■ **Output module:** The output module that receives messages from the congestion control module is the transmission module. Each message received from the congestion control module is affected to the corresponding channel to be effectively sent.

■ **Parameters modules:** The congestion control module interacts with different parameters modules. From the neighboring table module, the congestion control module evaluates the network and the neighborhood context, to recompute dynamically the messages priorities. Within the statistics module, are logged the different operations carried out by the congestion control module. And finally, the congestion control module interacts with QoS policies module to take into account the general quality of service policies, according to the priority of the transmitted messages.

Tests and measurements carried out to validate the presented implementation show that a fast and reliable communication architecture is ensured within the control channel and an acceptable communication architecture is established through the service channel.

8.5 Conclusions

We considered in this chapter the congestion control or avoidance issue within vehicular networks. We summarized existing research works on this topic, then we presented a congestion control approach, based on dynamic priority-scheduling transmissions. This approach takes into account the network load and the neighborhood context to adapt the priorities of the sent messages via the control and service communication channels. The efficiency of this congestion control technique was validated through formal verification, implementation, and real measurements.

References

[1] R. Allur and D. Dill, "A theory of timed automata," *Theoretical Computer Science* 126(1994):183–235.

[2] M. Chen, A. Abate, and S. Sastry, "New congestion control schemes over wireless networks: Delay sensitivity analysis and simulations," In *Proc. of the 16th IFAC World Congress*, Prague, CZ, July 2005.

[3] N. Bonmariage and G. Leduc, "A survey of optimal network congestion control for unicast and multicast transmission," *Computers Networks* 50(3) (2006):448–468.

[4] A. Cavalli, "Ingenierie des protocoles et qualite de service," *Reseaux et Telecoms*, Lavoisier Hermes, 2001.

[5] A. David, *UPPAAL2k: Small Tutorial*, Report version 3.2.11, October 2002.

[6] F. Kelly, "Charging and rate control for elastic traffic," In *European Transactions on Telecommunications*, (January 1997):33–37.

[7] E. Minack, Evaluation of the influence of channel conditions on Car2X communications. Diploma thesis, Chemnitz University, November 2005.

[8] V. Raghunathan and P. R. Kumar, "A counter example in congestion control of wireless networks," In *Proc. of the 8th ACM International Symposium on Modeling, Analysis and Simulation of Wireless and Mobile Systems*, New York, 2005. ACM:290–297.

[9] H. K. Rath and A. Sahoo, *Cross Layer Based Congestion Control in Wireless Networks*, Technical Report, April 2006.

[10] Y. Yi and S. Shakkottai, "Hop-by-hop congestion control over a wireless multi-hop network," *IEEE/ACM Transactions Networks* 15(1) (2007):133–144.

[11] G. Zhang, Y. Wu, and Y. Liu, "Stability and sensitivity for congestion control in wireless networks with time varying link capacities," In *Proc. of the 13TH IEEE International Conference on Network Protocols*, Washington, DC, 2005, IEEE Computer Society:401–412.

[12] Y. Zang, L. Stibor, X. Cheng, H.-J. Reumerman, A. Paruzel, and A. Barroso, "Congestion control in wireless networks for vehicular safety applications," In *Proceedings of the 8th European Wireless Conference*, Paris, France, 2007:7.

[13] L. Wischhof and H. Rohling, "Congestion control in vehicular ad hoc networks," In *Proc. of IEEE International Conference on Vehicular Electronics and Safety* (2005):58–63.

[14] D. Jiang, V. Taliwal, A. Meier, and W. Holfelder, "Design of 5.9 GHz DSRC-based vehicular safety communication," *IEEE Wireless Communications* 13(5) (October 2006):36–43.

[15] L. Stibor, Y. Zang, and H. J. Reumerman, "Neighborhood evaluation of vehicular ad-hoc network using IEEE 802.11p," In *Proc. of the 8th European Wireless Conference*, Paris, France, 2007:5.

[16] M. Torrent-Moreno, P. Santi, and H. Hartenstein, "Fair sharing of bandwidth in VANETs," In *Proc. of the 2nd ACM International Workshop on Vehicular Ad Hoc Networks*, New York, 2005:49–58.

[17] M.Sc. Yunpeng Zang, "Study on message dissemination algorithms for co-operative danger warning applications based on inter-vehicle communications," *ComNets 2008*. Seminar Paper.

Chapter 9

IP Address Autoconfiguration in Vehicular Networks

Maria Calderon, Hassnaa Moustafa,
Carlos J. Bernardos, and Roberto Baldessari

Contents

Many people in modern society spend a lot of time in their vehicles. Communication possibilities in vehicles have been restricted in the past mainly to cellular communication networks. Enabling broader communication facilities in vehicles is an important contribution to the global trend toward ubiquitous communications. Consequently, vehicles should provide access to the Internet and should be able to communicate among themselves, supporting new services and applications.

In the last years many mechanisms to enable vehicular communications based on the concept of vehicular ad hoc networks (VANETs) have been proposed. In order to enable the deployment and provision of IP-based (both vehicles-to-vehicles and vehicles-to-Internet) services in VANETs, every VANET node should be configured with at least one IP address. However, there is no standard mechanism to provide VANET nodes with IP configuration information, thus requiring nodes to be configured a priori and avoiding ad hoc networks to be spontaneously created.

Existing IP configuration protocols for traditional infrastructure-based networks assume the existence of a single multicast-capable link for signaling. Such a link does not exist in multihop infrastructureless networks, making it necessary to design new mechanisms that enable the autoconfiguration of IP addresses in a VANET.

This chapter will present and analyze the problem of IP autoconfiguration for VANETs, identifying the main issues that need to be tackled and providing a survey on existing proposals, as well as providing an overview of the work being done at the working group chartered at the IETF—called AUTOCONF—to work on the IPv6 autoconfiguration issue in mobile ad hoc networks (MANETs). The chapter starts by introducing the problem of IP autoconfiguration in ad hoc networks in general, identifying some key requirements and analyzing the solution space. The standardization activities within the IETF that tackle this problem are highlighted and the limitations of the conventional IP autoconfiguration protocols, when applied to ad hoc networks, are presented. Then, the chapter deals with vehicular networks particularities, presenting and analyzing the problem of IP autoconfiguration for VANETs. The IP autoconfiguration requirements in vehicular networks are also identified together with the main issues that need to be addressed. Finally, a survey on existing proposals is provided.

9.1 The Ad Hoc IP Address Autoconfiguration Problem

In order to allow wide deployment of ad hoc networks in which IP routing is the most candidate approach, IP configuration of nodes is a strong requirement that needs to be satisfied. IP address configuration is among the technical challenges in ad hoc networks, standing as an obstacle for the wide deployment of these networks. Consequently, such challenge exists in vehicular networks. Regarding ad hoc network characteristics, IP address configuration should be carried out in an automatic and distributed manner. So far, there is no standard IP autoconfiguration solution for ad hoc networks, and hence the problem becomes complex for vehicular networks. Indeed, the lack of standard IP address autoconfiguration solutions can threaten the services' quality and precisely the services' continuity in these networks.

A number of individual contributions exist in the literature and within the standardization activities, namely, the IETF. However, all these contributions are still in progress. Since the new version of IP (IPv6) is expected to be widely adopted to support the growth in a number of wireless devices, this chapter focuses on IPv6 autoconfiguration mechanisms.

We notice that the multihop nature of ad hoc networks and its lack of a single multicast-capable link for signaling prevents current IP address autoconfiguration-related protocol specifications (such as RFCs 4861 [1], 4862 [2], and so on) from being used as are in ad hoc networks. Moreover, the self-healing feature of ad hoc networks, where nodes partitioning from the network and merging to the network can take place in a dynamic manner, requires special functionalities in IP address autoconfiguration schemes. Indeed, duplicate address situations (i.e., two nodes using the same IP address) can frequently happen due to partitioning and merging. Consequently, IP address autoconfiguration mechanisms should guarantee that there is no address conflict in a dynamic manner.

To sum up, ad hoc networks have particular characteristics that should be taken into account when designing IP address autoconfiguration mechanisms and protocols. These unique characteristics make existing solutions for IP infrastructure–based networks difficult to be applied as is in ad hoc networks. The existing IP address autoconfiguration solutions in infrastructure-based networks have a number of limitations when applied in ad hoc networks (as discussed in the following section). Consequently, developing IP autoconfiguration mechanisms for ad hoc networks is needed, which satisfy the self-deployment and auto-healing requirements of these networks.

9.1.1 Main Limitations of Conventional IP Address Autoconfiguration Mechanisms

The simplest approach for IP address autoconfiguration is through the use of dynamic host configuration protocol (DHCP) [5] in which a DHCP server in place is responsible for allocating to each node an automatic IP address. Simply, a node (DHCP client) requiring an IP address contacts a DHCP server and requests an address. The DHCP server will dynamically assign an address from a certain pool of addresses and allocate it to the node for a certain lifetime (known as lease) that can be prolonged through a node request. In the case where the DHCP client and DHCP server do not share the same link (for instance, they are situated in different subnets), one or more DHCP relays are used to forward the messages between them. Indeed, the limitation of the DHCP solution in ad hoc networks concerns the direct communication with server (DHCP server), which is not always possible in ad hoc networks, and the communication through relay agents, which does not always work in ad hoc networks. The latter point constitutes the following problems: (i) relay agent definition, that is, which nodes are selected to be relay agents? (ii) relay agent configuration, that is, how should these relay agents be configured (should they be configured on MANET nodes or elsewhere), and (iii) how to configure IP addresses in the relay agents?

On the other hand, the stateless address autoconfiguration (SLAAC) [2] enables automatic configuration of an IP address to a node without needing to communicate with a server. First, the node (host) constructs a tentative IPv6 address by attaching its identifier (usually the MAC address) to the well-known link-local prefix. Then, the node (host) broadcasts neighbor discovery protocol (NDP) messages [1] in order to verify that the tentative address is not used by another host on the link. If the address is not unique, the autoconfiguration process will try another one. Otherwise, the node (host) may request a prefix from any router on the link by an exchange of NDP messages also. When obtaining the router prefix, the node (host) attaches its identifier to the prefix and carries out a uniqueness test as before. We notice that SLAAC relies on NDP signaling, where it works on the basic assumption that each node in the ad hoc network can communicate directly with every other node. This is a wrong assumption in ad hoc networks, since each node may see a different set of neighboring nodes. For the applicability of SLAAC in ad hoc networks, one should assume the existence of a single multicast-enabled link, which does not conform to the link properties in these networks. Maybe if an adaptation layer exists, the link properties problem can be solved and the SLAAC NDP approach can be applied in an adaptive manner. However, this latter approach is out of the *scope* of the AUTOCONF WG.

Prefix delegation (PD) using DHCP [6], also known as DHCP-PD, is a solution that allows for automatic allocation of IPv6 prefixes to MANET nodes using DHCP. In this approach, a MANET node contacts a DHCP server and requests a prefix by sending a DHCP request including the prefix delegation option. The server in its turn can delegate a subprefix (i.e., a subset of its address pool) to the MANET node. The DHCP message containing the prefix delegation option may also be relayed through one or more DHCP relays, similar to the case of DHCP [5]. We notice that DHCP-PD is based on DHCP and thus encounters the same previously discussed limitation of DHCP when applied in ad hoc network scenarios. This mainly concerns the server reachability and the configuration of relays. On the other hand, the prefix delegation might assume that the requesting MANET node has an IP address already configured, and hence the problems of DHCP are not shared.

9.1.2 The IETF AUTOCONF WG

The AUTOCONF WG (Ad Hoc Network Autoconfiguration Working Group[1]) is working toward standard specifications and solutions for IP address autoconfiguration within the different mobile ad hoc network (MANETs) environments. Indeed, the AUTOCONF WG lies under the Internet area at the IETF, and hence considers mobile ad hoc networks connected to the Internet, with the main goal of developing solutions for IPv6 address autoconfiguration. These addresses are (i) MANET-local addresses allowing communication of a node within the local MANET, and (ii) global addresses allowing a node to communicate with other nodes in the Internet or other external networks. Some inputs are acquired from the MANET WG (Mobile Ad hoc NETwork Working Group[2]). In this context, the AUTOCONF WG has developed an architecture document [3] aiming to serve the AUTOCONF WG in providing useful inputs for the IP autoconfiguration problem. This document defines the ad hoc networks' entities and architectural concepts, as well as the challenges for IP autoconfiguration protocols design and development and gives a model for IP addressing.

9.1.2.1 IP Addressing Architectural Model

The architectural model for IP addressing in ad hoc networks [3] assures the integrity of the conventional IP addressing architecture while considering

---

[1] http://www.ietf.org/html.charters/autoconf-charter.html.
[2] http://www.ietf.org/html.charters/manet-charter.html.

the characteristics of ad hoc networks' interfaces. In this context, a MANET router (MNR) is characterized by having one or more MANET interfaces and can also have zero or more non-MANET interfaces. Each MNR has nodes possibly attached and may be responsible for announcing the location of a particular address or set of addresses (i.e., a subnet prefix). An MNR can be also delegated zero or more prefixes. For example, if an MNR is delegated a prefix p::, then subnet prefixes derived from this prefix (e.g., p:1::/64, p:2::/64, and so on) may be assigned to the MNR's nonMANET interfaces(s), and nodes on these interfaces may be assigned addresses out of this prefix, and configured with this prefix according to the address autoconfiguration mechanisms governing these interfaces.

9.1.2.2 Scenarios Addressed by the AUTOCONF WG

Based on the MANET architecture model defined in [3], the AUTOCONF WG has identified two main scenarios that will be addressed by IP auto-configuration solutions [4]. These scenarios are as follows:

■ Autonomous MANETs: These networks, also known as standalone MANETs, are not connected to any external network. All traffic is generated by MANET nodes and destined to nodes in the same MANET. Hence, autonomous MANETs are MANETs upon which no external network imposes an addressing hierarchy. In such network scenarios, nodes may join or leave randomly. Besides, most likely no preestablished nonreliable address or prefix allocation agency will be present in the network. Typical examples of autonomous MANETs are networks setup in areas where infrastructure is un-available or inappropriate, for instance, vehicular networks, namely, vehicle-to-vehicle communication for sharing traffic and safety-related information. Other examples include file-sharing conference networks, battlefield networks, surveillance networks, and on-site emergency communication among rescue team members for disaster recovery.

■ Subordinate MANETs: These networks, also known as connected MANETs, have connectivity to one or more external networks, typically the Internet, by means of one or more gateways that are also known as MBRs (MANET border routers). Subordinate MANETs hence have at least one external network that imposes a specific addressing hierarchy on the MANET. This addressing hierarchy yields the use of specific prefixes for communications between nodes in the MANET and nodes in or across the external network. These prefixes need to be topologically correct, that is, allocated from out of a prefix p::, over which the point of attachment to the network has authority. A typical example of this scenario is the ad hoc network connected to the Internet, for instance, public wireless mesh networks in which scattered fixed WLAN access routers act as border routers and participate in an ad hoc network of mobile users. Also, vehicular networks with vehicle-to-road communication fall under this case. Another typical example is the coverage extension of a fixed wide-area wireless network, where one or more MNRs are connected to the Internet through technologies such as UMTS or WiMAX.

9.1.2.3 The AUTOCONF WG—Work in Progress

Based on the previously described architectural concepts, the AUTOCONF WG sorted some goals that should be addressed during the development of IP autoconfiguration solutions. These are as follows [4]: (i) Each MNR should configure, on its interfaces, IPv6 addresses that are unique within the ad hoc network, (ii) each MNR should be allocated IPv6 prefixes that are disjoint from prefixes allocated to other routers within the ad hoc network, (iii) each MNR should maintain the uniqueness of the configured addresses as well as the disjoint character of allocated prefixes (this uniqueness should be assured even in the case of network merging), and (iv) in the subordinate scenario, each MNR should be allocated topologically correct prefixes.

So far, there is no standard protocol for IP autoconfiguration in ad hoc networks. However, a number of propositions have been made for different ad hoc network scenarios. A survey on most of the proposed solutions is provided by Bernardos, Calderon, and Moustafa [7]. In this survey, a classification for the different proposals is given together with an analysis based on a number of evaluation criteria. Two major classification levels are considered in this context: (i) autonomous versus connected (or subordinate) scenarios, and (ii) partitioning and/or merging support.

Indeed, the dynamic and random nature of ad hoc networks together with the different environmental behavior require different functionalities in the IP autoconfiguration mechanisms. Consequently, it is important to keep in mind some related issues during the solutions development mainly concerning the signaling overhead, delay, different link type's applicability, interaction with the existing protocols, and security issues. Moustafa, Bernardos, and Calderon developed [8], some evaluation considerations for IP autoconfiguration mechanisms, serving as reference for the IP autoconfiguration solutions' space through illustrating some key features and behaviors for the different IP autoconfiguration mechanisms. These evaluation considerations can give some guidelines for solution developers during mechanisms' design and for implementers during the choice of the IP autoconfiguration mechanisms (during the decision of which protocol fits better the scenario requirements). This work refers to a previous study, carried out by Clausen [9].

As for the solution space, a survey by Bernardos, Calderon, and Moustafa [10] presents an analysis for the ad hoc IP autoconfiguration problem solution space, considering the different solution proposals and taking into account the developed evaluation considerations. The solution space analysis classifies, at a generic level, the solution space of the possible approaches that could be followed to solve the IPv6 autoconfiguration problem in ad hoc networks. The various approaches of IPv6 autoconfiguration for ad hoc networks are illustrated, showing their main key features and *scope*. Also, the benefits and trade-offs in different aspects of IPv6 autoconfiguration are explored.

9.1.3   IP Address Autoconfiguration Solution Requirements

This section highlights and describes the requirements that an IP autoconfiguration solution could need to meet. While this section provides an exhaustive list, depending on the particular scenario or use case, it might be the case that some requirements do not apply or have less relevance than others. The different requirements are classified into four different categories.

9.1.3.1   Node or Network Requirements

An IP autoconfiguration solution should be able to operate across a different range of network scenarios and support heterogeneous types of nodes.

As discussed previously in section 9.1.2.2, there exist two different ad hoc scenarios in which a network may operate: autonomous and subordinate. Different mechanisms' functionalities will be required according to the scenario type, where, for example, solutions designed for subordinate scenarios have to deal with the issue of getting global IPv6 addresses that allow nodes to get Internet connectivity. In fact, contributions designated for subordinate networks can mostly be general for both types of scenarios. Indeed, an autoconfiguration mechanism should take into account the case of scenario transition, where a subordinate network can converge to an autonomous one if it loses its attachment with the infrastructure, or a partially autonomous if subnetwork(s) exist(s) due to partitioning.

Another important requirement is that an IP address autoconfiguration solution should support nodes with different mobility patterns, from fixed (e.g., backbone mesh deployments) to highly mobile (e.g., vehicular ad hoc networks) scenarios. A high node mobility rate could lead to performance and stability problems if the solution is not properly designed to cope with these issues.

It is also important to avoid changing the operation of standard protocols and reuse them as much as possible.

9.1.3.2 Functional Requirements

Functional requirements are those functionalities that a mechanism must aim at featuring. An autoconfiguration mechanism can implement one or several of these functionalities.

The most important functional requirement is address uniqueness. This concerns two points: (i) duplicate address avoidance (preservice), and (ii) nonunique address detection (in-service). Duplicate address avoidance is a mandatory characteristic in any autoconfiguration mechanism. It consists in making all autoconfiguration mechanisms' functionalities assign addresses only after checking their uniqueness. Hence, this principle must be the core of any design of an autoconfiguration mechanism. On the other hand, nonunique address detection is the process used to detect address collisions and resolve them. This might be a heavy process in any IP autoconfiguration mechanism, requiring a considerable number of control messages.

Another important requirement is the support of network merging, that is, the ability of the autoconfiguration mechanism to detect MANETs' merging and provide functionalities to avoid IP address conflicts and connectivity problems in such cases. Merging support is important in the case where two previously disjointed ad hoc networks get connected, since there might be nodes with duplicated addresses in the merged network, and therefore it is needed to make some nodes change the IP addresses they are using to avoid the conflict.

Related to the previous one, there also exists the requirement of supporting network partitioning, that is, the ability of the autoconfiguration mechanism to detect MANETs' partitioning and provide functionalities to avoid connectivity problems in such cases (this is particularly relevant in subordinate scenarios).

A third functional requirement that might be needed in certain scenarios is the capability of delegating prefixes, since an ad hoc node (e.g., a MANET router) may have attached "traditional" hosts, and therefore it needs to acquire IPv6 prefixes instead of just single addresses.

9.1.3.3 Performance Requirements

Performance is always a critical issue in communication protocols. In the case of IP autoconfiguration, performance might be specially relevant in certain scenarios (for example, in constrained devices). The following are the most relevant performance requirements:

■ Low protocol overhead: An IP autoconfiguration solution may require additional signaling to work. Such signaling is considered as protocol overhead that might have a significant performance impact, and therefore should be kept, in general, as low as possible. The protocol overhead might have an impact on the convergence time and thus the scalability of the autoconfiguration mechanism.

◼ Robustness: An IP autoconfiguration should be as robust and reliable as possible, not assuming any strong property of the underlying physical layer.

◼ Low convergence time: Depending on the scenario and/or the application, we may define the convergence time as the time required by a single node to get a usable (and unique) IPv6 address or as the time required by the whole network to have all its nodes configured with correct addresses. For several scenarios, it might not be important for an IP autoconfiguration solution to take a long time to finish its operation (e.g., several minutes) because the IP autoconfiguration mechanism is only run once and then the nodes remain stable. However, for other scenarios, it might be required that the autoconfiguration solution should take less to converge than a given amount of time.

◼ Scalability: An IP autoconfiguration solution should be able to operate across a disparate range of network sizes, since an ad hoc network might comprise from just a few nodes to thousands. Therefore, the scalability of the solution in terms of size (this also has an impact on convergence time) is very important.

◼ Efficient IP address space utilization: An IP address autoconfiguration solution should make an efficient use of the available IP address space, especially in those scenarios in which IP addresses are scarce.

9.1.3.4 Security Requirements

Solutions should address security threats considered in existing IPv6 autoconfiguration mechanisms. In addition, solutions should address potential ad hoc network-specific threats, in particular, attacks enabled due to the multihop and unmanaged nature of these networks.

9.1.4 Analysis of the IP Address Autoconfiguration Solution Space

In this section, we analyze the vast solution space of MANET IP autoconfiguration by answering the following questions:

1. Which entities are involved?
2. What type of IP delegation: addresses or prefixes?
3. How are IP addresses obtained?
4. How is an IP address uniqueness guaranteed?
5. How is signaling performed?

6. What are the security considerations?
7. Are existing protocols modified?

9.1.4.1 Which Entities Are Involved?

There are several combinations of entities involved in IP autoconfiguration process:

- MANET routers (distributed approach): A MANET can be IP autoconfigured in a fully distributed way, without any nodes having special responsibilities in the IP autoconfiguration process. In this scenario, the responsibility of the task is shared among all the participant nodes.

  A possible approach is that each MANET router chooses randomly an IP address and then checks that there is no address conflict, by asking the other nodes in the MANET (e.g., Perkins [11] follows this approach). Additionally, the responsibility of detecting conflicts may be distributed, having all the nodes the potential to detect conflicts. This can be done, for example, by analyzing incoming routing protocol messages and looking for inconsistencies [12–14].

  Some solutions assign an IP address pool to every new node that enters into the MANET and as of that moment, this node has the potential to split their own IP pool and assign it to another new node [15,16] (e.g., all nodes collectively perform the functionality of a DHCP server).

  The main advantage of this distributed approach is that the existence of a single point of failure is avoided and, therefore, in general this kind of solution would be more robust and scalable than a centralized approach. On the other hand, the main concern with this approach is that the probability of an address conflict happening is higher, since there is no centralized server that can assure that two nodes are not assigned the same address.

- MANET routers and border routers: Some solutions may involve border router(s) (also known as Internet gateways) playing an active role in the IP autoconfiguration process (besides its role of gateways bounding the MANET and providing connectivity to other routing domains). The most common approach is that MANET border routers (MBRs) announce within the MANET the global IPv6 prefixes that can be used by MANET routers in the configuration of their IPv6 addresses [17]. MANET routers would still play an important role in the IP autoconfiguration process, and may, for example, be responsible for the detection and resolution of address conflicts.

■ MANET routers and distributed servers: Alternatively, the existence of some special nodes within the MANET that participate in the IP autoconfiguration process playing a predominant or special role (leader nodes) may be considered. These nodes are responsible for parts of the IP autoconfiguration of some other MANET routers [18] (e.g., by issuing router advertisements to nodes within their *scope*). In some solutions, a hierarchy is established by these special nodes.

The advantage of this approach is that it may be easier to avoid address conflicts than in a completely distributed approach, because there exist a set of servers in charge of assigning IP addresses. Furthermore, the reliability of the solution—when compared to a completely centralized solution (described next)—is improved, since there is no single point of failure. The main concern with this approach is the need for a mechanism to elect these leader nodes and to coordinate or synchronize them (in case this is required). If the leader node role cannot be played by every node (when requested to behave as leader node), then only specific ones can do it. In this case, the same issues pointed out about border routers also apply here.

■ MANET routers and centralized server(s) (centralized approach): In this case, a centralized server is in charge of the whole IP autoconfiguration process.

Centralized approaches may make use of DHCPv6 [5], for example, by deploying a DHCPv6 server (within the infrastructure—in case of connected MANETS—or within the MANET itself) and configuring all MANET routers as DHCP relays to get to the server when a new node joins the network [19].

Due to the centralized nature of these solutions (i.e., all the IP autoconfiguration information is managed and kept in one single entity), it becomes easier to ensure a correct IP configuration across the MANET (e.g., no duplicate addresses configured). The main concerns with this kind of approach are related to scalability and reliability (the server is a single point of failure). Besides, support of partitioning and merging becomes more complicated and the mobility management in general is not easy.

9.1.4.2 What Type of IP Delegation: Addresses or Prefixes?

One important aspect of an IP autoconfiguration mechanism, which usually has a very important impact on the mechanism operation, is the type of IP addressing resources that are delegated to MANET routers: addresses or prefixes.

Current MANET architecture model [3] basically defines MANET participant nodes as MANET routers. These MANET routers, beside having one or more MANET interfaces, may also have non-MANET interfaces, enabling legacy or non-MANET–enabled IPv6 hosts (i.e., hosts not running a MANET routing protocol) to attach to and obtain connectivity from a MANET router. In this particular scenario, allocating IPv6 prefixes to MANET routers appears as an important feature to be provided. Most of the first proposals for IP autoconfiguration mechanisms only tackled the address delegation problem, whereas lately some proposals support also prefix delegation.

It is usually harder to check prefix uniqueness within MANET than address uniqueness. Because of that, the most straightforward approach to provide prefix allocation is to do it in such a way that it is not needed to perform a prefix duplication check. Some ways of doing that is by using a centralized mechanism (e.g., based on DHCPv6 [19]) or by using a generation or delegation approach that guarantees the prefix uniqueness beforehand.

9.1.4.3   How Are IP Addresses Obtained?

This is an important question, since the way an IP address or prefix is obtained may also have an impact on other questions, for example, those related to the uniqueness of this address or prefix within the MANET.

One of the goals that IP autoconfiguration mechanisms try to achieve is the efficient provision of valid IP addresses to nodes, requiring as less time as possible. In IPv6, there are several mechanisms currently defined for infrastructure-based networks, and they can be classified basically into two main groups, depending on how the IP address is obtained: (i) the IP address is locally selected by the node (stateless autoconfiguration [2]), or (ii) the IP address is assigned by a DHCPv6 [5] server (stateful autoconfiguration). Additionally, the node is responsible for checking that the obtained candidate IP address is not being used in the subnet. To do that, a mechanism called duplicate address detection (DAD) has been also standardized [2].

Some of the autoconfiguration mechanisms used by non-MANET IPv6 nodes to choose or generate an IP address can also be considered for the MANET scenario. For example, a MANET router may generate its addresses based on the EUI-64 mechanism [2]. This approach has two main advantages. By reusing the same solution that has been defined for IPv6 infrastructure-based networks, the implementation of the mechanism becomes easier. Additionally, the EUI-64 procedure provides certain guarantees on the global uniqueness of the generated IPv6 address (basically, if the IEEE MAC addresses were globally unique—which is almost true in most cases—the EUI-64 generated IPv6 addresses would be globally unique).

An alternative approach, used by several ad hoc IP autoconfiguration mechanisms (e.g., draft by Perkins [11]), consists of generating a random IPv6 address out of a known prefix. This solution has the advantage of being quite simple, but special care should be taken in the implementation of the random generator, since a bad or limited one may lead to different nodes choosing the same IPv6 addresses. This could be an issue in resource-limited devices, where the implementation of a good random number generator could be hard or difficult.

Other solutions may make use of address pools, from which nodes may take IP addresses. These pools can also be distributed within the ad hoc network—for example, hierarchically, by using a binary split approach [15]—also providing certain address uniqueness guarantees. On the other hand, the management of these address pools may be complicated, especially in environments that present high mobility patterns.

Alternative ways of IP address generation can also be considered, for example, those that embed certain information in the IP address. This is the case, for example, of the cryptographic generated address (CGAs) [20] for which the interface identifier is generated by computing a cryptographic one-way hash function from the node's public key and the IPv6 prefix (among other additional information).

An additional aspect that might also be worthwhile being tackled is the address space distribution within the MANET (already briefly discussed when we described the address pool distribution). Basically, in IPv6 infrastructure-based networks, nodes are attached to subnets, where all the attached nodes share the same prefix(es). In ad hoc networks, given its multihop nature, this is not the only model that can be considered (that is, all the MANET routers within a MANET configuring their IPv6 addresses from the same prefix). We may consider, for example, the distribution of different IPv6 prefixes within the MANET, so different MANET routers may configure addresses from disjointed IPv6 prefixes. How these prefixes are distributed may be based on different aspects, such as the geographic location of the node, its relative position and distance from a MANET border router, its position within a particular hierarchy, and so on. The extreme case of this prefix distribution approach is the delegation of a different IPv6 prefix (or set of prefixes) per MANET router.

9.1.4.4 How Is an IP Address Uniqueness Guaranteed?

This question actually encompasses three different important ones.

How Is Address Uniqueness Detection Performed?

It should be noted that a nonunique address detection mechanism is not always needed, since some methods of obtaining or generating addresses ensure the uniqueness of the assigned addresses or prefixes. This is the case when a centralized approach is followed (e.g., a DHCPv6 server), which keeps an up-to-date list of assigned or available addresses, or when a set of coordinated servers is deployed, which collectively perform the DHCP functionality. For example, a new MANET router may take IP addresses from a set of address pools (a disjoint set of available IP addresses) distributed within the ad hoc network (e.g., hierarchically, by using a binary split approach [16]). Nodes owning a pool, which collectively perform the DHCP functionality, are somehow coordinated to assure the pools are collision-free. Additionally, there exist some methods of address generation, which ensure the uniqueness of assigned addresses (e.g., a special type of function is used to generate a series of random numbers, IPv6 addresses [21]).

On the other hand, some methods of obtaining or generating addresses do not ensure the uniqueness of the assigned addresses or prefixes (e.g., each MANET router generates a random IPv6 address out of a known prefix [11]). In these cases, mechanisms to detect and resolve address conflicts are needed.

A first approach to detect address conflicts is to check that the tentative address (e.g., randomly chosen) is not being used by another node in the network. A first possibility is that each MANET router, before using a tentative address, floods an address request message [11] in the network to query for the usage of this address. The node waits for a while after sending this query for the reception of a reply. The process is repeated if no answer is received, and if after a number of attempts no reply has been received, the node assumes that the tentatively chosen IPv6 address is unique and starts using it. A main concern with this approach is its scalability, which is strongly correlated with the organization of the network, that is, a flat structure or a hierarchical one. In the former case, every address acquisition results in extra traffic throughout the whole network; however, only group leaders or representatives need to take action in the latter case [18]. Additionally, this approach is not applicable in networks where message delays cannot be bounded (e.g., the use of timeouts cannot reliably detect the absence of a message).

Another possibility is that before choosing a tentative address a positive acknowledgment (ACK) is required from all known nodes in the network [22]. This approach requires each MANET router to maintain an up-to-date list of all the nodes in the network. This list can additionally be used to detect partitions in the network (e.g., if a set of nodes do not reply, it could be due to a partition). This approach has several significant drawbacks. Updating a list of participants in highly dynamic scenarios is expensive, and it is not applicable in networks where message delays cannot be bounded, which is a likely occurrence in dynamic ad hoc networks because of its use of timeouts (e.g., the absence of a message could be misinterpreted as a partition).

Another plausible approach is to relax the requirement of avoiding duplicate addresses and focus on preventing a packet from being routed to a wrong destination even if an address conflict exists [23]. For example, a unique per MANET router key is included in the routing control packets and in the routing table entries. Thus, every node is identified by a unique tuple [address, key] (e.g., virtual IP address). Following this approach, even if two nodes happen to have chosen the same IP address, they can still be identified by the use of their unique keys. The main concern with this approach is that it implies modifying current routing protocols.

Another way of addressing the detection of duplicate address events is looking for the consequences of a potential address conflict. This can be accomplished passively by continuously monitoring routing protocol traffic (e.g., looking for inconsistencies) [12]. The basic idea of this approach is to exploit the fact that some protocol events occur in case of a duplicate address, but (almost) never in case of a unique address. Most of the existing solutions following this approach work with proactive routing protocols (i.e., OLSR) but it can also be applied to on-demand routing protocols [13]. The main advantages of this approach are that it can work with current routing protocols (e.g., without any modifications), and it does not introduce any extra overhead to perform address uniqueness detection. On the other hand, the time needed to detect conflicts may be high and during this time, a MANET router may be experiencing deficiencies in its communications.

When Address Uniqueness Detection Is Performed Preservice and/or In-Service?

Address uniqueness detection may be needed at different times of the communication. When a MANET router has just chosen a tentative address, and before assigning it to the interface and using it, it is checked that there is not an address conflict (i.e., preservice detection).

Additionally address conflicts may occur at any time, mainly caused by mergers and partitions in the MANET. It may happen that nodes in different networks or partitions may independently obtain the same address, and duplicate addresses result if these networks merge later. Thus, the address uniqueness detection may be needed to take place in a continuous manner during the whole life of the MANET (in-service detection) [4].

Generally speaking, address uniqueness detection approaches commented above could be used both as preservice and as in-service mechanisms [18], [23], [24]. Nevertheless, some of them seem to be more appropriate for just one of the situations (preservice or in-service). For instance, querying the rest of MANET routers to check whether an IP address is available or not, seems to be more acceptable in the case of preservice detection (e.g., flooding is not repeated periodically over the time). In general, the overhead introduced by the mechanism is going to be a more critical issue in the case of in-service than in the case of preservice detection. So, approaches that analyze routing protocol messages looking for inconsistencies (e.g., [23]) or uniquely identify nodes in routing protocol messages (e.g., [21]) without adding extra messages seem to be more suitable for the case of in-service detection.

How Are Address Conflicts Resolved?

Whenever an address conflict is detected the most common approach is to use a heuristic to decide which MANET router keeps on using the duplicated address and which one has to look for a new IP address. In the case of preservice detection the solution is quite straightforward; the newcomer (e.g., trying to use an already-assigned address) has to look for a new address. However if the conflict has been caused by a merging (in-service detection) a different heuristic can be used (e.g., the node that detects the conflict keeps on using the duplicated address [24]).

An interesting issue to be addressed is what happens in the event of an address conflict while the node has an ongoing session. Session continuity should be guaranteed after an address duplication episode. One possible way of ensuring session continuity is IP tunneling of data packets to the new assigned address (e.g., the MANET router, keeping on using the duplicated address, tunnel packets to the other MANET router [24]).

9.1.4.5   How Is Signaling Performed?

In general, the IP configuration mechanism requires some extra signaling—in addition to the signaling introduced by the ad hoc routing—to reach its goal. The ways signaling is performed may have an impact on the scalability and convergence time of the IP autoconfiguration mechanism.

This extra signaling may be sent as separate messages or may be added or piggybacked to existing routing protocol messages (e.g., prefix or border router information). The size of this added overhead and its periodicity may vary on the different solutions. The main concern of adding or piggybacking signaling information to the existing routing protocol messages is that the IP autoconfiguration mechanism is routing protocol dependant. On the other hand, the main advantages of this approach are that the IP autoconfiguration mechanism may somehow take advantage of the routing discovery phase of the ad hoc routing protocol (e.g., discovering of available prefix and border routers) and do not introduce extra messages (e.g., MANET routers have to process less messages).

Flooding the MANET with signaling messages is required by some mechanisms, for example, asking for the approval of the rest of the nodes with each new address acquisition. There exist different ways of decreasing the effects of flooding such as limiting the *scope*, for example, organizing the network in a hierarchical structure, where only group leaders need to take action with each new address acquisition.

An alternative approach consists on relying—partially or totally—on ad hoc routing signaling to perform IP autoconfiguration. An example of this approach is passive address uniqueness detection [15], where conflicts are identified by analyzing received routing protocol messages and detecting inconsistencies. The main advantage of this approach is that no extra signaling is introduced in the network and the routing protocol is used as is (e.g., without modifications); on the other hand, this kind of mechanism is routing protocol dependent (e.g., the mechanism may be quite different for each particular routing protocol).

9.1.4.6 What Are the Security Considerations?

The wireless ad hoc environment attacks can lead to improper functioning of autoconfiguration mechanisms. Nevertheless, the IP autoconfiguration mechanisms proposed so far do not propose special mechanisms to secure the address autoconfiguration process.

Assuring reliable IP autoconfiguration mechanism signaling (i.e., secure transfer) is critical for the proper functionality of any IP autoconfiguration mechanism. In this sense, secure communication should be assured between MANET routers, where this problem can be differently tackled according to the approach used by the IP autoconfiguration mechanism. For IP autoconfiguration mechanisms depending on the routing protocol, this can be done through securing the routing protocol (especially, the control message transfer). However, IP autoconfiguration mechanisms that are routing protocol–independent need special security mechanisms. In spite of the type of IP autoconfiguration mechanism (routing protocol dependent or not), cooperation between nodes is an important factor to assure the proper message (signaling) forwarding during the autoconfiguration process.

Generally, security considerations can differ depending on different MANET scenarios, where connected MANETs allow to have a central authority that can play the role of a trusted third party to authenticate MANET routers, for example, or provide cryptographic keys.

9.1.4.7 Are Existing Protocols Modified?

IP autoconfiguration mechanisms should function in a compatible manner to the other underlying protocols; however, some of these protocols can be modified or extended in order to allow the proper IP autoconfiguration mechanisms' functioning and signaling transfer. Autoconfiguration mechanisms can extend the IPv6 neighbor discovery protocol (NDP) to work in multihop wireless networks (for instance, extending the NDP router solicitation and advertisement messages) or employ ICMPv6 messages in a modified manner. Also, some mechanisms can modify DHCP protocol, allowing MANET routers to act as modified DHCP proxies.

IP autoconfiguration mechanisms can use the routing protocol messages to transfer the IP autoconfiguration signaling. This takes place by simply encapsulating such signaling in routing protocol control messages with no routing protocol modification or through adding new control messages to the routing protocol ones. In the former, IP autoconfiguration mechanisms are mostly open to any existing routing protocol, proactive or reactive according to their functioning mode. While in the latter, IP autoconfiguration mechanisms extend the functioning of a given routing protocol to support IP autoconfiguration, which in turn limits their application *scope.*

9.2   IP Address Autoconfiguration in Vehicular Networks

The envisioned vehicular scenario is depicted in Figure 9.1, and it consists in a multihop vehicular network, where vehicles directly communicate to each other, without of relying on a centralized entity, thus making feasible time-critical applications such as safety applications. Intersection collision avoidance, emergency brake notification, and postcrash warning

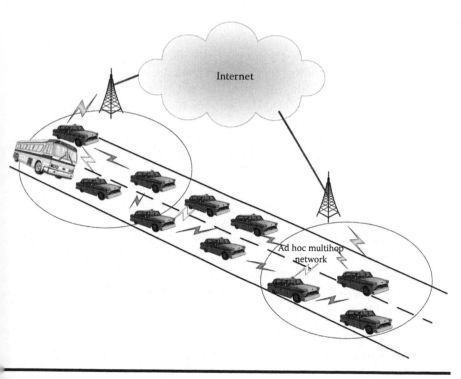

Figure 9.1   Vehicular scenario.

are only a few of the envisaged applications based on this technology. Additionally, by connecting the VANET to an infrastructure network (i.e., the Internet), other less critical safety applications and infotainment services can be implemented. For example, a dangerous situation can be reported to a service center and then notified to a much wider area, while centralized traffic management can also benefit extremely from this technology.

Automatic IP address configuration is a challenging and still almost unexplored issue in vehicular ad hoc network (VANETs) environments, where high mobility and variant density of vehicles impede the direct utilization of traditional networking techniques and protocols.

9.2.1 Solution Requirements

The vehicular applications put specific requirements for IP address configuration functionalities. These requirements are described in the following list:

- The address autoconfiguration solution must provide the capability to configure globally valid addresses. In fact, vehicles must be able to exchange packets with nodes in the infrastructure (i.e., a service center).
- A minimum amount of signaling messages should be used to save wireless resources.
- The address autoconfiguration solution must allow for usage of mobile IPv6 or network mobility basic support (i.e., the vehicle changes its point of attachment to the infrastructure while on the move and it may need session continuity). This implies that the solution must be suitable for movement detection procedures.
- Gateway selection must be provided by the address autoconfiguration solution in case multiple Internet gateways (MBRs) are simultaneously reachable. It is a desired functionality that the gateway selection algorithm is under control of the infrastructure operator (e.g., network operator, road administrator).
- Partitioning or merging support must be provided by the address autoconfiguration solution to cope with situations where address duplication occurs due to a merging or an Internet gateway becomes unreachable because of a partition.
- For network security reasons, the address autoconfiguration solution must not require selected vehicles to carry out more important tasks on which the whole VANET operability depends. In other words, the solution must be executed in a noncentralized fashion.
- The address autoconfiguration solution must protect the privacy of vehicle users. This implies that the solution must not reveal information that could potentially be used to track vehicles.

9.2.2 Proposed Solutions

For the time being, not much attention has been paid to the IP address configuration problem in vehicular networks. There exist two main proposals that target address autoconfiguration in this scenario. The first one, named vehicular address configuration (VAC) [25] proposes to use a distributed scheme based on an enhanced DHCP service with dynamically elected leaders. The second one, called geographically scoped stateless address configuration (GeoSAC) [26] consists of adapting the existing IPv6 stateless address autoconfiguration (SLAAC) mechanisms to geographic addressing and routing, where the concept of IPv6 link is extended to a specific geographic area associated with a (or several) point of attachment.

9.2.2.1 Vehicular Address Configuration

Vehicular address configuration (VAC) [25] assumes that the topology of a vehicular ad hoc network is linear (i.e., vehicles form a chain) and that a group of vehicles move almost together following a track with a low internal mobility with respect to each other. On the other hand, VAC assumes that vehicular communications are usually going to take place among users a few hops away, therefore nodes located very far from each other can utilize the same identifier (i.e., the same IP address).

VAC is a leader-based protocol. It organizes leaders in a connected chain such that every node (vehicle) lies in the communication range of at least one leader. The hierarchical organization of the network allows to limit the signal overhead for the address management tasks. Only leaders communicate with each other and maintain updated information on configured address management tasks. Normal nodes ask leaders for a valid IP address whenever they need to be configured. The communication that involves normal nodes is based on local broadcast within a short communication range (i.e., DSRC/802.11p). Each normal node communicates only with leaders in its transmission range, thus reducing flooding and multihop interactions among nodes in the network. Moreover, normal nodes snoop leaders' packets to catch information they need to maintain their IP addresses.

Leaders act as servers of a distributed DHCP protocol. Each of them manages a different subset of possible addresses to serve requests for address assignment coming from normal nodes located within their communication range.

VAC guarantees unique IP addresses within a defined *scope* around the leader, where the *scope* of the leader A is the set of leaders whose distance from A is less or equal to *scope* hops. Considering the normal node Y that received the IPy address from A, IPy will be unique as long as Y moves within the *scope* of A. If Y goes out of the *scope* of A to still ensure the address uniqueness, Y has to ask the new leader for another address. Considering that the relative speed between the nodes is low, changes in the address configuration due to having left the current leader's *scope* are not frequent.

The metric used for building the leader chain is the distance among nodes (e.g., vehicles are equipped with GPS). Periodically, a leader estimates the distance between itself and the closest among the other leaders in the *scope*. If the distance between these two leaders exceeds a threshold, a node in the middle has to become a new leader. On the contrary, if the distance falls under a threshold, one of these two leaders will become a normal node.

Leaders periodically exchange "hello" messages that include the list of the other leaders in their *scope*. These hello messages are also listened by normal nodes. In this way a normal node can detect when it has gone out of *scope* of its current leader and consequently, they should identify a new leader and ask it for a new address.

We next analyze the key features of VAC protocol:

■ Globally valid addresses: The proposed solution does not provide global addresses but only local addresses (i.e., it guarantees unique IP addresses within a defined *scope*). Nevertheless it could be extended with some mechanisms in order to provide global addresses.

■ Address uniqueness: The proposed solution does not employ any nonunique address detection mechanisms; however, it guarantees address uniqueness for each configured node.

■ Distributed or centralized approach: The proposed solution employs a partially distributed approach, where distributed DHCP is run by some mobile nodes (vehicles) that are elected as leaders in a dynamic manner to assign IP addresses to the requesting nodes.

■ Merging support: No special merging mechanisms are explained in this proposed solution, however, it could support merging. The *scope* principle together with the distributed DHCP permit the nodes to join or leave different scopes while acquiring a new address from the *scope* leader.

■ Routing protocols' dependency: Apparently VAC does not depend on any special routing protocol. However no clear definition is given on how and what control messages are exchanged to configure each node requiring an IP address.

■ Protocol overhead: The hierarchical organization in this solution limits the signaling overhead and avoids flooding. The overhead in this solution mainly concerns the signaling messages for communication between leaders nodes, the request messages sent by mobile nodes requesting an IP address from their leaders (this takes place in a limited *scope*), and the reply messages from the leaders to the requesting mobile nodes for assigning IP addresses (this also takes place in a limited *scope*). It is noticed that this solution does not use nonunique address detection mechanisms due to the distributed DHCP functionality among leader nodes, which helps in limiting the signaling overhead.

9.2.2.2 Geographically Scoped Stateless Address Configuration

Geographically scoped stateless address configuration (GeoSAC) [26] uses the car-to-car consortium (C2C-CC) architecture [27] as its reference architecture. In the C2C-CC system architecture, vehicles are equipped with devices termed on-board units (OBU), which implement the C2C-CC protocol stack. Units of different vehicles can communicate with each other or with fixed stations installed along roads termed road-side units (RSUs). OBUs and RSUs implement the same network layer functionalities and form a self-organizing network. RSUs can be connected to a network infrastructure, most presumably an IP-based network. Also, it is reasonable to assume that RSUs will act as IPv6 access routers (AR) or as network bridges connected to an AR. Passenger or driver devices attached to the vehicle on-board system are called application units (AU). AUs are assumed to have a standard IPv6 protocol stack, OBUs act as gateways for the in-vehicle network optionally enhanced with the network mobility basic support protocol [28]. Figure 9.2 depicts the resulting set of communicating devices and their protocols with respect to IPv6.

GeoSAC adapts the existing IPv6 stateless address autoconfiguration (SLAAC) mechanisms to geographic networking, where the concept of IPv6 link is extended to a specific geographic area associated with a (or several) point of attachment. For a better understanding of the solution and its potential applications, in the following text we refer to the C2C-CC protocol architecture already introduced, though the proposed method can be applied in principle to different protocol stacks, as described below.

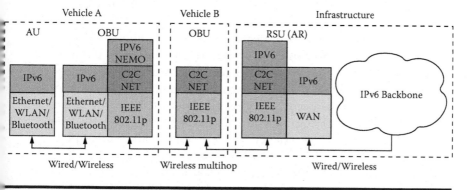

Figure 9.2 IPv6 deployment in the C2C-CC system architecture.

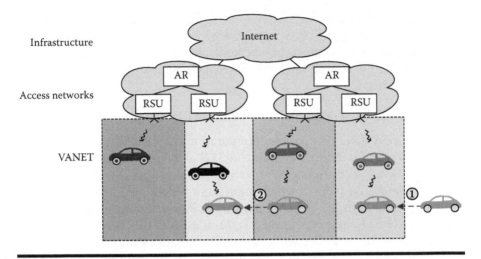

Figure 9.3 VANET splitting achieved with the proposed solution.

In the considered protocol architecture (Figure 9.2), routing and addressing functionalities are split and assigned to two different protocol layers. A lower layer (C2C NET) deals with ad hoc routing by applying geographic networking and presents to the upper layer a flat network topology. The upper layer can run standard IPv6 procedures, as the concept of *link* is emulated by the lower layer. In this approach, at the reasonable cost of some additional overhead due to encapsulation into a sub-IP protocol header (i.e., C2C NET header), IPv6 packets can be routed from a point of attachment to an arbitrary node that is reachable via multihop. Furthermore, dividing of the VANET into different clusters is also provided by the sub-IP layer (Figure 9.3), which in fact enables to limit the link *scope* to a certain geographic area. This results in a broadcast domain that, as seen by IPv6, is flexible and can be configured on a per packet basis according to geographic parameters, instead of pure topological ones like the traditional "time to live."

Exploiting the efficient multihop distribution and network splitting offered by the sub-IP geographic routing, a point-of-attachment in the proposed solution sends out standard IPv6 router advertisement (RA) messages, which reach all the nodes currently located within a well-defined area. In particular, the access point specifies as target of the sub-IP protocol header a preassigned geographical area, which is served by this gateway. Upon reception of this packet, a node applies the geographic filtering before delivering the RA to the IPv6 layer and forwards the message according to the geocast forwarding procedure. As a result, if a multihop path exists, all the nodes within the area receive the RA and the IPv6 instance running above geo networking processes the message as if the node was directly connected to the access router that issued the message.

According to IPv6 SLAAC, at this point a host generates an address appending its network identifier derived from the MAC address to the received IPv6 prefix and performs the IPv6 duplicate address detection (DAD) procedure. This consists in sending a set of neighbor solicitation (NS) messages to the just configured address to make sure that is is not already in use. For this purpose, we propose that the same geographic area specified by the access point is again set as broadcast domain, which allows for uniqueness of the addresses within this area. This implies two possible behaviors of an RSU, according to whether the RSU acts as network bridge (Figure 9.3) or access router. In the first case, as the IPv6 prefix is shared among different RSUs, the DAD messages should be forwarded among the RSUs to assure uniqueness in the entire IP subnet. In the latter, a DAD restricted to the geographical area served by the RSU is enough to assure global uniqueness, because the IPv6 prefix is exclusively associated to this access point.

Regarding detection of duplicate addresses, we argue that the execution of DAD might be unnecessary because MAC addresses in VANET might be required to be unique, at least within macroregions where vehicles are sold and can potentially communicate with each other (e.g., a continent). This property, in fact, is highly desirable for security and liability reasons, as it would allow (i) forensic teams to rely on vehicular communications to reconstruct accident scenes or other critical situations and (ii) to detect malicious nodes and reduce considerably effects of network attacks. Despite uniqueness of addresses, privacy of users can be protected by equipping vehicles with sets of addresses to be used for limited intervals as *pseudonyms* [29]. These addresses could be assigned by authorities and, when coupled with the usage of digital certificates and cryptographic protection [30], this mechanism can accomplish support for liability as well as privacy protection from malicious users.

A technique that maximizes the benefits of the proposed mechanism consists of shaping the geographical areas assigned to the access point in an adjacent and nonoverlapping fashion, as depicted in Figure 9.3. By doing so, the following key advantages are obtained: (i) Unique gateway selection is achieved efficiently and with the infrastructure having full control of it, as only one RSU is assigned per geographical area; (ii) Network splitting is obtained that considerably improves movement detection procedures of IPv6 mobility and also allows for location-based services. In particular, a vehicle moving across regions served by different RSUs (case 2 in Figure 9.3) experiences a sharp subnet change, without traversing *gray areas* where router advertisements are received from multiple access points.

In addition to the already mentioned benefits, the VANET splitting obtained with the proposed solution implements a matching between geographical area and IPv6 prefix assigned to an access point. For the purpose

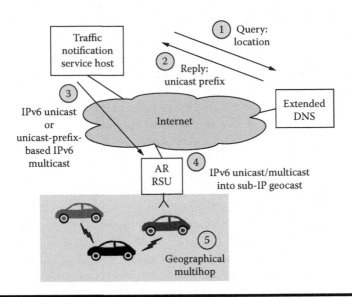

Figure 9.4 Proposed solution coupled with extended DNS (usage example).

of deploying location-based applications, this matching can be efficiently coupled with already proposed techniques for geographical routing in the infrastructure, like extended DNS [31] or geographic IPv6 prefix format [32]. In the first approach, DNS servers are extended with the capability to resolve geographical locations into IP addresses, without requiring changes in the routing behavior of today's Internet (Figure 9.4). The second approach consists of encoding the geographic position as part of the IPv6 prefix and performing the actual routing accordingly.

We next analyze the key features of GeoSAC protocol:

■ Globally valid addresses: The proposed solution fulfills this requirement because a globally valid prefix can be included into the RA broadcast within each geographical area.

■ Address uniqueness: The proposed solution does not require a DAD mechanism.

■ Distributed or centralized approach: The proposed mechanism does not rely on any particular node on the VANET playing a special role in the IP address autoconfiguration process, but only on some nodes located at the infrastructure side (i.e., the RSUs or access routers attached to them).

■ Routing protocols' dependency: GeoSAC does not require to implement a sub-IP geographic routing, but can be also applied as an IPv6 extension. This requires first to allow an IPv6 host to rebroadcast router advertisements, and second, to enforce at IP layer a filtering mechanism based on the geographic position somehow specified in the RA message itself (e.g., in an IPv6 extension header or as part of the prefix itself). However, a quick comparison of these possible approaches in VANETs shows how sub-IP ad hoc routing results in lower complexity (unmodified standard IPv6 behavior), clean separation of roles and better integration with safety communication systems that are currently being designed by the automotive industry [27].

■ Protocol overhead: The solution has low complexity in terms of per node requested functionality. Each node only needs to perform the geographically scoped broadcast forwarding to ensure that RA messages reach all the intended destinations and to process them in the usual way [2]. On the other hand, due to the use of geographical routing, RA messages can reach all nodes within a given geographical area. These geographically distributed RAs are the only signaling messages required by the solution to work.

9.3 Conclusions

Besides vehicle-to-vehicle communication, the potential in the vehicle-to-infrastructure architecture is promising in allowing vehicular Internet access as well as provision of Internet-related services to drivers and passengers. However, IP configuration is an important challenge that can threaten the services' quality and precisely the services' continuity in vehicular networks. Regarding the vehicular network characteristics, inherited from ad hoc networks, IP address configuration should be carried out in an automatic and distributed manner. So far, there is no standard for IP autoconfiguration in ad hoc networks, and hence the problem becomes complex for vehicular networks. This problem is attracting the standardization attention, where the AUTOCONF WG has been created within the IETF, with the main objective to develop and standardize IP autoconfiguration protocols for ad hoc networks, focusing on IPv6. In this context, vehicular networks have been defined among the ad hoc scenarios to be considered by the AUTOCONF WG work.

We notice that the conventional IP autoconfiguration protocols (DHCP, SLAAC/NDP, and DHCP-PD) provide only a partial solution with respect to the required goals for achieving appropriate IP autoconfiguration. Hence, these protocols cannot be applied as is generally in ad hoc networks and particularly in vehicular networks. There is a need for special mechanisms and functionalities that can deal with the dynamic and distributed nature of these networks, as well as the multihop communication. Moreover, IP autoconfiguration solutions in vehicular networks should specially consider the strong mobility requirements.

References

[1] T. Narten, E. Nordmark, W. Simpson, and H. Soliman, Neighbor Discovery for IP version 6 (IPv6). Internet Engineering Task Force, RFC 4861 (Draft Standard), September 2007.

[2] S. Thomson, T. Narten, and T. Jinmei, IPv6 Stateless Address Autoconfiguration. Internet Engineering Task Force, RFC 4862 (Draft Standard), September 2007.

[3] I. Chakeres, J. Macker, and T. Clausen, Mobile Ad hoc Network Architecture. draft-ietf-autoconf-manetarch-07 (work in progress), November 2007.

[4] E. Bacceli (ed.), Address Autoconfiguration for MANET: Terminology and Problem Statement. Draft-IETF-autoconf-statement-04 (work in progress), February 2008.

[5] Ralph Droms, Jim Bound, Bernie Volz, Ted Lemon, Charles E. Perkins, and Mike Carney, Dynamic Host Configuration Protocol for IPv6 (DHCPv6). Internet Engineering Task Force, RFC 3315 (Proposed Standard), July 2003.

[6] O. Troan and R. Droms, IPv6 Prefix Options for DHCPv6. Internet Engineering Task Force, RFC 3633 (Proposed Standard), December 2003.

[7] C. Bernardos, M. Calderon, and H. Moustafa, Survey of IP address autoconfiguration mechanisms for MANETs. draft-bernardos-manet-autoconf-survey-03 (work in progress), November 2008.

[8] C. H. Moustafa, C. Bernardos, and M. Calderon, Evaluation Considerations for IP Autoconfiguration Mechanisms in MANETs. draft-bernardos-autoconf-evaluation-considerations-03 (work in progress), November 2008.

[9] T. Clausen, Evaluation Criteria for MANET Autoconf Mechanisms. draft-clausen autoconf-criteria-00 (work in progress), July 2005.

[10] C. Bernardos, M. Calderon, and H. Moustafa, Ad-Hoc IP Autoconfiguration Solution Space Analysis. draft-bernardos-autoconf-solution-space-02 (work in progress), November 2008.

[11] Charles E. Perkins, IP Address Autoconfiguration for Ad Hoc Networks. draft-perkins-manet-autoconf-01 (work in progress), November 2001.

[12] K. Weniger, "PACMAN: Passive auto configuration for mobile ad hoc networks," *IEEE Journal on Selected Areas in Communications*, 23(3)(March 2005):507–519.

[13] H. Jeong, Passive Duplicate Address Detection for On-demand Routing Protocols. draft-jeong-autoconf-pdad-on-demand-01 (work in progress), April 2007.

[14] K. Mase and C. Adjih, No Overhead Autoconfiguration OLSR. draft-mase-manetautoconf-noaolsr-01 (work in progress), April 2006.

[15] M. Mohsin and R. Prakash, "IP address assignment in a mobile ad hoc network," In *Proceedings of the IEEE Military Communications Conference (MILCOM)*, Anaheim, CA, 2002.

[16] A. Tayal and L. Patnaik, An address assignment for the automatic configuration of mobile ad hoc networks. Personal Ubiquitous Computing, London: Springer, 2004.

[17] S. Ruffino and P. Stupar, Automatic configuration of IPv6 addresses for MANET with multiple gateways (AMG). draft-ruffino-manet-autoconf-multigw-03 (work in progress), June 2006.

[18] K. Weniger and M. Zitterbart, "IPv6 autoconfiguration in large scale mobile ad-hoc networks," In *Proc. of European Wireless*, Florence, Italy, 2002.

[19] D. Templin, S. Russert, and S. Yi, The MANET Virtual Ethernet (VET) Abstraction. draft-templin-autoconf-dhcp-14 (work in progress), April 2008.

[20] Tuomas Aura, Cryptographically Generated Addresses (CGA). Internet Engineering Task Force, RFC 3972 (Proposed Standard), March 2005.

[21] H. Zhou, L. Ni, and M. Mutka, "Prophet address allocation for large scale MANETs," In *Proc. of INFOCOM*, San Francisco, CA, 2003.

[22] S. Nesargi and R. Prakash, "MANETconf: Configuration of hosts in a mobile ad hoc network," In *Proc. of INFOCOM*, New York, 2002.

[23] N. Vaidya, "Weak duplicate address detection in mobile ad hoc networks." In *Proc. of MOBIHOC'02*, Lausanne Switzerland, 2002.

[24] J. Jeong, Ad Hoc IP Address Autoconfiguration. draft-jeong-adhoc-ip-addr-autoconf-06 (work in progress), January 2006.

[25] M. Fazio, C. Palazzi, S. Das, and M. Gerla, "Facilitating real-time applications in VANETs through fast address auto-configuration," *IEEE CCNC International Workshop on Networking Issues in Multimedia Entertainment (CCNC/NIME 2007)*, Las Vegas, NV, January 2007.

[26] R. Baldessari, C. Bernardos, and M. Calderon, "GeoSAC—scalable address autoconfiguration for VANET using geographic networking concepts," In *Proc. of IEEE International Symposium on Personal, Indoor and Mobile Radio Communications (PIMRC'2008)—Intelligent Transportation Special Session*, Cannes, France, September 2008.

[27] Car-to-Car Communication Consortium Manifesto Version 1.1. Available at: http://www.car-to-car.org, August 2007.

[28] Vijay Devarapalli, Ryuji Wakikawa, Alexandru Petrescu, and Pascal Thubert, Network Mobility (NEMO) Basic Support Protocol. Internet Engineering Task Force, RFC 3963 (Proposed Standard), January 2005.

[29] E. Fonseca, A. Festag, R. Baldessari, and R. Aguiar, "Support of anonymity in VANETs—putting pseudonymity into practice," In *Proc. of IEEE Wireless Communications and Networking Conference (WCNC)*, Hong Kong, March 2007.

[30] C. Harsch, A. Festag, and P. Papadimitratos, "Secure position-based routing for VANETs," In *Proc. of VTC Fall*, Baltimore, MD, October 2007.

[31] T. Imielinski and J. Navas, GPS-Based Addressing and Routing. Internet Engineering Task Force, RFC 2009 (Experimental), November 1996.

[32] T. Hain, An IPv6 Provider-Independent Global Unicast Address Format draft-hainipv6-pi-addr-10 (work in progress), August 2006.

Chapter 10

Network Mobility in Vehicular Networks

Ignacio Soto, Roberto Baldessari, Carlos J. Bernardos, and Maria Calderon

Contents

Accelerated by the success of cellular technologies, mobility has changed the way people communicate. As Internet access becomes more and more ubiquitous, demands for mobility are not restricted to single terminals anymore. It is also needed to support the movement of a complete network that changes its point of attachment to the fixed infrastructure, maintaining the sessions of every device of the network. It is known as network mobility in IP networks. In this scenario, the mobile network has, at least, a (mobile) router that connects to the fixed infrastructure, and the devices of the mobile network connect to the outside through this mobile router.

Since the vehicular communications scenario involves a group of devices that move together, both the car-to-Internet (also known as vehicle-to-infrastructure [V2I]) and the car-to-car (also known as vehicle-to-vehicle [V2V]) cases may be addressed by assuming that there is a mobile router deployed in each vehicle, managing the mobility of the group of devices within the moving vehicle.

The car-to-Internet scenario fits quite well into the general network mobility paradigm. Therefore, the applicability of a generic network mobility framework to the car-to-Internet scenario solution should be analyzed. NEMO route optimization solutions may be applied to improve the performance. Actually, this is a very good example of a scenario where a route optimization solution for NEMO is needed.

The car-to-car scenario may also be addressed by using a generic NEMO approach. However, this kind of solution does not perform well in a car-to-car communication, even when a generic route optimization for NEMO solution is used. There is an opportunity for optimization that current research efforts within the field of intervehicular communication (IVC) systems are looking at. This optimization is based on the use of vehicular ad hoc networks (VANETs) to exploit connectivity between neighboring cars and set up a multihop network to support car-to-car services.

This chapter provides a detailed description of the network mobility problem, describing the currently standardized solution (NEMO basic support protocol), as well as identifying open issues currently being addressed by the research community—such as route optimization. The applicability of network mobility solutions in combination with ad hoc networks in the vehicular communications scenario is analyzed, providing an overview of two existing solutions.

10.1 Why Network Mobility: The Network Mobility Problem

Driven by the success of cellular technologies, mobility has changed the way users communicate. Ubiquity and heterogeneity [1,2] will be two key concepts of forthcoming 4G [3] networks, which are expected to enable users to communicate almost anytime, anywhere.

Triggered by these needs and the fact that deployed Internet protocols did not support mobility of any kind, the technical community designed several solutions that addressed the problem of mobility [4]. The problem of terminal *mobility* in IP networks has been studied for a long time within the IETF[1], and there exist IP-layer solutions for both IPv4 [5] and IPv6 [6] that enable the movement of terminals without stopping their ongoing sessions.

As Internet access becomes more and more ubiquitous, demands for mobility are not restricted to single terminals anymore. There exists also the need of supporting the movement of a complete network that changes its point of attachment to the fixed infrastructure, maintaining the sessions of every device of the network, which is known as *network mobility* in IP networks. In this case, the mobile network will have at least a (mobile) router that connects to the fixed infrastructure, and the devices of the mobile network will obtain connectivity to the exterior through this mobile router.

Supporting the roaming of networks that move as a whole is required to enable the transparent provision of Internet access in mobile platforms [7], in particular in vehicular scenarios such as the following:

■ Public transportation systems: That would enable passengers in trains, buses, planes, ships, and the like to travel with their own terminals (e.g., laptops, cellular phones, and PDAs) and obtain Internet access through a mobile router located at the transport vehicle, which connects to the fixed infrastructure.

■ Car scenarios: Future cars will benefit from having Internet connectivity, not only to enhance safety (e.g., by using sensors that could control multiple aspects of the vehicle operation, interacting with the environment, and communicating with the exterior), but also to provide personal communication and Internet-based entertainment services to passengers.

10.2   NEMO Basic Support Protocol

The IP terminal mobility solution (Mobile IPv6 [6]) does not support, as it is now defined, the movement of networks. As a result, the IETF NEMO (network mobility) Working Group (WG) was created to standardize a solution enabling network mobility at the IPv6 layer.[2] The current solution called network mobility basic support protocol is defined in the RFC 3963 [10].

---

[1] http://www.ietf.org/.

[2] At the end of 2007, NEMO-related work has been moved to MEXT (Mobility EXTensions for IPv6) WG of the IETF.

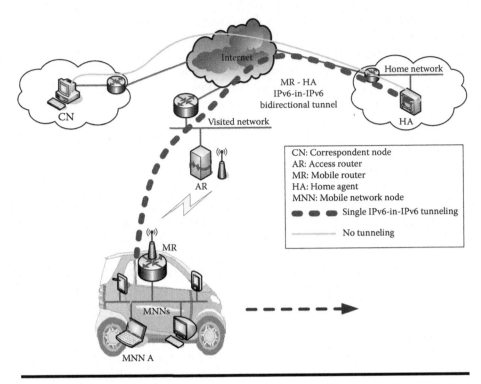

Figure 10.1   Mobile network scenario.

A mobile network (also known as NEtwork that MOves or NEMO) is defined as a network whose attachment point to the Internet varies with time. Figure 10.1 depicts an example of a network mobility scenario. The router in the NEMO that connects to the Internet is called mobile router (MR). It is assumed that the NEMO is assigned to a particular network, known as its home network, where it resides when it is not moving. Since NEMO is part of the home network, the mobile network has configured addresses belonging to one or more address blocks assigned to the home network—the mobile network prefixes (MNPs). These addresses remain assigned to NEMO when it is away from home. Of course, these addresses only have topological meaning when the NEMO is at home. When the NEMO is away from home, packets addressed to the nodes of the NEMO, known as mobile network nodes (MNNs), will still be routed to the home network. Additionally, when the NEMO is away from home, its MR acquires an address from the visited network, called care-of address (CoA), where the routing architecture can deliver packets without additional mechanisms.

When any node located at the Internet, known as a correspondent node (CN), exchanges IP datagrams with mobile network node (MNN A in Figure 10.1), the following operations are involved in the communication:

1. The CN transmits an IP datagram destined for MNN A. This datagram carries as its destination address MNN A's IPv6 address, which belongs to the NEMO's mobile network prefix.
2. This IP datagram is routed to the home network of the NEMO, where it is encapsulated inside a new IP datagram by a special node located on the home network of the NEMO, called the home agent (HA). The new datagram is sent to the mobile router's care-of address, with the home agent's IP address as source address. This encapsulation preserves mobility transparency (i.e., neither MNN A nor the CN are aware of the mobility of the NEMO) while maintaining Internet MNNs' established connections.
3. The encapsulated IP datagram is received by the mobile router, which removes the outer IPv6 header and delivers the original datagram to MNN A.
4. In the opposite direction, the operation is analogous. The mobile router encapsulates the IP datagrams sent by MNN A toward its home agent, which then forwards the original datagram toward its destination (i.e., the correspondent node). This encapsulation is required to avoid problems with ingress filtering, since many routers implement security policies that do not allow the forwarding of packets that have a source address that is topologically incorrect.

There are different types of mobile network nodes:

■ Local fixed node (LFN): It is a node attached to a NEMO that has no mobility-specific software. Therefore, an LFN is unable to move by itself, changing its point of attachment while maintaining its ongoing sessions. Its IPv6 address is taken from an MNP of the NEMO.
■ Local mobile node (LMN): It is a node that implements the mobile IPv6 protocol and whose home network is located in the mobile network. Its home address (HoA) is taken from an MNP.
■ Visiting mobile node (VMN): It is a node that implements the mobile IP protocol (and therefore, it is able to change its point of attachment while maintaining ongoing sessions), it has its home network outside the mobile network, and it is visiting the mobile network. A VMN that is temporarily attached to a mobile subnet (used as a foreign link) obtains an address on that subnet (i.e., its CoA is taken from an MNP).

Additionally, mobile networks may be nested. A mobile network is said to be nested when it attaches to another mobile network and obtains connectivity through it (Figure 10.2). An example is a user who gets into a vehicle with his personal area network (mobile network 2) that connects

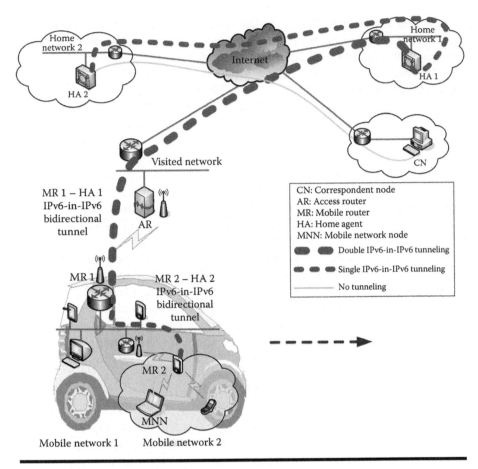

Figure 10.2 **Abstract view of a vehicular area network.**

through an MR—like a WLAN-enabled PDA—to the car's network (mobile network 1), which is connected to fixed infrastructure.

The NEMO basic solution protocol enables the mobility of an entire network, but this is just the first step to allow the deployment of new ubiquitous connectivity configurations, solving only the very basic problem, and raising some other issues that need to be carefully looked at. Among the issues that are still open, it is worth mentioning the following:

■ Route optimization support: When the NEMO basic support protocol is used, all communications to and from a node attached to the mobile network go through the MR-HA bidirectional tunnel when the mobile network is away. As a result, the packet overhead and the length of the route followed by packets are increased, thus resulting in an increment of the packet delay in most cases. This issue may have a serious impact on the performance of applications running on nodes within the NEMO and may even prevent communications from taking place.

- Multihoming support: The support of multihoming has shown to be very important in future 4G networks to fully exploit the heterogeneity in the network access. This is even more relevant for mobile networks, since a loss of connectivity or a failure to connect to the Internet has a more significant impact than on a single node. Furthermore, typical deployment scenarios, such as the provision of Internet access from moving vehicles, will typically require the use of several interfaces (using different access technologies), since the mobile network may be moving within distant geographical locations, where different access technologies are provided and governed by distinct access control policies [22]. Although there exist several works published regarding multihoming support for NEMO [13–21], there is no mechanism that fulfils all the requirements of a multihoming solution for mobile network environments. The applicability of the SHIM6 protocol [22] to provide NEMO multihoming support is one of the approaches that should be further investigated (an early attempt can be found in [23]).

- Multicast support: Current network mobility basic specification does not support multicast traffic transmission to or from a mobile network. With some broadcast technologies becoming popular, such as DVB, the support of multicast-like application would be required in future 4G platforms. Early attempts to provide such a support to mobile networks can be found in Simon et al. and Hugo et al. [24,25].

- Seamless handover support: To support real-time applications, not only the end-to-end delay should be kept under certain values [26], but also the interruption time due to handovers. Owing to the additional complexity of the NEMO scenario, the handover delay during handovers may be higher than for a single terminal. The applicability of some of the solutions for mobile IPv6, such as fast handovers for mobile IPv6 [27], to alleviate the increase in handover delay or the design of new ones should be investigated [28,29].

- QoS support: Mobile networks, because of their dynamic nature, pose additional challenges to the inherent difficulty of providing QoS over wireless links. Indeed, QoS provisioning in a NEMO involves additional mechanisms besides providing QoS to the various wireless links of the mobile network. Statistical analyses are required to guarantee the desired performance resulting from traversing several wireless links, each of which provides only statistical guarantees. In addition, novel signaling mechanisms need to be devised to perform QoS signaling over such a dynamic environment. An early attempt of reservation protocol adapted to NEMO can be found in the work of Tlais and Labiod [30].

■ Authentication, authorization, and accounting (AAA) support: The NEMO scenario poses some challenges to classical authentication, authorization, and accounting (AAA) schemes [31]. This issue has to be carefully analyzed, paying attention to real NEMO AAA deployment scenarios [32].

Although all the previously described topics are relevant, the route optimization issue is the most critical one, since it may even prevent mobile networks from being deployed in real scenarios. Therefore, it is very important to address this issue.

10.3 NEMO Route Optimization

10.3.1 Problem Description

By using a bidirectional tunnel between the mobile router and the home agent, the NEMO basic support protocol [10] enables mobile network nodes to reach, and be reachable by, any node in the Internet. However, such a solution also presents important performance limitations [33], which will be described in this section.

The network mobility basic solution forces—when a mobile network is not at home—all the traffic addressed to an MNN, to traverse the HA and to be forwarded to the mobile network through the tunnel established between the MR and the HA. The inverse path is followed by packets sent by an MNN. This phenomenon (see Figure 10.1) raises some inefficiency, both in terms of latency and effective throughput, and can be unacceptable for certain applications. More precisely, we can highlight the following limitations of the basic solution [10]:

■ It forces suboptimal routing (known as angular or triangular routing), that is, packets are always forwarded through the HA following a suboptimal path and therefore adding a delay in the packet delivery. This delay can be negligible if the mobile network or the correspondent node are close to the home agent (that is, close to the home network). On the other hand, when the mobile network or the correspondent node are far away from the home agent, the increase in the delay could be very large. This may have a strong impact on real-time applications where delay constraints are very important. In general, an increase in the delay may also impact the performance of transport protocols such as TCP, since the sending rate of TCP is partly determined by the round-trip time (RTT) perceived by the communication peers. A representative example of how large the impact on the delay can be found on aircraft communications, where a tunneled mobile IP communication takes almost 2 seconds to complete a TCP three-way handshake [8,9].

■ It introduces nonnegligible packet overhead, reducing the path MTU (PMTU) and the bandwidth efficiency. Specifically, an additional IPv6 header (40 bytes) is added to every packet because of the MR-HA bidirectional tunnel.

The effect of this overhead can be analyzed, for example, by looking at a VoIP communication using the widely utilized Skype[3] application. Skype [34] uses the iLBC (internet low bitrate codec) [35] codec, which is a free speech codec suitable for robust voice communication over IP. If an encoding frame length of 20 ms (as in RFC 3550 [36]) is used, it results in a payload bit rate of 15.20 kbps. Because of the additional IPv6 header (that is, 320 extra bits per packet, 50 packets per second with this codec) the bit rate used by the voice communication is increased in 16 kbps (which is more than the actual VoIP payload).

■ The HA becomes a bottleneck of the communication, as well as a potential single point of failure. Even if a direct path is available between an MNN and a CN, if the HA (or the path between the CN and the HA or between the HA and the MR) is not available, the communication is disrupted. Congestion at the HA or at home network may lead to additional packet delay or even packet loss. The effect of congestion is twofold. On the one hand, it affects data packets by making them be delayed or even discarded. On the other hand, delayed or discarded signaling packets (e.g., binding updates) may affect the set-up of the bidirectional tunnels, causing disruption of the data traffic through these tunnels.

Ng et al. [33] also describe additional limitations, such as increased processing delay, increased chances of packet fragmentation, and increased susceptibility to link failures.

Most of these concerns also exist in terminal mobility when using mobile IPv6 [6]. In order to solve them, a *route optimization* mechanism was developed and included as a part of the base protocol. In mobile IPv6, route optimization is achieved by allowing the mobile node (MN) to send binding update messages also to the CNs. In this way the CN is also aware of the CoA where the MN's home address (HoA) is currently reachable. The return routability (RR) procedure is defined to prove that the mobile node

---

[3] http://www.skype.com/.

has been assigned (that is, *owns*) both the home address and the care-of address at a particular moment in time [37].

The network mobility scenario brings a number of additional issues, making the problem more complex and difficult to solve.[4]

The aforementioned problems are exacerbated when considering nested mobility. The NEMO WG defined some useful terminology [11] related to the nested scenario. The mobile network at the top of the hierarchy connecting the aggregated nested mobile network to the Internet is called *root-NEMO* (e.g., mobile network 1 in Figure 10.2). Likewise, the mobile router of that root-NEMO is called *root-MR*[5] (e.g., MR 1 in Figure 10.2). In a mobile network hierarchy, the upstream mobile network providing Internet access to another mobile network further down in the hierarchy is named *parent-NEMO* and the downstream mobile network is called *sub-NEMO* (in Figure 10.2, mobile network 1 is a parent-NEMO of mobile network 2—which is therefore a sub-NEMO of the former). Similarly, the MRs of the parent-NEMO and the sub-NEMO are called, *parent-MR* and *sub-MR*, respectively (e.g., MR 1 and MR 2 in Figure 10.2).

The use of the NEMO basic support protocol in nested configurations amplifies the suboptimality of the routing and decreases the performance of the solution, since in these scenarios packets are forwarded through all the HAs of all the upper level mobile networks involved (known as multiangular or pinball routing, see Figure 10.2). This is because each sub-NEMO obtains a CoA that belongs to the mobile network prefix of its parent NEMO. Such a CoA is not topologically meaningful in the current location, since the parent-NEMO is also away from home, and packets addressed to the CoA are tunneled—thus increasing packet overhead—to the HA of the parent-NEMO.

There is an additional particular NEMO scenario that needs to be addressed, namely, when a mobile IPv6 host attaches to a mobile network (becoming a visiting mobile node [VMN]). Traffic sent to and from a VMN has to be routed not only via the home agent of the VMN, but also via the HA of the MR of the mobile network, therefore suffering from the same performance problems as in a 1-level nested mobile network.[6] Even if the VMN performs the mobile IPv6 route optimization procedure, this will only avoid traversing the VMN's HA, but the resulting route will not be optimal at all, since traffic will still have to be routed through the MR's HA.

---

[4] This situation made the IETF decide to address the route optimization problem in network mobility separately, not including the development of an RO solution as an item of the NEMO WG charter, but the analysis of the problem and solution space.

[5] Some authors alternatively use "top level mobile router" (TLMR) to refer to the root-MR.

[6] Some authors [33,38] consider this case as a particular one of nested mobility.

Because of all the limitations identified in this section, it is highly desirable to provide route optimization support for NEMO [33,38,39], enabling direct packet exchange between a CN and an MNN without passing through any HA and without inserting extra IPv6 headers.

10.3.2 Solution Space

Since the very beginning of research on network mobility, even before the IETF NEMO WG had been created, route optimization was a hot topic.[7] A plethora of solutions trying to enable network mobility support in an optimal way has been proposed since the beginning of the NEMO research. Most relevant proposals are identified and briefly described in various texts [38–40].

10.4 NEMO in the Vehicular Scenario

Using a NEMO solution, vehicles can communicate with the infrastructure (the Internet). Nevertheless, there exist several vehicular applications, such as multiplayer gaming, instant messaging, traffic information or emergency services, which might involve communications among vehicles that are relatively close to each other (i.e., car-to-car communications) and may even move together (e.g., military convoys). These applications are currently not well supported in vehicular scenarios.

Although automobiles can communicate with other vehicles through the infrastructure (the Internet), they could benefit from better bandwidth, delay, and, most probably, cheaper communication, by forming vehicular ad hoc networks (VANETs) and making use of the resulting multihop network to directly communicate with each other. The VANET can also be used to reach the infrastructure from a vehicle that is not within the coverage of the infrastructure.

This sections analyses the requirements on solutions for providing communications in vehicle scenarios. Then, it presents two different approaches combining the use of vehicular ad hoc network (VANET) and NEMO mechanisms. The first approach—called *vehicular ad hoc route optimization for NEMO (VARON)*—allows local car-to-car communications to be optimized,

---

[7] Before the IETF NEMO WG was finally created, it was thought that the working group would be chartered to work on route optimization issues. However, given the complexity of this topic (the design of a secure but still deployable route optimization solution for mobile IPv6 delayed the standardization process several years), the IETF considered that it was too early to standardize a route optimization protocol, so it focused the NEMO WG charter on the base specification.

by enabling—in a secure way—the use of VANET for local communications among cars. The second approach is a solution for applying NEMO in a VANET that is based on geographic routing.

10.4.1 Requirements Analysis

Applications for vehicular communications can be roughly grouped into safety (e.g., hazard warning and work-zone warning) and nonsafety (e.g., point-of-interest notification and Internet access). These application types put different and partially conflicting requirements on the system design.

Typically, nonsafety applications establish communication sessions with their peer entities. Data is transmitted as packets from source to destination, using unicast or multicast. In contrast, safety applications data is commonly regarded as spatial and temporal *state information* that needs to be disseminated in geographical areas. This implies *in-network processing* that allows to aggregate, modify, and invalidate the information to be forwarded. The fundamentally different information dissemination strategy of safety applications results in unique protocol mechanisms for geographically based data forwarding, congestion control, and reliable data transfer with strong cross-layer dependencies [43]. Clearly, these mechanisms are not part of the standard TCP/IP protocol suite.

To reach a considerable number of equipped vehicles after market introduction, safety and nonsafety applications must be integrated into a single system. In particular, a number of safety applications need a minimum share of equipped vehicles for vehicle-to-vehicle communication. The support for nonsafety applications is commonly regarded as a catalyst for successful market introduction of a safety communication system. Use cases are being currently specified [44] and comprehended, for example, notification services (traffic, weather, and news), peer-to-peer applications, and generic file transfers from the Internet. Also, vehicular nonsafety applications have not found wide deployment in the past—closed telematic platforms of vehicle vendors, high costs for the telematic hardware, and service fees are some of the reasons. It is expected that convenience (nonsafety) applications will boom when integrated with a communication-based safety system [45–47].

We classify the requirements of nonsafety applications into *economic, functional, performance,* and *deployability* requirements.

Economic Requirements: Costs represent an important factor for a vehicular communication system. Primarily, hardware and software of vehicular equipment must be inexpensive. Attractive nonsafety applications, such as Internet-based applications, would point out a visible and immediate added value to customers. Two more aspects can promote a vehicular communication system: (i) An investment in fixed communication units at designated locations along the road—by public authorities or private road operators—helps to overcome the market introduction barrier. (ii) To provide a large customer basis, vehicular communication must provide business opportunities for Internet service providers to generate revenue.

Functional requirements: A fundamental functionality of vehicular communication is the support of vehicle-to-infrastructure (V2I) and vehicle-to-vehicle (V2V) communication. Clearly, V2I communication can only work when connectivity to an RSU (road-side unit[8]) exists, possibly via multiple wireless hops. Conversely, direct V2V communication must work without an RSU being available since intermittent RSU access is a basic assumption for the vehicular communication system. Moreover, if V2V and V2I are feasible simultaneously, policies should determine which communication mode to use. The following nonexhausting list points out additional functional requirements for nonsafety applications:

- Vehicles carry unique identifiers for reachability via V2I and V2V communication.
- As a minimum, inexpensive solution, a vehicle equipped with only IEEE 802.11 technology can use V2I and V2V communication.
- Applications can utilize capabilities for geographic addressing specific to safety applications. This implies a mapping between IP addresses and geographical positions and areas.
- Data security for safety applications (authentication, integrity, and nonrepudiation) is a mandatory function since attacks by malicious nodes, as well as misconfiguration and malfunction can have disastrous effects [48]. Nonsafety applications must not introduce new security leaks for safety applications or render the security measures useless.
- The privacy of drivers and passengers[9] is a strong concern. To protect privacy, the use of preassigned, quasirandom, and changing identifiers—referred to as pseudonyms [49]—is considered for so-called revocable privacy in safety applications. Nonsafety applications should not reveal additional personal information when being used, nor allow for linking changed pseudonyms by sending constant identifiers as clear text.

Performance requirements: The dominant factor that limits the performance of vehicular communication is the available bandwidth. Considering the potentially high relative velocity of vehicles, control traffic for network

---

[8] RSUs are fixed stations installed along the roads.
[9] For example, communications that utilize geographical data for routing publicly disclose position, speed, and driving direction.

organization needs to be minimized. For networking, two aspects are important: First, for ad hoc routing in a vehicular environment a reactive scheme has significantly less signaling overhead than a proactive scheme.[10] Second, IP mobility support for handover among road-side units or hot spots and for global reachability must cope with the nodes' high velocity.

Deployability Requirements: A NEMO solution must be asymmetrically deployable. This means that communication between nodes must be possible, where only one node is equipped with a NEMO solution and the other with a standard IPv6 protocol stack.

10.4.2   Vehicular Ad Hoc Route Optimization Solution for NEMO

In this section, we describe the operation of a solution [42] that provides route optimization for NEMO in vehicular environments, where a vehicular ad hoc network (VANET) may be created and securely used to optimize local communications among vehicles.

It is assumed that the mobile router (MR) deployed in each vehicle will have at least three network interfaces: one *ingress* interface to communicate with the nodes inside the vehicle that belong to the NEMO (e.g., WLAN and Bluetooth), one or more *egress* interfaces to connect to the Internet (e.g., UMTS, WiMAX and even WLAN in some cases), and an additional ad hoc interface (e.g., WLAN) to communicate with neighboring cars and set up multihop networks. Compared to a normal mobile router (without any ad hoc optimization), only one (ad hoc) additional interface is required. It is important to notice that mobile routers deployed in vehicles will not be much concerned about energy constraints, as opposed to personal mobile devices or other ad hoc scenarios (such as sensor networks).

It is also assumed that vehicles' devices will always be able to communicate with other vehicles' devices through the Internet, by using the NEMO basic support protocol. On the other hand, there may exist the possibility of enabling these devices to directly communicate if a multihop vehicular ad hoc network could be set up by the involved vehicles and other neighboring cars. VARON aims making it possible to benefit from this optimization opportunity in a secure way.

In VARON, the MR is the node in charge of performing the optimization of the communications. The steps for carrying out this procedure are the following:

---

[10] Proactive schemes attempt to maintain at all times up-to-date routing information from each node to every other node in the network. Reactive routing protocols initiate a route discovery process on demand.

1. *Discovery of reachable MNPs*: The MR needs to find out which other MRs are available within the VANET, that is, which mobile network prefixes are reachable through its ad hoc interface.
2. *Creation of a secure ad hoc route*: The route is created between the MRs of the mobile networks that want to optimize the route they are using to exchange traffic. The ad hoc routing protocol used to create this route should provide certain security guarantees. The mechanism used by VARON to set up and maintain a secure ad hoc route is based on ARAN (authenticated routing for ad hoc networks) [41], modified and extended to fulfill the requirements of the network mobility–based vehicular scenario.

Next, we describe in detail each of these two steps.

10.4.2.1 Discovery of Reachable MNPs

Every MR announces its mobile network prefix (MNP) by periodically broadcasting—through the ad hoc interface—a message, called *home address advertisement* (HoAA), that contains its home address and an associated lifetime, to allow this information to expire. These messages are announced through the ad hoc interface, by using a hop-limited flooding, so every MR becomes aware of the MNPs that can be reached through the VANET.

The MR's HoA is chosen to belong to the NEMO's mobile network prefix. The length of the MNP is fixed to 64 bits (/64) due to security reasons that will be explained in Section 10.4.2.2. Hence, the MNP can be inferred directly from the HoA (it is the network part of it). With the MRs' announcements, every MR is aware of all the MRs' HoAs (and associated mobile network prefixes) that are available within the ad hoc network.

10.4.2.2 Creation of a Secure Ad Hoc Route

In case a mobile router detects that there is an ongoing communication between a node attached to it and a node attached to another MR that is available through the VANET and this communication is decided to be optimized (how this decision is taken is out of the scope of this work), the MR needs to build a multihop route to send packets directly through the ad hoc network.

An example (Figure 10.3) is used to illustrate in more detail the proposed mechanism. *A* device (e.g., a back-seat embedded video game system) in car *A* is communicating with another device in car *B*.[11] This communication is initially being forwarded through the Internet, following the suboptimal

---

[11] Another example could be car A from an emergency service convoy communicating with another emergency car B.

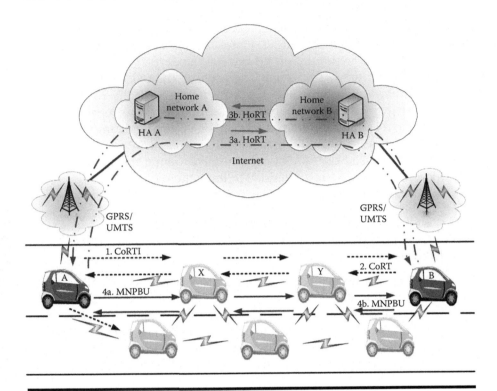

Figure 10.3 **Nested bidirectional tunneling.**

path determined by the NEMO basic support protocol, thus traversing home networks *A* and *B* before being delivered to the destination. We call this route *home route*. By listening to the announcements (i.e., HoA_A messages) received in the ad hoc interface, MR *A* becomes aware that the destination of such communication may also be reachable through a VANET formed by neighboring VARON-enabled vehicles. Then, MR *A* may decide to start using the vehicular ad hoc network to route this traffic, instead of sending it through the Internet.

The first step in this optimization process is that MR *A* must learn and set up a bidirectional route through the vehicular ad hoc network to MR *B* (the MR claiming to manage MNP *B*). We call this route *care-of route*. For doing this, MR *A* (the *originator MR*) sends—through its ad hoc interface—a *care-of route test Init* (CoRTI) message (Table 10.1 summarizes the notation) to its one-hop neighbors:

$$A \rightarrow \text{one-hop neighbors} : [\text{CoRTI}, \text{HoA}_B, N_A, \text{HoA}_A, K_{A+}]_{K_{A-}}$$

This message includes, besides the identifier of the message (CoRTI), the destination MR's HoA (HoA_B), a nonce N_A (to uniquely identify a CoRTI

Table 10.1   Table of Variables and Notation

K_{A+}	Public key (and CGA-related information) of MR A
K_{A-}	Private key of MR A
$[d]_{K_{A-}}$	Data d digitally signed by MR A
N_A	Nonce issued by MR A
HoA_A	Home address of MR A
CoRTI	Care-of route test Init message type
CoRT	Care-of route test message type
CoRE	Care-of route error message type

message coming from a source; every time an MR initiates a route discovery, it increases the nonce), the IP address of MR A (HoA_A), and its public key (K_{A+}), all signed with the MR A's private key (K_{A-}). When an MR receives through its ad hoc interface a CoRTI message, it sets up a reverse route back to HoA_A (MR A's HoA), by recording the MR from which it received the message (so it knows how to send a reply in case it receives a message that has to be sent back to HoA_A). To authenticate the message, a mechanism that securely binds the IP address of MR A (HoA_A) with K_{A+} is needed. One possibility is to use certificates issued by a third trusted party, as proposed in ARAN [41], but VARON adopts a more generic solution that does not impose that requirement. Instead, this secure binding is obtained by using special type of addresses—cryptographically generated addresses (CGAs) [43].

Cryptographically generated addresses are basically IPv6 addresses for which the interface identifier is generated by computing a cryptographic one-way hash function from a public key and the IPv6 prefix.[12] The binding between the public key and the address can be verified by recomputing the hash function and comparing the result with the interface identifier (Figure 10.4). In this way, if the HoA used by MRs is a CGA, a secure binding between the MR's HoA and the MR's public key is provided, without requiring any public key infrastructure (PKI) to be available. Notice that by itself, CGAs do not provide any guarantee of prefix ownership, since any node can create a CGA from any particular mobile network prefix by using its own public–private key pair. But a node cannot spoof the CGA that another node is legitimately using, because it does not have the private key associated with the public key of that IP address.

A receiving MR (e.g., MR X in Figure 10.3) uses MR A's public key (included in the message) to validate the signature, then appends its own public key (K_{X+}) to the message, and signs it using its private key (K_{X-}).

---

[12] There are additional parameters that are also used to build a CGA to enhance privacy, recover from address collision, and make brute-force attacks unfeasible. We intentionally skip these details. The interested reader may refer to Cryptographically Generated Addresses RFC [43] for the complete procedure of CGA generation.

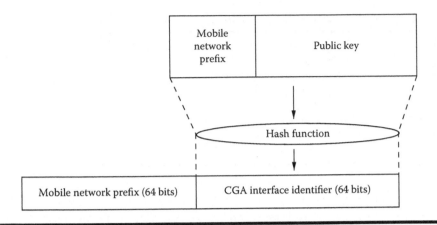

Figure 10.4 Simplified overview of CGA creation and structure.

The signature prevents spoofing or message modification attacks that may alter the route or form loops. Then it forwards the CoRTI message:

$$X \rightarrow \text{ one-hop neighbors} : [[\text{CoRTI}, \text{HoA}_B, N_A, \text{HoA}_A, K_{A+}]_{K_{A-}}, K_{X+}]_{K_{X-}}$$

Upon receiving this CoRTI message from neighbor MR X, MR Y verifies the signatures from the originator MR A and the neighbor MR X stores the received nonce to avoid reply attacks and adds a route to HoA_A through HoA_X (MR X). Then, the signature and public key of the neighbor MR X are removed, and MR Y appends its own public key, signs the message, and forwards it:

$$Y \rightarrow \text{ one-hop neighbors} : \left[\left[\text{CoRTI}, \text{HoA}_B, N_A, \text{HoA}_A, K_{A+}\right]_{K_{A-}}, K_{Y+}\right]_{K_{Y-}}$$

This last step is repeated by any intermediate node along the path until the CoRTI message reaches the destination (the *target* MR, MR B) or the allowed hop limit expires. Notice that MR B, after receiving the CoRTI message, has the guarantee that only the node that has the private key associated with HoA_A (K_{A-}) could have sent the CoRTI message.

Once MR B receives the CoRTI message, it generates a reply message (including the received nonce N_A), called *care-of route test* (CoRT), and unicasts it back following the previously learnt reverse path to the originator (MR A):

$$B \rightarrow Y : [\text{CoRT}, \text{HoA}_A, N_A, \text{HoA}_B, K_{B+}]_{K_{B-}}$$

Each node in the reverse path performs a similar procedure to when forwarding the CoRTI. The first MR in the reverse path that forwards the message (i.e., MR Y) verifies the signature and, if correct, adds its public key K_{Y+}, signs the message, and sends it to the next MR in the path:

$$Y \to X: \left[\left[\text{CoRT, HoA}_A, N_A, \text{HoA}_B, K_{B+} \right]_{K_{B-}}, K_{Y+} \right]_{K_{Y-}}$$

The MR X also sets up a reverse route back to MR B's HoA by recording the MR from which it received the message.

The remaining MRs in the reverse multihop route, when receiving the CoRT message, verify the signature of the previous MR, remove it and the associated public key, add their public key, sign the message, forward it to the next MR, and set up the reverse route. In the example, when MR X receives the message from MR Y, it sends the following to MR A:

$$X \to A: \left[\left[\text{CoRT, HoA}_A, N_A, \text{HoA}_B, K_{B+} \right]_{K_{B-}}, K_{X+} \right]_{K_{X-}}$$

When the originator MR (MR A in the example) receives the CoRT message, it verifies the signature and nonce returned by the destination MR (MR B). Once this procedure is completed, MR B has successfully established a route with MR A within the multihop vehicular ad hoc network. This route is basically a temporal path (care-of route) to reach MR B's HoA, additional to the default route that MR A may always use to send packets toward MR B (through the Internet, using the home route), and vice-versa.

The care-of route cannot be used to forward packets between NEMO A and NEMO B yet, since it has not been proved neither that MR A manages MNP A, nor that MR B manages MNP B. So far, only the validity of a route to a node (B and A) with an address (HoA$_B$ and HoA$_A$) for which the node has the respective private key has been proved to MR A and MR B. It has not been verified that MR A and MR B are actually the routers authorized to manage MNP A and MNP B, respectively. Without further verification, nothing could prevent MR from stealing a mobile network's traffic. For example, a malicious node could be able to claim the ownership of a given IP address (an address belonging to MNP A) and steal packets addressed to that prefix (MNP A). This issue is similar to that of route optimization in mobile IPv6, where a mechanism is required to enable the mobile node to prove that it *owns* both the care-of address and the home address.

The return routability procedure defined for mobile IPv6 is based on two messages sent by the CN, one sent to the mobile node's home address and another to mobile node's care-of address. Based on the content of the received messages, the mobile node sends a message to the correspondent node [6], [37]. By properly authenticating the message, this procedure is enough to prove that the mobile node has received both messages and therefore it has been assigned (that is, *owns*) both the home address and the care-of address at that time.

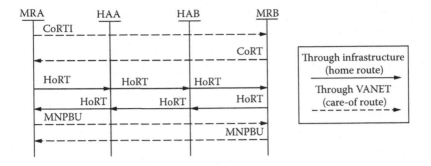

Figure 10.5 Care-of route authentication signalling.

In VARON, we borrow from the return routability (RR) procedure some of the underlying security concepts. With the RR, the correspondent node is provided with a mechanism to verify that a mobile node is able to send and receive packets from two different addresses. In VARON, what is needed is to provide a pair of end-point MRs (which are communicating with each other through the home route) with a mechanism to verify that the multihop route within the VANET connects each of them with the same node that is reachable through the infrastructure when addressing the respective HoA. In this way, the two end-point MRs may choose to use that care-of route instead of the home route.

The essence of the care-of route authentication procedure in VARON is that the two end-point MRs involved in a particular route optimization procedure request each other to verify that the VANET care-of route may be used to send traffic between the two NEMOs. This is done (Figure 10.5) as follows:

- Each mobile router generates a key K_{mr}, which can be used with any other MR. In addition, the MR generates nonces at regular intervals. These nonces[13] and K_{mr} will be used to generate a security association between the two end-point MRs.
- Each MR creates two tokens and sends each of them through one of the possible routes (care-of and home routes). Tokens are generated from K_{mr} and a particular nonce.
- The first part of the care-of route authentication procedure is done at the same time—and using the same messages—as the care-of route setup. The first token, called *care-of keygen token*, is sent piggy-backed in the CoRTI message, plus a *care-of cookie*, and the index of the nonce used to generate the token. The correspondent MR replies

---

[13] Note that these nonces are different from the ones used during the ad hoc route discovery and setup procedure.

in the CoRT message, including its own care-of keygen token, its nonce index and copying the cookie received in the CoRTI message.

■ The second token, called *home keygen token*, is sent, plus a *home cookie* and a nonce index, in a separate message, called *home route test* (HoRT), through the MR-HA tunnel (protected by IPsec ESP in tunnel mode) configured by the NEMO basic support, using the routing infrastructure. To verify that the correspondent MR is actually managing the IPv6 network prefix it claims to, that is, the mobile network prefix assigned to the NEMO, the HoRT message is sent to a random address within the MNP. The MR that manages the prefix has to intercept[14] that message therefore showing that it actually manages the MNP.[15] The mobile network prefix length used by VARON MRs is fixed to 64 bits (/64) to avoid a malicious node to "steal" a prefix. Otherwise, for instance, if an MR was assigned a /64 prefix, then with probability 1/2 it could try to spoof a /63 prefix (and steal its *"neighbor's"* packets). By fixing the MNP length, this attack is no longer feasible.

As in the case of the care-of route test, the correspondent MR replies to this message with another HoRT message, including its own home keygen token and nonce index and copying the received cookie.

■ Each MR uses the received home and care-of keygen tokens to create a key, K_{bm} that can be used to authenticate a *mobile network prefix binding update* (MNPBU) message[16]—sent along the care-of route— that enables other MRs to check that the mobile network (MNP) reachable through the VANET (care-of route) is the one reachable through the infrastructure. This verification can be done because each MR has the information required to produce the key when the MNPBU is received and therefore authenticate the message.

At this point VARON signaling has finished. MR A has found out that MR B—which owns HoA$_B$ and its associated private key—that is reachable through the VANET, is also capable of receiving and sending packets sent to any address from the mobile network prefix (MNP) B through the

---

[14] It is not required for the MR to continuously examine every received packet in order to intercept HoRT messages. The MR may start inspecting packets after sending (or receiving) a CoRT message.

[15] This test does not guarantee that a node manages a certain prefix, but that this node is at least in the path toward that prefix. This provides the solution with a similar security level that today's IPv4 Internet has.

[16] The generation of this key (K_{bm}) and the keygen tokens and the authentication of the message follow the same mechanism that the return routability procedure [13] [47] and the proposal to extend it to support network prefixes [54].

infrastructure. This only happens if the HA responsible for routing packets addressed to this MNP (that is, HA *B*) is forwarding to MR *B* those packets, which are addressed to MNP *B*. HA *B* would only be doing that if proper authentication has taken place, and MR *B* is authorized to manage MNP *B*. The same guarantee also holds for MR *B* regarding MNP *A* and MR *A*.

The care-of route authentication mechanism performed in VARON, as the return routability procedure defined in mobile IPv6, implicitly assumes that the routing infrastructure is secure and trusted. As long as this is true, the mechanism defined is appropriate to secure the mobile network prefix binding update, since it does not introduce any new vulnerability that was not possible in today's IPv4 Internet.

After this process is completed, the end-point MRs (MR *A* and MR *B*) may exchange traffic using the set-up care-of route within the VANET.

10.4.2.3 Optimized Routing Using the VANET

Once the care-of route authentication procedure has finished, all MRs involved in the creation of the ad hoc route can forward packets to the HoAs of the end-point MRs (see an example in Figure 10.6). However, only the end-point MRs have verified the association of the corresponding MRs' HoA and the respective MNP. Intermediate MRs (i.e., MR *X* and MR *Y* in the example) have only learnt host routes toward the home addresses of the two end-point MRs (i.e., HoA_A and HoA_B). In order to route data traffic between cars' nodes with addresses belonging to MNP *A* and MNP *B*, each end-point has to tunnel the packets toward the HoA of the other MRs through the VANET route. In this way, intermediate MRs in the ad hoc route just forward the packets based on the host routes (with the end-point MR's HoAs as destination) added to their routing tables during the ad hoc care-of route creation process (see Figure 10.6).

The care-of route discovery and validation signaling is repeated periodically, both to refresh the ad hoc routes and to avoid time-shifting attacks. If an ad hoc route becomes invalid (e.g., because it expires) or it is broken, and traffic is received through this route, a *care-of route error* (CoRE) message is sent (and forwarded) by each MR in the path to the source MR. For example, if intermediate MR *Y* in Figure 10.6 receives data traffic from MR *A* addressed to MR *B* and the link between MR *Y* and the next hop toward MR *B* (in this case, MR *B* itself) is broken, then MR *Y* sends a CoRE message to the next MR along the path toward the source MR (MR *A*), which is MR *X*, indicating that there is a problem with this care-of route:

$$Y \rightarrow X : [\text{CoRE}, \text{HoA}_A, \text{HoA}_B, N_Y, K_{Y+}]_{K_{Y-}}$$

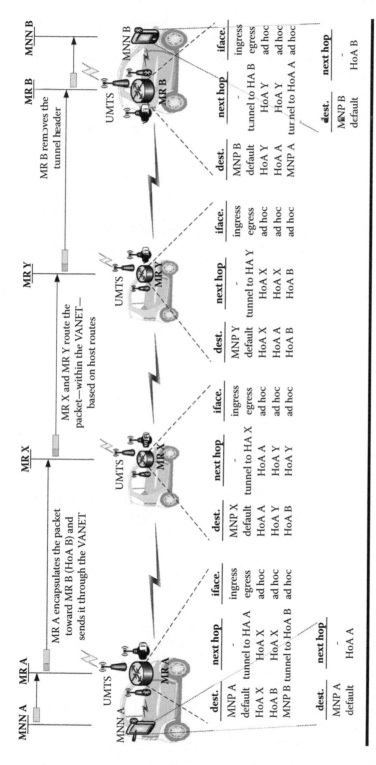

Figure 10.6 Overview of packet routing within the VANET.

This message is received by MR X, which after verifying the authenticity of the received CoRE, signs the message, adds its public key K_{X+} and the signature to the message (as performed by intermediate MRs when processing and forwarding CoRTI and CoRT messages) and sends it to the next hop towards MR A.

$$Y \to X: \left[\left[\text{CoRE}, \text{HoA}_A, \text{HoA}_B, N_Y, K_{Y+} \right]_{K_{Y-}}, K_{X+} \right]_{K_{X-}}$$

Upon reception of this error message, the source MR (MR A in the example) switches to use the home route for sending packets, and it may start a new route discovery procedure to set up a new optimized care-of route within the VANET. To avoid DoS attacks, a CoRE message indicating that a route has become invalid is only processed by an MR if the neighbor that is forwarding the message is the next hop of this route. Otherwise, malicious nodes would be able to set as invalid any care-of route.

There exist several possible mechanisms that can be used to detect that a care-of route is no longer working. As an example, mobile routers may check if the data packets forwarded within the VANET have been correctly received by the next hop making use of link layer acknowledgment frames (if the MAC layer supports that). If several data frames have not been acknowledged, this may be used as an indication that the next hop is no longer reachable and therefore the care-of route is broken.

An exhaustive performance analysis and a security validation of VARON can be found in [42].

10.4.3 A Solution for Application of NEMO in VANET Using Geographic Routing

In this section, we provide an overview of another proposed approach for the application of NEMO to VANET [50]. This approach suggests a way to run the NEMO basic support protocol in VANET based on geographical routing, where the packet routing and forwarding in the ad hoc domain is taken care of by a position-based routing protocol.

10.4.3.1 Geographic Routing in VANET

Recently, protocols for ad hoc routing in vehicular networks making use of geographic positions to forward packets [55] have gained a remarkable interest in research projects [56–58] and automotive industry consortia [54,59], and are currently moving from the status of pure research to being deployed in the first field tests [60,61].

The basic principles of geographic-based routing and forwarding (also called *geonetworking*) were originally proposed as alternative to pure topology-based Internetworking [51] and in mobile ad hoc networks [52].

Geonetworking assumes that vehicles acquire information about their position (i.e., geodetic coordinates) via GPS or any other positioning system. Every vehicle periodically advertises this information to its neighbor vehicles and hence, a vehicle is informed about all other vehicles located within its direct communication range. If a vehicle intends to send data to a known target geographic location, it chooses another vehicle as message relay, which is located in the direction toward the target position. The same procedure is executed by every vehicle on the multihop path until the destination is reached. This approach does not require establishment and maintenance of routes. Instead, packets are forwarded "on the fly" based on most recent geographic positions.

Principles of geographic routing and forwarding have recently been adopted by the Car-2-Car Communication Consortium (C2C-CC) [54] in the definition of the C2C-CC network (NET) layer, a non-IP network layer protocol that is adopted here as part of the proposed solution. C2C-CC is an industry consortium initiated by major European vehicle manufacturers in 1992. To date, C2C-CC counts about 50 partners from industry and research. This consortium is defining protocols and mechanisms that will guarantee interoperability between different European car manufacturers. The recommendations produced by C2C-CC are adopted by the newly created ETSI technical committee for *intelligent transport systems* [62].

10.4.3.2 Application of NEMO in VANETs with Geographic Routing

The first required functionality for usage of NEMO in ad hoc networks is that a NEMO mobile router gains infrastructure connectivity via multihop access. The solution proposed here achieves this functionality by (i) extending IPv6 address autoconfiguration so that the access router appears to the mobile router as if it was directly attached, and (ii) by providing sub-IPv6 geographical ad hoc routing to deliver IPv6 packets over multihop. Before illustrating the details of the two mechanisms, we describe the assumed protocol stack.

In Figure 10.7, one road side unit (RSU) and two on-board units (OBUs)[17] with different configurations are shown. The C2C-CC NET layer is common to every node, whereas a standard IPv6 network layer is put on top of the C2C-CC NET in Vehicle *A* and RSU. The C2C-CC NET layer provides geographic routing as currently considered by C2C-CC and is able to transport and forward IPv6 datagrams, such that OBUs can potentially act as relays even without the IPv6 layer (Vehicle *B*).

According to C2C-CC preliminary studies [54], the C2C-CC NET protocol should provide routing of unicast and multicast packets based on the geographical position of source, destination, and intermediate forwarders. In

---

[17] An OBU is a vehicle node implementing the C2C-CC protocol stack.

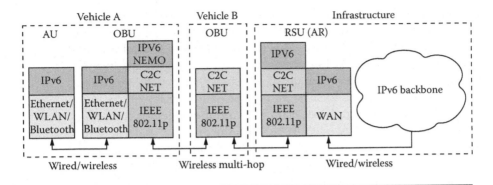

Figure 10.7 Protocol stack of involved entities in the proposed solution.

our proposed solution for unicast routing, we adopt the basic *greedy* strategy [52]. In this strategy, after a *location service* signaling for the resolution of the destination node's position has taken place, packets are routed hop-by-hop selecting the closest direct neighbor to the destination location. The *greedy* strategy is enhanced with multicast support based on position, also called *geocast*, where the packet's destination is not a single node but a target area. Geocast is used in the distribution of IPv6 multicast packets.

As depicted in Figure 10.7, IPv6 is put on top of geographic routing. To avoid changes in standard IPv6 behavior, the sub-IP layer presents Ethernet-like characteristics so that standard mechanisms for datagram transport can be used [53]. Moreover, by applying geocast to IP multicast packets, the sub-IP layer provides a geographic partitioning of the VANET, which implies that a subset of the VANET nodes is presented as a single broadcast domain.

In this approach, at the reasonable cost of additional packet overhead due to encapsulation into a sub-IP protocol header, IPv6 packets can be routed from a point-of-attachment to an arbitrary node that is reachable via multihop. Furthermore, mapping partitioning of the VANET into different clusters is also provided by the sub-IP layer, which in fact enables the addressing of nodes within a limited geographic area and therefore can filter packets before delivering them to IPv6. This results in a broadcast domain that, as seen by IPv6, is flexible and can be configured on a per packet basis according to geographic parameters, instead of pure topological ones like the traditional time-to-live.

For address autoconfiguration, by relying on the multihop distribution and network partitioning offered by the sub-IP geographic routing, in the proposed solution [63] a point-of-attachment sends out standard IPv6 router advertisement (RA) messages that reach all and only the nodes currently located within a well-defined area. In particular, the attachment point specifies as target of the sub-IP protocol header a preassigned geographical area, which is served by this gateway. Upon reception of this packet, a node applies the geographic filtering before delivering the RA to the IPv6 layer and forwards the message according to the *geocast* forwarding procedure. As a result, if a multihop path exists, all the nodes within the area receive the RA, and the IPv6 instance running above geonetworking processes the message as if the node was directly connected to the access router that issued the message.

10.5 Conclusions

In this chapter, we have reviewed the application of network mobility solutions in vehicular networks. Vehicular communications involve a group of devices moving together, so having a mobile router with network mobility functionality is an efficient way of managing the mobility of the set of devices.

Network mobility solutions can provide both car-to-Internet and car-to-car communications. The latter case opens an opportunity for optimizing communications using ad hoc network solutions. Thus, the combination of NEMO and ad hoc solutions brings a strong set of possibilities to manage communications in vehicle scenarios. In this chapter, we have described two solutions that combine NEMO support mechanisms and ad hoc networks to enable efficient communications in vehicular scenarios.

References

[1] Antonio Cuevas, Pablo Serrano, Jose Ignacio Moreno, Carlos J. Bernardos, Juergen Jaehnert, Rui L. Aguiar, and Victor Marques, "Usability and Evaluation of a Deployed 4G Network Prototype," *Journal of Communications and Networks* 7(2) (June 2005):222–230.

[2] Rui Aguiar, Albert Banchs, Carlos J. Bernardos, Maria Calderon, Marco Liebsch, Telemaco Melia, Piotr Pacyna, Susana Sargento, and Ignacio Soto, "Scalable QoS-aware Mobility for Future Mobile Operators," *IEEE Communications Magazine* 44(6) (June 2006).

[3] Suk Yu Hui and Kai Hau Yeung, "Challenges in the migration to 4G mobile systems," *IEEE Communications Magazine* 41(12) (December 2003):54–59.

[4] Thomas R. Henderson, "Host mobility for IP networks: A comparison," *IEEE Network* 17 (November/December 2003):18–26.

[5] Charles E. Perkins, IP Mobility Support for IPv4. Internet Engineering Task Force, RFC 3344 (Proposed Standard), August 2002. http://www.ietf.org.

[6] David B. Johnson, Charles E. Perkins, and Jari Arkko, Mobility Support in IPv6. Internet Engineering Task Force, RFC 3775 (Proposed Standard), June 2004.

[7] Hong-Yon Lach, Christophe Janneteau, and Alexandru Petrescu, "Network mobility in beyond-3G systems," *IEEE Communications Magazine* 41(7) (July 2003):52–57.

[8] Andrew L. Dul, Global IP Network Mobility Using Border Gateway Protocol (BGP). Presented at 62nd IETF, Minneapolis, MN, March 2005.

[9] John Bender and Don Bowman, Global Network Mobility. Presented at RIPE 48, May 2004.

[10] Vijay Devarapalli, Ryuji Wakikawa, Alexandru Petrescu, and Pascal Thubert, Network Mobility (NEMO) Basic Support Protocol. Internet Engineering Task Force, RFC 3963 (Proposed Standard), January 2005. http://www.ietf.org.

[11] Thierry Ernst and Hong-Yon Lach, Network Mobility Support Terminology. Internet Engineering Task Force, RFC 4885 (Informational), July 2007. http://www.ietf.org.

[12] Chan-Wah Ng, Thierry Ernst, Eun Kyoung Paik, and Marcelo Bagnulo, Analysis of Multihoming in Network Mobility Support. Internet Engineering Task Force, RFC 4980 (Informational), October 2007. http://www.ietf.org.

[13] Eun Kyoung Paik, Hosik Cho, Taekyoung Kwon, and Yanghee Choi, "Mobility-aware mobile router selection and address management for IPv6 network mobility," *Journal of Network and Systems Management* 12(4) (December 2004):485–505.

[14] Eun Kyoung Paik, Hosik Cho, Thierry Ernst, and Yanghee Choi, "Load sharing and session preservation with multiple mobile routers for large scale network mobility," In *Proc. of the 18th International Conference on Advanced Information Networking and Applications (AINA)* Fukuoka, Japan 1(2004):393–398.

[15] Chan-Wah Ng and Thierry Ernst, "Multiple access interfaces for mobile nodes and networks," In *Proc. of the 12th IEEE International Conference on Networks, 2004. (ICON 2004)* Singapore 2 (November 2004):774–779.

[16] Nicolas Montavont, Thierry Ernst, and Thomas Noel, "Multihoming in nested mobile networking," In *Proc. of the 2004 International Symposium on the Applications and the Internet Workshops, 2004 (SAINT 2004 Workshops)* Tokyo, Japan (January 2004):184–189.

[17] Masayuki Kumazawa, Taisuke Matsumoto, Shinkichi Ikeda, Makoto Funabiki, Hirokazu Kobayashi, and Toyoki Kawahara, "Router selection for moving networks," In *Proc. of the 1st IEEE Consumer Communications and Networking Conference, 2004, (CCNC 2004)*, Las Vegas, NV (January 2004):99–104.

[18] Thierry Ernst and Julien Charbon, "Multihoming with NEMO basic support," In *Proc. of the First International Conference on Mobile Computing and Ubiquitous Computing (ICMU)*, Yokosuka, Japan, January 2004.

[19] Lucian Suciu, Jean-Marie Bonnin, Karine Guillouard, and Thierry Ernst., "Multiple network interfaces management for mobile routers," In *Proc. of the 5th International Conference on ITS Telecommunications (ITST)*, Brest, Japan, June 2005.

[20] Koshiro Mitsuya, Manabu Isomura, Keisuke Uehara, and Jun Murai, "Adaptive application for mobile network environment," In *Proc. of the 5th International Conference on ITS Telecommunications (ITST)*, Brest, Japan, (June 2005):211–214.

[21] Hiroshi Esaki, "Multi-homing and multi-path architecture using mobile IP and NEMO framework," In *Proc. of the 2004 International Symposium on Applications and the Internet:* (SAINT 2004), Tokyo, Japan, (January 2004):6.

[22] Marcelo Bagnulo and Erik Nordmark, Level 3 multihoming shim protocol. Internet Engineering Task Force, draft-ietf-shim6-proto-10.txt (work-in-progress), February 2008. http://www.ietf.org.

[23] Marcelo Bagnulo, Application of a multi6 protocol to nemo. Internet Engineering Task Force, draft-bagnulo-nemo-multi6-00.txt (work in progress), November 2004. http://www.ietf.org.

[24] Csaba Simon, Rolland Vida, Peter Kersch, Christophe Janneteau, and Gosta Leijonhufvud, "Seamless IP multicast handovers in overdrive," In *Proc. of the 13th IST Mobile and Wireless Communications Summit*, Lyon, France 2 (June 2004):606–610.

[25] Dirk V. Hugo, Holger Kahle, Carlos J. Bernardos, and Maria Calderon, "Efficient multicast support within moving IP sub-networks," In *Proc. of the IST Mobile and Wireless Communications Summit 2006*, Myconos, Greece, June 2006.

[26] Mansour J. Karam and Fouad A. Tobagi, "Analysis of the delay and jitter of voice traffic over the Internet," In *Proc. of the 20th Annual Joint Conference of the IEEE Computer and Communications Societies (INFOCOM)* Anchorage, AK, 2 (April 2001):824–833.

[27] Rajeev Koodli, Fast Handovers for Mobile IPv6. Internet Engineering Task Force, RFC 5268 (Proposed Standard), June 2008. http://www.ietf.org.

[28] Henrik Petander, Eranga Perera, Kun-Chan Lan, and Aruna Seneviratne, "Measuring and improving the performance of network mobility management in IPv6 networks," *IEEE Journal on Selected Areas in Communications* 24(9) (September 2006):1671–1681.

[29] Youn-Hee Han, JinHyeock Choi, and Seung-Hee Hwang, "Reactive handover optimization in IPv6-based mobile networks," *IEEE Journal on Selected Areas in Communications* 24(9) (September 2006):1758–1772.

[30] Mazen Tlais and Houda Labiod, "Resource reservation for NEMO networks," In *Proc. of the International Conference on Wireless Networks, Communications, and Mobile Computing*, Maui, HI, (June 2005):232–237.

[31] Saber Zrelliand, Thierry Ernst, Julien Bournelle, Guillaume Valadon, and David Binet, "Access control architecture for nested mobile environments in IPv6," In *Proc. of the 4th Conference on Security and Network Architecture (SAR)*, Batz sur Mer, France, June 2005.

[32] Hanane Fathi, SeongHan Shin, Kazukuni Kobara, Shyam S. Chakraborty, Hideki Imai, and Ramjee Prasad, "LR-AKE-based AAA for network mobility (NEMO) over wireless links," *IEEE Journal on Selected Areas in Communications* 24(9) (September 2006):1725–1737.

[33] Chan-Wah Ng, Pascal Thubert, Masafumi Watari, and Fan Zhao, Network Mobility Route Optimization Problem Statement. Internet Engineering Task Force, RFC 4888 (Informational), July 2007. http://ietf.org.

[34] Salman A. Baset and Henning Schulzrinne, "An analysis of the skype peer-to-peer internel telephony protocol," In *Proc. the 25th IEEE International Conference on Computer Communications (INFOCOM 2006)*, Barcelona, Spain, April, 2006.

[35] Soren Vang Andersen, Alan Duric, Henrik Astrom, Roar Hagen, W. Bastiaan Kleijn, and Jan Linden, Internet Low Bit Rate Codec (iLBC).

Internet Engineering Task Force, RFC 3951 (Experimental), December 2004. http://www.ietf.org.

[36] Henning Schulzrinne, Stephen L. Casner, Ron Frederick, and Van Jacobson, RTP: A Transport Protocol for Real-Time Applications. Internet Engineering Task Force, RFC 3550 (Standard), July 2003. http://www.ietf.org.

[37] Pekka Nikander, Jaru Arkko, Tuomas Aura, Gabriel Montenegro, and Erik Nordmark, Mobile IP Version 6 Route Optimization Security Design Background. Internet Engineering Task Force, RFC 4225 (Informational), December 2005. http://www.ietf.org.

[38] Chan-Wah Ng, Fan Zhao, Masafumi Watari, and Pascal Thubert, Network Mobility Route Optimization Solution Space Analysis. Internet Engineering Task Force, RFC 4889 (Informational), July 2007. http://www.ietf.org.

[39] Eranga Perera, Vijay Sivaraman, and Aruna Seneviratne, "Survey on network mobility support," *ACM SIGMOBILE Mobile Computing and Communications Review* 8(2) (April 2004):7–19.

[40] Carlos J. Bernardos, Route Optimisation for Mobile Networks in IPv6 Heterogeneous Environments. PhD Thesis, defended in Madrid on November 28, 2006.

[41] Kimaya Sanzgiri, Daniel LaFlamme, Bridget Dahill, Brian Neil Levine, Clay Shields, and Elizabeth M. Belding-Royer, "Authenticated routing for ad hoc networks," *IEEE Journal on Selected Areas in Communications* 23(3) (March 2005):598–610.

[42] Carlos J. Bernardos, Maria Calderon, Ignacio Soto, Fernando Boavida, and Arturo Azcorra, "VARON: Vehicular Ad-hoc Route Optimisation for NEMO." *Computer Communications* 30(8) (June 2007): 1765–1784.

[43] Tuomas Aura, Cryptographically Generated Addresses (CGA). Internet Engineering Task Force, RFC 3972 (Proposed Standard), March 2005. http://www.ietf.org.

[44] Chan-Wah Ng and Jun Hirano, Extending Return Routability Procedure for Network Prefix (RRNP). Internet Engineering Task Force, draft-ng-nemo-rrnp-00.txt (work-inprogress), October 2004. http://www.ietf.org.

[45] ABI Research, Dedicated Short Range Communications (DSRC) The Emerging Wi-Fi and RFID Market for Advanced Automotive Identification, Commerce and Communications. Research Report, May 2005.

[46] Karl Matheus, Rudolf Morich and Andreas Luebke. "Economic background of car-to-car communications," In *Proceedings of IMA,* Braunschweig, Germany, October 2004.

[47] Karl Matheus, Rudolf Morich, Igor Paulus, Cornelius Menig, Andreas Luebke, Bernd Rech, and Wolf Specks, "Car-to-car communication—market introduction and success factors," In *Proc. of the 5th European ITS Congress and Exhibition,* Hannover, Germany, June 2005.

[48] Ahmid Aijaz, Bernd Bochow, Fred Doetzer, Andreas Festag, Matthias Gerlach, Tim Leinmueller, and Robert Kroh," "Attacks on inter vehicle communication systems—an analysis," In *Proc. of WIT,* Hamburg, Germany, March 2006:189–194.

[49] Emanuel Fonseca, Andreas Festag, Roberto Baldessari, and Rui Aguiar, "Support of anonymity in VANETs—putting pseudonymity into practice," In *Proc. of WCNC,* Hong Kong, March 2007.

[50] Roberto Baldessari, Andreas Festag, Wenhui Zhang, and Long Le, "A MANET-centric solution for the application of NEMO in VANET using geographic routing," In *Proc. of TridentCom*, Innsbruck, Austria, March 2008.

[51] Gregory Finn, Routing and Addressing Problems in Large Metropolitan-Scale Internetworks. ISI Research Report, ISI/EE-87-180, March 1987.

[52] Brad Karp and H. T. Kung. "GPSR: Greedy perimeter stateless routing for wireless networks," In *Proc. of MobiCom*, Boston, USA, August 2000.

[53] Matt Crawford, Transmission of IPv6 Packets over Ethernet Networks. Internet Engineering Task Force, RFC 2464 (Draft Standard), December 1998. http://www.ietf.org.

[54] Car2Car Communication Consortium. C2C-CC Manifesto. Version 1.1, August 2007.

[55] Martin Mauve, Joerg Widmer, and Hannes Hartenstein, "A survey on position-based routing in mobile ad hoc networks," *IEEE Network* 15(6) (November 2001):30–39.

[56] NoW—Network on Wheels, http://www.network-on-wheels.de.

[57] FleetNet—Internet on the Road, http://www.et2.tu-harburg.de/fleetnet.

[58] The GeoNet Project, http://www.geonet-project.eu.

[59] ISO TC204 WG16 (CALM), http://www.isotc204.com.

[60] SIM—TD—Sichere Intelligente Mobilitaet Testfeld Deutschland, http://www.aktivonline.org/.

[61] PRE-DRIVE C2X Project. http://cordis.europa.eu/fetch?CALLER=FP7_PROJ_EN&ACTION=D&DOC=10&CAT=PROJ&QUERY=011cb00077ee:06a9:689b95a0&RCN=87604.

[62] ETSI technical committee for Intelligent Transport Systems, http://www.etsi.org.

[63] Roberto Baldessari, Carlos J. Bernardos, and Maria Calderon, "Scalable address autoconfiguration for VANET using geographic networking concepts," In *Proc. of IEEE International Symposium on Personal, Indoor and Mobile Radio Communications (PIMRC)*, Cannes, France, September 2008.

Chapter 11

Mobile Ad Hoc NEMO

Ryuji Wakikawa

Contents

Mobile ad hoc NEMO (MANEMO) is a technology introduced to integrate the capabilities of NEMO [2] and MANET. It has been discussed recently at IETF. MANEMO is defined in many ways as "Mobile ad hoc NEMO," "MANET and NEMO," "management of nested NEMO," and "MANET for NEMO." NEMO provides movement transparency to a network, while MANET provides ad hoc routing with neighboring nodes. Several industries such as transportation and the military look for both capabilities in their network systems. MANEMO introduces a new concept known as the wireless fringe stub—a cloud of NEMO-capable mobile routers connected by wireless or wired links and a stub at the edge of the Internet, interconnecting various types of devices, which discover each other and form a network in an ad hoc fashion to provide global connectivity to each other. One example of MANEMO networks is a vehicular network. The concept of MANEMO, possible issues, and proposed solutions are briefly explained in this chapter.

11.1 Introduction

When the network mobility support protocol is widely deployed to vehicles, public transportation, and even personal area networks (PANs), it is expected that the impact on existing network environments will be considerable. Current mobility protocols rely on the availability of well-managed, fixed network infrastructures. From the mobile node point of view, once the node acquires connectivity, it is assured of reachability and communication to the Internet. However, as network mobility support becomes available, a mobile node no longer assumes the presence of such fixed infrastructures. The mobile node may attach to the Internet through mobile routers providing the Internet connectivity. There is no guarantee that the mobile node always gets reachability to the Internet over the mobile router since the mobile router is also moving and may lose connectivity to the infrastructure. Since a mobile network provided by a mobile router can be viewed as a regular IPv6 network, the mobile node cannot tell whether it is connected to a mobile network or a fixed network. This causes considerable change to the connectivity assumption of the mobile node. Moreover, by nesting mobile routers, multiple wireless hops appear on the path between end nodes. Most wireless communication today consists of the last-one-hop wireless path(s) and the fixed infrastructure.

11.2   MANEMO Wireless Fringe Stub

When mobile routers and mobile nodes converge at the edge of the Internet using wireless interfaces, they can form a wireless network cloud. This type of network is called a MANEMO fringe stub (MFS) as shown in Figure 11.1. An MFS is a stub at the edge of the Internet, interconnecting various types of devices, which discover each other and form a network in an ad hoc fashion to provide Internet connectivity to one another. Participants in an MFS are not only mobile routers but also mobile hosts, fixed hosts, and fixed routers. The fixed nodes are located either within one of the mobile networks or within the fixed infrastructure.

The exit router is a router that provides Internet connectivity in MFS. In Figure 11.1, the exit router is either a mobile router connected directly to the Internet (Exit Router1) or a fixed access router supporting MANEMO

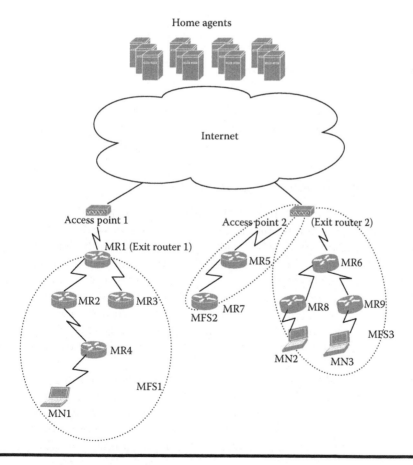

Figure 11.1   Mobile fringe stub.

(Exit Router2). Multiple exit routers may be available in an MFS, as shown in MFS2 in Figure 11.1.

Different types of links are used to form an MFS, including WiFi, Bluetooth, 802.15.4, and meshed wireless technology (802.11s, 802.11 in ad hoc mode, etc.). Exit routers are connected to the infrastructure by wired link, WiFi, WiMAX, and cellular technology (LTE, HSDPA, EvDo, GPRS, etc.).

An MFS can also be disconnected from the infrastructure. In such a disconnected MFS, mobile routers communicate only with nodes inside the same MFS.

Any node requiring Internet connectivity has to select the best exit router toward the Internet. Therefore, it is necessary for mobile nodes to maintain a local topology in the MFS. MANEMO provides the necessary additions to existing protocols (IPv6, NDP [13], NEMO [2]), for the MFS to find the most suitable exit router for the infrastructure. MANEMO enables basic internal connectivity within the MFS whether the infrastructure is reachable or not.

11.3   MANEMO Characteristics and Requirements

When we consider MFS and MANEMO, several new features are introduced to the mobility environment [4] [5]. This section explains the characteristics and requirements of MANEMO by comparing existing solutions such as NEMO [15] and MANET.

11.3.1   Supporting Flexible Path Selection

Figure 11.2 compares the MANET and NEMO communication models inside an MFS. In both scenarios, the mobile node (MN1) attached to MR4 communicates with the mobile node (MN2) attached to MR8. When the NEMO protocol is employed, the packets are routed toward the infrastructure to reach the home agents and return to the MFS. Since there are multiple mobile routers on the path between MN1 and MN2, multiple encapsulation occurs. On the other hand, in the MANET scenario, mobile routers maintain the ad hoc links and manage local routing information. The path between MN1 and MN2 is directly established without relying on the fixed infrastructure.

One of the goals of MANEMO is flexible path selection depending on the MFS environment and communication conditions. A problem found within the MANET communication model is that the majority of the MANET routing protocols select the shortest path in all cases. This causes congestion at some links when many nodes generate traffic. For example, the MR7

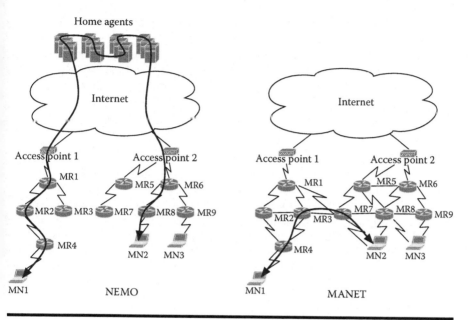

Figure 11.2 NEMO and MANET communication models.

link may become a possible bottleneck when the shortest path is taken in Figure 11.2.

In an MFS, infrastructure is available. Therefore, the mobile router transmits packets to the infrastructure even if the destination node is located nearby. Even though the path length may increase, the path over the Internet is often more reliable than the shortest ad hoc link. The additional overhead associated with transmitting packets to the infrastructure is a trade-off of several aspects such as latency, congestion, reliability, and cost of local routing management over wireless links.

Each mobile router should be able to decide whether packets are routed to the infrastructure or to the destination directly over the ad hoc links. Other issues involving the use of fixed infrastructure by mobile routers are described in Section 11.3.2.

11.3.2 Avoiding Redundant Tunnels and Paths

When multiple mobile routers are connected in a nested formation, redundant tunnels and unoptimized paths are often found. This is a well-known constraint of the NEMO basic support protocol [10,13]. The NEMO basic support protocol does not provide any route optimization mechanism and mandates all traffic from and to a mobile router going through a home agent over a bidirectional tunnel. The route optimization has been investigated and summarized by Ng et al. [11] [12].

For example, in Figure 11.2, the path from a mobile network node (MN1) to another (MN2) becomes MN1->MR4->MR2->MR1->HA1->HA2->HA4->HA8->HA6->MR6->MR8->MN2. Note that if MN1 and MN2 are visiting mobile nodes, two home agents are added to the path above. The path is obviously redundant compared to the MANET approach (figure on the right of Figure 11.2).

In addition, whenever packets are routed over a mobile router, an additional IPv6 header is inserted in the packet for tunneling. Therefore, fragmentation may occur due to the availability of multiple tunnels between end nodes. Since the MFS is formed with resource-scarce wireless technology, larger packets hinder communication performance.

MANEMO is expected to reduce the overhead of the bidirectional tunnels caused by the nested NEMO.

11.3.3 Forming Loop-Free MFS

A network loop occurs when two mobile routers are connected to each other. In Figure 11.2, MR2 can connect to the mobile network of MR4. The loop occurs between MR2 and MR4.

Because the mobile network is seen as a regular IPv6 network, nodes connecting to the mobile network are unable to distinguish whether the attached network is mobile or not. The Internet reachability of a mobile router is not always guaranteed due to the mobile router's movement. If the mobile router does not register the binding to its home agent, the mobile network is equivalent to a disconnected link.

MANEMO provides useful information on access link conditions to nodes attached to the MFS. In addition, MANEMO provides a mechanism to form the MFS loop-free.

11.3.4 Supporting Movement Transparency

When a mobile router changes its point of attachment, it must hide the changes from any nodes located within its mobile network. Since nodes in the mobile network move together, sets of mobile routers can move at once in an MFS.

For instance, in Figure 11.2, MR2 moves its point of attachment from MR1 to MR3. The movement has minimum impact on MR4 and MN1 under the MR2. On the other hand, under most MANET and AUTOCONF schemes, the change of MR4's attachment affects the neighboring nodes (and possibly the entire network). Most routing protocols require route recalculation or route rediscovery (route maintenance) when topology changes take place. This should be avoided as it breaks the nature of the NEMO basic support protocol.

MANEMO inherits from NEMO and supports movement transparency when a mobile router changes its attachment point.

11.3.5 Supporting Diversity of Wireless Access

An MFS is formed with multiple wireless access technology such as WiMAX, WiFi infrastructure mode, LTE, and even WiFi ad hoc mode (802.11p). This is possible as each mobile router has at least two interfaces such as egress and ingress interfaces. Moreover, a mobile router (e.g., in a vehicle) might have more egress interfaces for the high capability of the network connectivity. On the other hand, in most MANET scenarios, a MANET router uses the same wireless access media (e.g., 802.11b ad hoc mode) in the same MANET because of the flooding capability.

MANEMO should support the diversity of wireless access media.

11.3.6 Supporting Home Agent-Independent Communication

Mobile IPv6 [3] and the NEMO basic support protocol [2] rely on an entity called a home agent. Without registering the binding to the home agent, mobile nodes and routers cannot send or receive packets from foreign links. However, Internet connectivity is not always available in mobile scenarios and is often intermittent. In an MFS, there are multiple mobile routers (i.e., wireless links) along the path toward the Internet.

Specifically when two mobile routers in the same MFS communicate, packets are not necessarily routed to the infrastructure. If local routing is available in the MFS, packets can be routed directly to the destination without involving home agents.

MANEMO should support local routing inside the MFS and decrease the dependency of home agents on Mobile IPv6 and NEMO basic support protocols.

11.3.7 Supporting Local Routing

In MANET, each router can route the packet received at the MANET interface [16]. A route can receive a packet from a MANET interface and can send the packet from the same MANET interface according to its routing table. However, under NEMO basic support, a mobile router can route only the tunneled packet to and from its mobile network. Without the bidirectional tunnel, the mobile router never routes the nontunneled packet. The packet sent from the mobile network is always routed to the mobile router's home agent by using IP encapsulation. Incoming packets must be always tunneled from the mobile router's home agent except for packets meant for the mobile node itself.

11.4 MANEMO Scenarios: Vehicular, Disaster, and Public Safety Networks

Since MANEMO supports network connectivity, ad hoc communication (infrastructureless), self-forming, movement transparency, and diversity of wireless access media, real deployment scenarios for mobile computing naturally can expect to receive benefit from MANEMO technology. Examples of such possible scenarios are mesh networks, sensor networks, vehicle networks, personal area networks, disaster networks, and ship networks.

11.4.1 Vehicular Network

Once mobile routers are well deployed in vehicles, personal devices, and the like, it is expected that we will begin to see access networks that are on the move. The best access network for users might depend on more than layer 2 information and location knowledge. For instance, a passenger in a vehicle (e.g., bus or train) should connect to the access network provided by that vehicle while a stationary passenger located in the station should get access from a fixed resource. Some of the required information to make the proper decision should be delivered to users. MANEMO is a mechanism employed to discover and select the best access network for users. The MANEMO scenarios are very close to our daily life and related to human movement patterns.

The vehicle network for vehicle to vehicle (V2V) and vehicle to road (V2R) communications is another possible MANEMO scenario. While a mobile router will be deployed on a vehicle and provide network connectivity to nodes inside the vehicle, the vehicle needs to communicate locally to the vehicle driving in front or to road-side units. These local communications are not always served by the mobile router because of the cost of communications, the ad hoc nature of communications, and the existence of delay sensitive communications (e.g., safety applications). MANEMO may be a good solution candidate for future vehicular networks.

11.4.2 Disaster and Public Safety Network

Disaster and public safety scenarios are another hot topic related to MANEMO. The MetroNet6 project [15] was introduced as an example of a possible MANEMO scenario by Jim Bound (HP Fellow) in IETF. The MetroNet6 is developing dynamic, secure wireless networks formed with both wireless and wireline access media in an ad hoc manner for first responders to disaster cases. Its aim is to connect police, firefighters, and hospitals to a command center (e.g., National Homeland Security Office) over the MetroNet6 and the Internet infrastructure in the event of a disaster. All personnel involved in the disaster recovery are equipped with wireless handheld devices for voice, video, medical, and any data communication over the MetroNet6 infrastructure. The project began in the state of California (Sacramento). The network expected to be deployed under the MetroNet6 project is very close to the characteristics of a MANEMO. In Europe, there is a similar activity called U-2010 project [18] developing a ubiquitous IP infrastructure for effective and flexible communication in disaster and public safety arenas. In Asia, after the Indian Ocean tsunami struck in 2004, several projects were started for disaster recovery networking. The Digital Ubiquitous Mobile Broadband OLSR (DUMBO) project [19] aims at providing dynamic wireless ad hoc networks for disaster scenarios in Thailand.

A common feature of these projects is the quick recovery of communication infrastructure and flexible connectivity management by combining dynamic wireless networks and existing wireline infrastructure (e.g., Internet). For MANEMO, several mobile routers on emergency vehicles, military vehicles, and rescue crews are deployed in the disaster area and form an MFS to restore connectivity. The MFS might be disconnected initially due to the complete breakdown of the infrastructure, but it can be extended later to the Internet over wireless connectivity (satellite) from emergency vehicles. The MFS can be used for local recovery actions and also for remote recovery actions from remote command centers.

11.4.3 Scenario Analysis

When MANEMO scenarios are analyzed, there are several characteristics such as

■ Group mobility: Nodes in an MFS move as a group. Nodes do not move randomly in MANEMO scenarios. Nodes often move within a set of groups. For instance, passenger nodes are moved together with a vehicle. Alternatively, MANET routing protocols address the random movement of nodes. Vehicles driving on a highway are sometimes moving together for a while.
■ Less mobility: The topology of an MFS does not change frequently in MANEMO scenarios. The majority of scenarios deal with vehicle networks. Once nodes board a vehicle, the topology inside the vehicle does not change until the nodes disembark from the vehicle. On the other hand, topology might be frequently changed in vehicle to vehicle networks depending on wireless medias and the moving speed of vehicles. There are less topology changes when vehicles are in a traffic jam.

    In disaster scenarios, a wireless mesh backbone formed with MANEMO is relatively stable, while the nodes of rescue members move under the backbone. Therefore, mobility is definitely involved in MANEMO scenarios, but topology changes may be less frequent than when compared to MANET scenarios.

■ Internet-oriented communication: Communications tend to be established between nodes in an MFS and in an infrastructure. Local communication in the same MFS occurs rarely, as participants in an MFS do not have a strong relationship with each other in MANEMO scenarios. For example, in a train, all passenger communications are established with nodes in the infrastructure (e.g., Internet). Therefore, a solution should be designed to support global communications to and from an MFS.

11.5   MANEMO Architecture

This section outlines the MANEMO architecture including envisioned topology configuration and addressing assignments.

11.5.1   *The NEMO Basic Support Protocol*

Before explaining the MANEMO architecture, we briefly explain the architecture of the NEMO basic support protocol. A mobile network is defined in RFC 3753 [1] as, "An entire network, moving as a unit, which dynamically changes its point of attachment to the Internet and thus its reachability in the topology. The mobile network is composed of one or more IP-subnets and is connected to the global Internet via one or more Mobile Routers (MR). The internal configuration of the mobile network is assumed to be relatively stable with respect to the MR." A mobile network is seen as an IPv6 subnet by any nodes other than mobile routers.

According to RFC 3963, Figure 11.3 is the common interpretation of a mobile router. Each mobile router has an egress interface(s) to reach the home agent through the Internet and also an ingress interface(s) attaching to the mobile network. A mobile network node is a node attached to the mobile network such as fixed or mobile routers and fixed or mobile hosts. A mobile router obtains its care-of address at the egress interface and establishes a bidirectional tunnel to the home agent. It routes all packets intercepted at the ingress interface to the bidirectional tunnel. A packet's source address must belong to the mobile network prefix. Only packets sent to and from a mobile network are routed to the tunnel by the mobile router. Some known remarks from this NEMO basic support are

■ Unless a mobile network node (host or router) is connected to a mobile network, NEMO guarantees session continuity to the node.

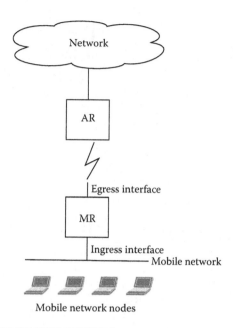

Figure 11.3   The mobile router's configuration.

- Mobile network nodes are unaware of the mobile router's changing attachment point. The mobile network can be seen as just an IPv6 network.
- An access router at a visiting network is not aware of the existence of a mobile network behind the mobile router.

11.5.2   MANEMO Topologies

The NEMO basic support protocol introduces two different interfaces on a mobile router known as the egress interface and the ingress interface as explained in Section 11.5.1. These interfaces are not necessarily physically available. If we interpret egress and ingress interfaces as conceptually defined interfaces, a mobile router can be operated with a single physical interface defining both the ingress and egress functions. In MANEMO, several topologies can be realized through the attachment of a combination of these two interfaces.

When considering MANEMO, the following topology can be logically possible. Figure 11.2 shows all the possible topologies of MANEMO. The MFS is formed either by one of these topologies or a combination of the topologies listed above.

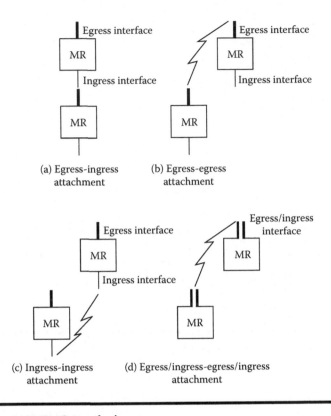

Figure 11.4 MANEMO topologies.

- Egress and ingress (E-I) attachment (Figure 11.4a): The E-I attachment is the common configuration of the NEMO basic support. A mobile router connects to the other mobile routers' mobile network by its egress interface. A mobile router of a personal area network of a driver can connect to a mobile router of a vehicle by E-I attachment.

- Egress and egress (E-E) attachment (Figure 11.4b): The E-E attachment is found when a mobile router uses an ad hoc type of interface such as 802.11b ad hoc mode, 802.11s, or 802.11p as its egress interface. Mobile routers connect to each other by egress interfaces. This configuration is similar to MANET topology. In MANEMO, this scenario is also considered for intervehicle networks (VANET).

- Ingress and ingress (I-I) attachment (Figure 11.4c): Although this configuration is logically possible, it is slightly unrealistic. When ingress interfaces of different mobile routers are connected, two different mobile networks are merged into a single mobile network. A mobile network node will obtain multiple IP addresses from different mobile routers. Whenever a mobile router leaves the MFS, addresses generated from the mobile network prefix of the departing mobile router become unreachable. This configuration may break the fundamental features of the NEMO basic support protocol due to the lack of movement transparency. This configuration is not considered in MANEMO.

- Egress/ingress and Egress/ingress (EI-EI) attachment (Figure 11.4d): EI-EI is a similar configuration to the E-E attachment. A mobile router is equipped only with a single wireless interface and utilizes it conceptually as both the egress and ingress interface. Under NEMO basic support, a mobile router is assumed to have two physically different interfaces. However, in this context, the ingress and egress interfaces are provided over the same physical interface. A mobile router exposes its mobile network prefix to the interface and also obtains a care-of address at the same interface.

11.5.3   Addressing Architecture

The addressing architecture is a key factor in the design of a MANEMO solution(s). It brings several constraints to communication and routing. Each mobile router needs to obtain an address as a care-of address for the egress interface at visiting networks. This care-of address is used to exchange signaling and also to tunnel packets to the home agent. There are two different addressing assignment approaches today in IETF: the NEMO addressing approach and the AUTOCONF addressing approach. The main difference is where the address originates from. The arrows in Figure 11.5 show how each mobile router and node obtains an address in MFS.

In the NEMO addressing approach, the address is obtained from the upper router. If the upper router is a mobile router, the address is retrieved from the mobile network prefix of the mobile router. This is the basic concept of the NEMO basic support protocol. The node in the MFS is not aware of routers other than the upper router to which the node is attached. The movements of other routers are hidden by the upper mobile router. The address assigned to the node is not a topologically correct address, but it is defined at the home link of the upper mobile router. Therefore, all packets sent from the address are routed to the home agent by the upper mobile router. This is the origin of the nested NEMO problem in MFS.

The Ad hoc Network Autoconfiguration (AUTOCONF) working group in IETF has discussed the MANET architecture [19] and an address assignment mechanism for MANET. In the context of the MANET architecture, the egress interface of each mobile router can be treated as a MANET interface. The MANET architecture suggests assigning a unique address or prefix to the MANET interface. Although no solution has been defined for the MANET address assignment mechanism yet, the MANET addressing approach assumes that the address of each router in the MFS is derived from

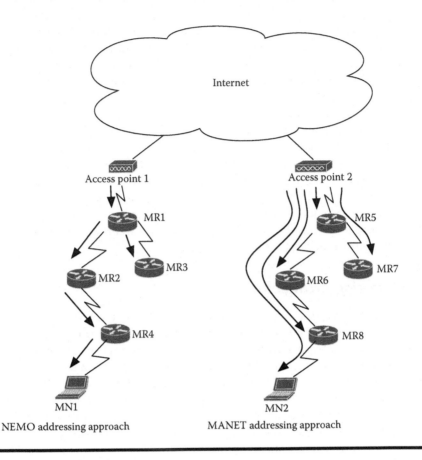

Figure 11.5 MANEMO topologies.

the exit router over multihops. To deliver addressing configuration information from the exit router, it is mandatory to maintain the local path between each router and the exit router. This is actually the natural behavior of MANET and its routing protocols. Unlike the NEMO addressing approach, whenever the MFS topology is altered due to the movement of one or more routers, this movement affects all nodes (or nodes behind the moving routers). Each node may update local routes by a MANET routing protocol. It is also required to update its address whenever the associating exit router is changed. The address is a topologically correct address that the exit router is serving. Packets are forwarded locally inside the MFS and routed to the destination by the exit router as required. There might not be a constraint that requires packets to travel all the way to the home network.

However, there are several unclear and undefined areas to address toward achieving the MANEMO goals. Each mobile router obtains multiple, varying addresses on its egress interface such as one from the access router and one from its upper mobile router. Each mobile router runs IPv6 address autoconfiguration [16] at the attached link (i.e., mobile network of the attached mobile router). As an example, MR8 obtains an address from the exit router (Access Point2) and another address from the upper mobile router, MR6. The mobile router selects the address assigned from the access router as a care-of address and registers it to the home agent. By doing so, tunnel overhead issues raised in the MANEMO problem statement are generally avoided. The packet is directly routed to and from the MFS via only the mobile router's home agent even if multiple mobile routers are located in the path of end nodes. Once the packet arrives at the MFS, it is routed to the target mobile router by MANET local routing.

An additional consideration is that the mobile router has its own mobile network prefixes and additional addresses on its ingress interface. Therefore, the mobile router must carefully check the source address of all packets to decide whether the packets should be routed to the bidirectional tunnel or not. If the packet is sent from the address of the mobile network, it is tunneled to the home agent. Otherwise, the packet is locally routed to the destination. If the destination is not in the same MFS, the packet is first delivered to the exit router and then routed to the destination in the infrastructure behind the exit router.

The loop-free topology formation is essential of MANET routing protocols. The multiple exits issue may be solved if the MANET routing protocol can be extended to carry additional information of exit routers along with the route information. Due to the local routing, if the destination and the source nodes are located in the same MFS, they can communicate without reaching the home agent.

11.6 Solution Candidates

Several solutions are proposed for MANEMO in IETF. The existing solutions can be classified into a NEMO-centric and MANET-centric approach.

11.6.1 MANEMO Solution Requirements

Before discussing and comparing the MANEMO solutions, we discuss the MANEMO solution requirements. Here are the lists of requirements proposed by Wakikawa et al. [4]:

1. The MANEMO protocol must enable the discovery of multihop topologies at layer 3 from mere reachability and elaborate links for IPv6 usage, regardless of the wired or wireless media.
2. The MANEMO protocol must enable packets transmitted from mobile routers visiting the MFS to reach the Internet via an optimized path toward the nearest exit router and back.

3. MANEMO must enable IP connectivity within the nested NEMO whether the infrastructure is reachable or not.
4. The MANEMO protocol must enable packets transmitted from mobile routers visiting the MFS to reach the Internet with a topologically correct address.
5. The MANEMO protocol should aim at minimizing radio interference with itself as the control messages get propagated in the MFS.
6. The MANEMO protocol must enable inner movements within MFS to occur and ensure details of this movement are not propagated beyond the MFS.
7. An MFS may split to become two separate MFSs, in this case MANEMO will continue to maintain local connectivity within the separate MFSs and connectivity between the MFSs will be restored once a NEMO connection becomes available.
8. The MANEMO protocol should enable and optimize the trade-off between ensuring some reciprocity between MFS peers and maintaining a safe degree of CIA properties between the peer mobile routers.
9. The MANEMO protocol should enable the mobile routers to be deployed in restoring connectivity if parts of an MFS went isolated or extend the connectivity in the areas that are not covered.
10. The solution must not require modifications to any node other than nodes that participates in the MFS. It must support fixed nodes, mobile hosts, and mobile routers in the NEMO that form the MFS and ensure backward compatibility with other standards defined by the IETF.
11. The MANEMO protocol shall enable multicast communication for nodes within the MFS and on the Internet. Translation of MANEMO multicast signaling and multicast signaling on the Internet shall take place on the exit router.
12. The MANEMO protocol shall optimize the path to the Internet using cross-layer metrics.

As we discussed in Section 11.4, the characteristics of "less mobility," "group mobility," and "Internet-oriented communication" are reasons for developing new solutions for MANEMO. Moreover, the following design decisions should be considered when developing a MANEMO solution.

In the MANET-centric approach, the assumption is one-to-all communication as the flooding mechanism is used for protocol operations. Flooding is used to disseminate packets to the entire MANET. Packets are delivered to all nodes in the MANET over multihop. The flooding mechanism is not employed in the NEMO-centric approach. The proposed MANEMO protocols are operated one-to-one. A mobile router exchanges packets only with one-hop neighbors in the MFS. A packet is sometimes delivered beyond the one-hop neighbors, but it is always processed and routed by the intermediate nodes in the MFS. The one-to-all assumption brings with it various security issues. It is extremely difficult to establish secure relationships among all nodes in an MFS. Nodes in the MFS do not always share a common relationship in all MANEMO scenarios. MANET technology has not addressed security issues to this point in IETF. If security is not guaranteed in solutions, the MFS cannot be deployed and operated in real scenarios.

The MANEMO solution is to merge several features of several different protocols. As it is assumed MFSs will appear everywhere in the world, the MANEMO solution should be installed in all mobile devices similar to neighbor discovery protocol. The MANEMO solution should be one of the core networking protocols in a device. The solution should not introduce considerable overhead and modifications to the mobile device, since most mobile devices have limited computing and networking resources. Therefore, the trade-off between MANEMO solutions and MANEMO functions should be carefully considered. The MANET protocol may achieve most of the MANEMO goals, however, at the same time it might also introduce considerable overhead and modification to mobile devices.

As we discussed in Section 11.4, group mobility is a MANEMO-specific movement pattern. By using the NEMO basic support protocol, the impact of a mobile router changing attachment is hidden by that mobile router. Even if a mobile router changes its point of attachment, this change is perfectly hidden from nodes behind the mobile router. The topology behind the mobile router stays the same. On the other hand, in the MANET and AUTOCONF approach, the impact is propagated to all or some mobile routers that are located behind the mobile router which changes the attachment point. In the AUTOCONF addressing architecture, if a mobile router changes its attachment to a new access router, all nodes behind the mobile router should obtain a new address from the new access router.

11.6.2 Solution Classification

Possible MANEMO solutions can be classified into two. The difference of these two approaches is which addressing architecture solution is selected (i.e. NEMO addressing architecture versus MANET addressing architecture).

■ NEMO centric approach: The nested NEMO is the initial root problem MANEMO seeks to address. Mobile routers attach to one another, and MANEMO should optimize the resulting topology for access to the infrastructure, provide a safe model for mobile routers to help one another, and offer some degree of inner routing. For this approach, tree discovery [6] has been proposed to form a loop-free tree in MANEMO, and NINA [7] provides some routing in that space. Reverse routing header (RRH) [8] is a solution toward bypassing the number of home agents when mobile routers form a nested NEMO.

■ Mobile ad hoc (MANET centric): NEMO mobile routers form a MANET in an MFS. If MANET approaches are taken, loop-free and local routing is somehow guaranteed. However, there are several optimized levels to be found by using different MANET mechanisms. The decision on which optimization level is ideal may depend on the MANEMO scenario under consideration. More details can be found in Section 11.6.6.

11.6.3 Tree Discovery

The tree discovery protocol [6] is a distance vector protocol used to find the nearest exit toward the Internet and form a tree structure in the form of directed acyclic graphs. It is an extension to neighbor discovery protocol to carry information and metrics to form a loop-free tree structure in the MFS in an autonomous fashion. The tree information option is newly defined and carries the depth of the tree, the status of network connectivity, and so on by router advertisement down the tree.

Although each mobile router sends a router advertisement for its mobile network prefix, the tree information option is propagated down the tree. The value of the tree information option is updated by each intermediate mobile router in the tree structure.

As explained in Section 11.3.3, the risk of having loops in the MFS with the NEMO basic support protocol exists. With the tree discovery protocol, the router can avoid the loop by carefully checking the tree-depth value advertised in the tree information option. A mobile router should not connect to another mobile router advertising a higher tree depth in order to avoid the loop.

In the tree information option, the status of connectivity to the infrastructure can be stored. By using such status, a mobile router can decide on the best tree with which to reach the infrastructure in an MFS. The tree depth is also one metric used to decide the tree head.

11.6.4 Network in Node Advertisement

The network in node advertisement (NINA) protocol [7] enables local routing between mobile routers in MFS. Local routing can prevent packets from going through multiple home agents between mobile routers in the same MFS. NINA exposes the mobile network prefixes up to the tree after tree discovery forms the loop-free tree structure. NINA defines a new option named network in node (NINO) option to carry the mobile network prefix in the neighbor advertisement message. By exchanging the NINO options by neighbor advertisement messages up to the tree, a mobile router learns the mobile network prefix of all other mobile routers down its tree. When two mobile routers in the same tree communicate, the upper mobile router of those two mobile routers can direct packets without going to the Internet. Since the cost of Internet connectivity in the mobile environment is expensive and unstable, local routing is beneficial in optimizing communication performance inside the MFS. The mobile router can also be released from a dependency on home agents for communication to other nodes inside the MFS.

11.6.5 Reverse Routing Header

The reverse routing header (RRH) [8] is a source routing protocol used to avoid multiple tunnels and redundant routes in a nested NEMO. It introduces a new routing header called the reverse routing header and records the sequences of traversed mobile routers on the way out of an MFS.

While the packet is forwarded along the tree formed by tree discovery, all the mobile router's care-of addresses are recorded in the reverse routing header of the packet. The reverse path is then recorded in each packet. Once the packet arrives at the MFS, it forwards the packet to a node in the MFS along the path recorded in the reverse routing header of the packet. It avoids multiple IP-in-IP tunnel encapsulation (40-byte header), while adding 16 bytes in the reverse routing header.

11.6.6 MANET and AUTOCONF Solutions

MANET and AUTOCONF solutions are not specifically designed for MANEMO goals. However, there are several candidate solutions. The main goal of MANET and AUTOCONF solutions is to add local routing capability to mobile routers of the NEMO basic support protocol in an MFS.

For a simple solution, a mobile router only discovers the neighbor mobile router(s) and communicates only with it directly. Multihop capability is not available in this case. To add one-hop local routing capability, NANO [9] provides a very simple solution to exchange the NEMO prefix between neighboring mobile routers. Likewise, NHDP [17] is a candidate protocol for discovering two-hop neighbors. Although NHDP is designed to run with a MANET routing protocol, it is a protocol used to maintain two-hop neighbors in the MANET. A mobile router manages routes for two-hop neighbor mobile routers and can optimize the path only for two-hop neighbors. No MANET routing protocols are required for one-hop and two-hop optimization of local routing.

If a MANET routing protocol such as the optimized link state routing protocol [18] is run in the MFS, a mobile router can be made aware of the local route in the entire MFS and can form a loop-free topology in the MFS.

MANEMO could still help in several fashions, for instance, providing scalability by splitting the larger network into a number of more manageable islands, interconnected by NEMO over the infrastructure.

11.7 Conclusions

This chapter presented the MANEMO architecture and its goals. MANEMO has just begun the discussions in IETF and research community. As such the MANEMO concept is not well understood and has not been sufficiently addressed within IETF. However, several projects and individuals share a definite interest in the MANEMO concept and have begun work on solutions in IETF. Pascal Thubert and I began MANEMO discussions originally as a possible new solution for the nested NEMO in 2004. An unofficial BOF was held at IETF-68 in Prague and presentations have continued at the meetings of several working groups. During IETF-69, MANEMO was discussed at the AUTOCONF working group. Through the course of discussions on MANEMO at IETF, we have realized that MANEMO contains more issues than the nested NEMO problem. As was introduced at the outset, MANEMO is defined in several ways, MANET for NEMO, mobile ad hoc NEMO, and management of nested NEMO. All definitions of MANEMO seem to be valid. Several MANEMO solutions have been proposed and discussed, though a working group for MANET has not yet been formed within IETF. Following deployment of IP mobility technology, several mobile scenarios will surely move closer to what we have defined as MANEMO. The MANEMO will be the technology to mesh several features of different protocols.

References

[1] J. Manner and M. Kojo, "Mobility Related Terminology," RFC 3753, IETF, June 2004. http://www.ietf.org/rfc/rfc3753.txt.

[2] V. Devarapalli, R. Wakikawa, A. Petrescu, and P. Thubert, "Network Mobility (NEMO) Basic Support Protocol," RFC 3963, IETF, January 2005. http://www.ietf.org/rfc/rfc3963.txt.

[3] D. Johnson, C. Perkins, and J. Arkko, "Mobility Support in IPv6," RFC 3775, IETF, June 2004. http://www.ietf.org/rfc/rfc3775.txt.

[4] R. Wakikawa, P. Thubert, T. Boot, J. Bound, and B. McCarthy, "Problem Statement and Requirements for MANEMO," draft-wakikawa-manemo-problemstatement-01.txt, IETF, Internet Draft, (work in progress), July 2007. http://tools.ietf.org/id/draft-wakikawa-manemo-problem-statement-01.txt.

[5] R. Wakikawa, T. Clausen, B. McCarthy, and A. Petrescu, "MANEMO Topology and Addressing Architecture," draft-wakikawa-manemoarch-00.txt, IETF, Internet Draft, (work in progress), July 2007. http://tools.ietf.org/id/draft-wakikawa-manemoarch-00.txt.

[6] P. Thubert, C. Bontoux, and N. Montavont, "Nested Nemo Tree Discovery," draftthubert-tree-discovery-06.txt, Internet Draft, (work in progress), July 2007. http://tools.ietf.org/id/draft-thubert-tree-discovery-06.txt.

[7] P. Thubert, "Network In Node Advertisement," draft-thubert-nina-00 (work in progress), February 2007. http://tools.ietf.org/id/draft-thubert-nina-00.txt.

[8] P. Thubert, "IPv6 Reverse Routing Header and its application to Mobile Networks," draft-thubert-nemo-reverse-routing-header 06.txt, Internet Draft, IETF (work in progress), September 2006. http://tools.ietf.org/id/draft-thubert-nemo-reverse-routing-header-06.txt.

[9] A. Petrescu and C. Janneteau, "The NANO Draft (Scene Scenario for Mobile Routers and MNP in RA)," draft-petrescu-manemo-nano-00 (work in progress), March 2007. http://tools.ietf.org/id/draft-petrescu-manemo-nano-00.txt.

[10] T. Clausen, Baccelli, E., and R. Wakikawa, "Network Mobility Route Optimisation Problem Statement," draft-clausen-nemo-ro-problem-statement-00 (work in progress), October 2004. http://tools.ietf.org/id/draft-clausen-nemo-ro-problem-statement-00.txt.

[11] C. Ng, P. Thubert, M. Watari, and F. Zhao, "Route Optimization Problem Statement," RFC 4888, IETF, July 2007. http://tools.ietf.org/rfc/rfc4888.txt.

[12] C. Ng et. al, "Network Mobility Route Optimization Solution Space Analysis," RFC 4889, IETF, July 2007. http://www.isi.edu/in-notes/rfc4889.txt.

[13] T. Narten, Nordmark, E., and W. Simpson, "Neighbor Discovery for IP Version 6 (IPv6)," RFC 2461, December 1998. http://www.ietf.org/rfc/rfc2461.txt.

[14] MetroNet6, http://www.metronet6.org/. (accessed December 10, 2008).

[15] U-2010, http://www.u-2010.eu/. (accessed December 10, 2008).

[16] I. Chakeres, "Mobile Ad hoc Network Architecture," draft-ietf-autoconf-manetarch-01 (work in progress), March 2007. http://tools.ietf.org/id/draft-ietf-autoconf-manetarch-01.txt.

[17] T. Clausen, "MANET Neighborhood Discovery Protocol (NHDP)," draft-ietf-manetnhdp-00 (work in progress), June 2006. http://tools.ietf.org/id/draft-ietf-manet-nhdp-00.txt.

[18] T. Clausen, "The Optimized Link-State Routing Protocol Version 2," draft-ietf-manetolsrv2-02 (work in progress), June 2006. http://tools.ietf.org/id/draft-ietf-manet-olsrv2-02.txt.

[19] DUMBO, http://www.interlab.ait.ac.th/dumbo/what is dumbo.php. http://tools.ietf.org/id/draft-ietf-manet-olsrv2-02.txt.

Chapter 12

Security in Vehicular Networks

Christian Tchepnda, Hassnaa Moustafa,
Houda Labiod, and Gilles Bourdon

Contents

Vehicular networks are emerging as a new class of wireless networks, spontaneously formed between moving vehicles equipped with wireless interfaces that could be of homogeneous or heterogeneous technologies. Indeed, the recent advances in wireless technologies and the current trends in ad hoc network scenarios allow a number of possible architectures for vehicular networks' deployment. However, the security of vehicular networks constitutes a major challenge, that can impact their future deployment and applications. Beside drivers and passengers' safety, it is important to have reliable solutions for secure communication between participants as well as authorized and secure service access. Consequently, appropriate security architectures should be in place providing secure communications between vehicles and allowing different service access. Moreover, there is a need for security mechanisms suitable for any vehicular network environment and aiming to provide trust, authentication, access control, authorized, and secure service access.

This chapter focuses on the problem of security in vehicular networks. The chapter starts by presenting the special characteristics of vehicular networks and their impact on the security of these networks. The different types of attacks in such networks are then illustrated, and some relevant attacks examples are introduced. Consequently, some key security requirements are derived, considering different service types (services related to intelligent transportation system [ITS] as well as non-ITS related services). Finally, the chapter presents a panorama of a number of existing security contributions in vehicular networks, analyzing such contributions and discussing their deployment feasibility.

12.1 Introduction on Vehicular Networks Security

Although security is an important challenge in vehicular networks, little attention has been devoted to such problem. Currently, most of the research in vehicular networks is being focused on the development of a suitable MAC and PHY layers, deployment architectures, routing and dissemination protocols, and potential applications ranging from collision avoidance to onboard infotainment services. However, the problem of security is being postponed to later research and development phases by academics and industry [1]. Since the deployment of vehicular networks is expected soon, their success and safety will depend on viable security solutions acceptable to consumers, cars manufacturers, network operators or service providers, and governments. Indeed, vehicular network security is a crucial challenge. On one hand, it is essential to make sure that life-critical information cannot be inserted, modified, or truncated by an attacker, where only authorized users should be able to manipulate the transmitted information. On the other hand, only authorized users should be able to access the offered non-ITS services, thus allowing service access legitimacy. One should notice that providing strong security in vehicular networks raises important privacy concerns that must be also considered.

Security and privacy are major concerns in the development and acceptance of services in vehicular networks and should not be compromised by ease-of-use of service discovery protocols. As the demand for service discovery is growing, drivers and passengers may use services in foreign networks and create security problems for themselves and for other network users (e.g., theft of services due to nonauthorized service access or impersonation during service access, drivers getting false information due to consulting an illegitimate entity—discovered through a nonsecure service discovery protocol—on traffic information, a wearable medical device can benefit from the service discovery architecture to communicate with the doctor's office—following an accident—while leaking sensitive data to outsiders). Moreover, nodes behavior is an important issue that can threaten the security of communication and service delivery in vehicular networks and hence is worth consideration. The dynamic and open environment of vehicular networks makes nodes cooperation an important aspect that should be satisfied to allow successful communication between vehicles. In fact, nodes may behave selfishly by not forwarding messages for others to save power, bandwidth, or just because of security and privacy concerns. Consequently, appropriate security architectures should be in place providing secure communication between vehicles as well as secure service access in a cooperative manner. The security mechanisms, in this context, need to be suitable for any vehicular network environment, and should provide trust, authentication, access control, and authorized and secure service access. Appropriate mechanisms should also be developed to detect selfishness and enforce nodes cooperation in vehicular network environment.

To enhance the service access ubiquity in vehicular networks, security solutions should take advantage of (i) the ad hoc multihop communication concept and (ii) the distribution based approach. Since much of the impetus for the development of vehicular network applications comes from vehicle manufacturers as well as integrated operators, there is a need for developing security techniques that operate in a distributed, ad hoc fashion. This approach will speed vehicular network deployment, since it allows car manufacturers to incorporate these techniques with no dependency on particular infrastructures, and allows network operators or service providers to offer advanced services for drivers and passengers without deploying additional infrastructure or deviating from their core business model.

To sum up, security and privacy must be two primary concerns in the design of vehicular networks. Poorly designed vehicular networks that permit serious attacks on the network can jeopardize the goal of increased driving safety. Also, a vehicular network design that enables unauthorized and illegitimate third parties to collect private information about drivers, for example, by making tracking vehicles a possibility, will certainly be refused by drivers. Moreover, vehicular networks constituted of nodes behaving in a selfish and noncooperative manner would lead to unreliable service access and could also threaten peoples' safety. Indeed, the specific characteristics of vehicular networks result in hard to address security issues, making the field of secure intervehicular communication an interesting research topic that is still very open and active.

12.2 Attacks in Vehicular Networks

Vehicular networks special characteristics make them susceptible to a wide range of attacks. In this section we discuss the most relevant attacks threatening vehicular networks and intervehicular communication. A classification of these attack is provided and some concrete attack examples are illustrated.

12.2.1 Attack Classification

An important key in securing vehicular networks is determining the types of attacks that threaten the communication in such networks. Different types of attacks may exist according to the type of environment as well as the usage scenario. The following list presents a classification of the possible attacks types. These are based on the taxonomy provided by Raya and Hubaux [1], while introducing a little extension to consider the cooperative malicious behavior.

- Internal or external attacks: An internal attack can be mounted by an authenticated member of the vehicular network. In other words, this member is identified by other members as a legitimate member. This type of attack is probably the most critical one because it is mounted by legitimate network members. On the other hand, an external attack can be mounted by a nonauthenticated entity that is hence considered as an intruder by legitimate members. Unlike an internal attacker, an external attacker is limited in the diversity of attacks that they can mount.
- Intentional or unintentional attacks: An intentional attack is mounted by an entity aiming voluntarily to disrupt the network operation. Conversely an unintentional attack is mostly due to potential transmission or network operation errors.
- Active or passive attacks: An active attack is mounted by an attacker who generates or modifies the network traffic. In contrast, a passive attack is mounted by an attacker who will only eavesdrop the wireless channel for later unauthorized use.
- Independent or coordinated attacks: An independent attack is caused by a unique attacker whereas a coordinated attack is caused by a group of attackers sharing the same interest.

12.2.2 Attack Examples

Indeed, it is hardly possible to give an exhaustive list encompassing all attacks in vehicular networks. This subsection only introduces some important and relevant attack examples that have been identified, where an analysis of these attacks is illustrated based on the attack classification presented above.

■ Privacy disclosure: This type of attack is considered as the big brother case where an attacker can actively monitor the network traffic to disclose vehicles identities and trace them. Other examples of traceable identifiers are IP addresses, MAC addresses, certificates IDs, and so on. Even if individual messages do not contain such identifiers, a particular string that does not in itself identify the vehicle could appear in a series of messages. If an attacker ever, unambiguously, observes the vehicle emitting that string, they can use this string as an identifier for the vehicle. Moreover, using radio fingerprinting is another way to identify and trace vehicles. This latter is a physical layer attack with the main goal to detect the signal (device) features that form a valid device fingerprint and based on which associations can be made between observed messages and their senders. Figure 12.1 illustrates an identity disclosure example. Considering the attack classification given in the previous subsection, this type of attack is characterized by being internal or external, intentional, passive, or independent.

■ Information inconsistency: This type of attack enables the attacker to inject wrong information in the network to affect the behavior of other vehicles. An example scenario is a vehicle cheating with positioning information to alter its perceived position, speed, direction, itinerary, and the like. By doing so, the attacker can divert the traffic from a given road and hence free that road for himself. Figure 12.2 and Figure 12.3 illustrate this case. In Figure 12.2, the attacker diffuses bogus traffic information, where such type of attack is characterized by being *internal, intentional, active,* and *independent.* Whereas in Figure 12.3, some attackers are cheating with positioning information, where such type of attack is characterized by being internal, intentional, active, or coordinated.

■ Impersonation or masquerading: In this type of attack, the attacker uses a false identity pretending to be another vehicle. More generally, the attacker uses a false credential to be granted another vehicle privileges. Figure 12.4 shows an example of such attack. This specific attack example is characterized by being internal or external, intentional, active, or independent.

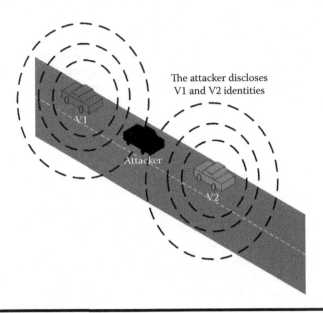

Figure 12.1 Identity disclosure.

- Denial of service (DoS): In this type of attack, the attacker prevents legitimate vehicles to access the network services. This can be done by jamming the wireless channel, overloading the network or having a noncooperative behavior (e.g., dropping packets). Figure 12.5 illustrates an example of such attack, where the attacker causes a

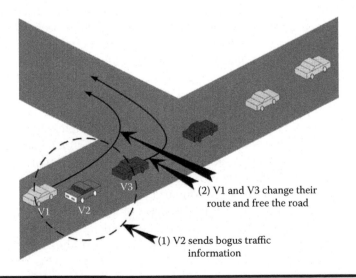

Figure 12.2 Bogus traffic information injection.

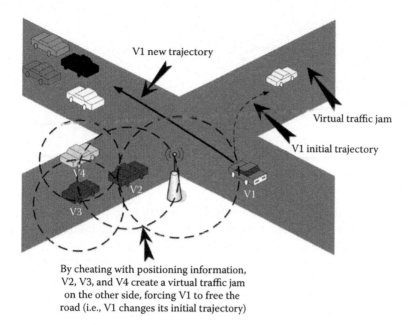

By cheating with positioning information,
V2, V3, and V4 create a virtual traffic jam
on the other side, forcing V1 to free the
road (i.e., V1 changes its initial trajectory)

Figure 12.3   Cheating with positioning information.

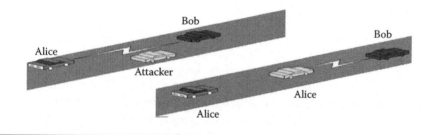

Figure 12.4   Masquerading.

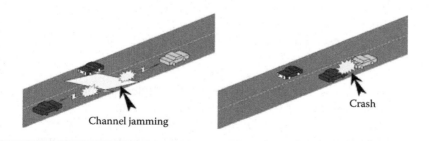

Figure 12.5   DoS—channel jamming.

Figure 12.6   Eavesdropping a commercial transaction.

crash by preventing legitimate vehicles to exchange critical traffic information. This attack type is characterized by being internal or external, intentional, active, independent or coordinated.

■ Eavesdropping: In this type of attack, the attacker monitors the network traffic to extract any sensitive information. Figure 12.6 shows the case where an attacker eavesdrops and extracts a commercial transaction password. This attack type is characterized by being internal or external, intentional, passive, or independent.

12.3   Security Challenges

Vehicular networks are expected to provide a wide set of useful services to drivers and passengers. These services can be generally classified into two main classes: (i) ITS also including safety services, and (ii) non-ITS services. ITS was the main objective in the emergence of vehicular communications with the main target to minimize accidents and improve traffic conditions, while non-ITS services aim at providing commercial, leisure, and convenience services. In this section, we show the main security challenges in vehicular networks considering both ITS and non-ITS services. These security challenges should be considered during the design of vehicular network architecture, security protocols, cryptographic algorithms, and software and hardware architectures.

The following list presents the main security challenges in vehicular networks [2,3]:

■ Real-time constraints: A critical feature in ITS services is their time sensitiveness, where ITS data are mostly real-time data with about 100 ms critical transmission delay [4]. In fact, ITS services are expected to carry out much more signature verification than signature generation, due to their broadcast nature. So, an important challenge, achieving real-time constraints, is choosing the fastest public key cryptosystem in signature verification which at the same time performs well in signature generation. Also, the selected cryptosystem should be as compact as possible. In any case, any security mechanism for ITS services should take into consideration these real-time constraints.

■ Data consistency liability: This is also an important security issue for ITS services as even authenticated vehicles could become malicious by sending bogus information to gain an undue advantage and thus can cause accidents or disturb the network operation. As a mechanism example ensuring data consistency or liability, we can quote verification by correlation which consists of correlating, through a reputation-based or a recommendation-based system, data received from a given source with those received from other sources. Some other approaches consist of enabling any node to search for possible explanations for the data it has collected based on the fact that malicious nodes may be present. Explanations that are consistent with the node model of vehicular networks are scored, and the node accepts the data as dictated by the highest scoring explanations [5].

■ Low tolerance for errors: Some applications use protocols that rely on probabilistic schemes to provide security (e.g., the traffic reporting application by Picconi et al. [6] and the vehicle safety application by Guo, Baugh, and Wang [7]. However, given the life-or-death nature of ITS applications, a small probability of error will be unacceptable. Consequently, such applications will have to rely on deterministic schemes or probabilistic schemes with security parameters large enough to make the probability of failure minor.

■ Key distribution: Generally, key distribution is a fundamental building block for security protocols. Particularly in vehicular networks, key distribution poses several significant challenges. First, vehicles are manufactured by many different companies, so installing keys at the factory would require coordination and interoperability between manufacturers, as well as a general agreement on standards for key distribution. Second, relying on certification authority (CA) to certify each vehicle's public key is a challenging approach, since vehicles from different countries may not be able to authenticate each other unless vehicles trust all CAs, which reduces security. Also, certificate-based key establishment might have the danger of violating driver privacy, as the vehicle's identity might be revealed during each key establishment

■ Incentives: Successful deployment of vehicular networks will require incentives for vehicle manufacturers, network operators, consumers, and the government, and reconciling their often conflicting interests will prove challenging. For example, law enforcement agencies would quickly embrace a system in which speed-limit systems broadcast the mandated speed and vehicles automatically report any violations. Obviously, consumers would reject such intrusive monitoring, giving vehicle manufacturers little incentive to include such a feature. Conversely, consumers might appreciate an application that provides an early warning of a police speed trap. Manufacturers might be willing to meet this demand, but law enforcement is likely to object.

■ High mobility: This is a crucial challenge in designing vehicular network security systems. We assume that the vehicle's computing platforms have the same computational capability and energy supply as wired clients such as desktop PCs. However, they significantly differ in their mobility support and their resulting throughput capability. These factors result in a mismatch between security protocols execution time in vehicular networks and their execution time in wired networks. This execution gap for security protocols is an important issue that must be faced by vehicular network designers. An attempt to lower this gap is making vehicular security protocols and their inherent cryptographic algorithms lightweight and fast without loosing security robustness. For instance, by implementing only a subset of security protocols features, it is possible to reduce the overhead and execution time. This goal can also be achieved by selecting optimal software or hardware implementations for cryptographic algorithms and adapting the encryption policies based on the content of the data that is being encrypted (e.g., video encryption). Two approaches for reducing the execution time and adapting to high mobility are as follows:

1. Low complexity security algorithms: Indeed, the current security protocols such as SSL/TLS [8,9], DTLS [10], and WTLS [11] generally use RSA-based public key cryptography [12] for authentication. The security of the basic RSA algorithm is derived from the integer factorization which is NP-hard. Hence, RSA can provide high security if the modulus is a large integer (e.g., 1024 or 2048 bits) whose factoring is extremely complex. This means that the basic computation for decrypting data is performed using large keys, making it expensive computationally and time wise. We can take advantage of alternative public key cryptography standards (PKCS) that provide security robustness while requiring less execution time. Elliptic curve cryptosystems (ECC) [13] and lattice-based cryptosystem NTRU [14] are examples of such alternative public key cryptosystems that are increasingly being used in wireless security software toolkits. For bulk data encryption or decryption, a protocol such as AES [15] is preferred since older ones like DES [16] or 3DES [17] appear less attractive due to security limitations or computational and time expensiveness.

2. Transport protocol choice: When securing transactions over IP, the transport layer over which security protocols are implemented should be carefully chosen. For example, transport layer security (TLS), which secures application-layer traffic over TCP/IP is discouraged for mobile use as it operates over TCP. Conversely, datagram TLS (DTLS), which secures application-layer traffic over UDP/IP is better accepted as it operates over a connectionless transport layer (i.e., UDP). A protocol such as Internet protocol security (IPSec) [18], which secures IP traffic should be avoided as its secure connections (known as security associations [SAs]) are cumbersome to set up, requiring too many messages. However, when vehicles are not in motion (e.g., in a parking space), protocols like IPsec or TLS might become appropriate.

12.4   Security Requirements

As a countermeasure against most of the attacks mentioned above, a number of requirements should be satisfied. We notice that the message transmission model in vehicular networks depends on the type of provided services (ITS or non-ITS), where broadcast transmission is mostly relevant for ITS/safety services and unicast or multicast transmission is used for non-ITS services. Consequently, the security requirements can be different according to the service class.

The following list presents a number of security requirements that are worth consideration in vehicular network environments [2]:

- Availability: Denial of service (DoS) attacks due to channel jamming, network overloading, or noncooperative behaviors may result in network unavailability. It is also noticed that the availability problem becomes worse if the communication layer is less reliable. Hence, a continuous network operation should be supported by alternative means. The nonavailability risk can be mitigated by exploring the numerous security mechanisms (e.g., monitoring, reputation-based systems, etc.), which can be used for noncooperative node detection or by exploring channel and technology switching and cognitive radio techniques that in turn can be used against jamming attacks. Indeed, this security requirement is required for any class of services and is particularly important for ITS services.
- Message integrity: This security requirement applies to ITS and non-ITS services as it protects against altering a message in transit. In practice, authenticity and integrity could not be separated, since there are no means to correctly identify a message's origin if the message content is altered.

- Confidentiality: We believe that ITS services are directly related to people's physical safety. If a vehicle that is not authenticated (i.e., an illegitimate vehicle) is in accident, authenticated vehicles (i.e., legitimate vehicles) may also be affected. Consequently, to efficiently achieve the main goal of ITS services, which is mainly peoples' safety, every vehicle (legitimate and illegitimate) should be capable of receiving and processing ITS data. Unlike non-ITS services, which may require confidentiality due to their commercial nature, ITS services should not require protection against eavesdropping. However, ITS message transmission origin should be verified through source authentication.

- Source authentication: In ITS services, the only restriction lies in avoiding illegitimate vehicles to generate ITS data. This could simply be implemented through discarding ITS data not having a proof of authenticity of their sources (e.g., a digital signature). Indeed, this security requirement appears more relevant for ITS services since mostly broadcast communication is considered for these services, and because almost all vehicles (legitimate or illegitimate) are expected to receive safety messages.

- Mutual authentication, authorization, and access control: The commercial or transactional nature of non-ITS services requires them to support mutual authentication between each client (vehicle) and the service provider (or the network operator) on one hand and between each communicating vehicles on the other hand. In this context, mutual authentication is useful, where it aims at preventing attacks such as man-in-the-middle (MITM) attack. A simple solution to carryout authentication in such environment is to employ a symmetric key shared by all nodes in the network. Although this mechanism is considered as a *plug-and-play* solution and requires less processing and communication overhead, it is limited to closed scenarios of small number of vehicles, mostly belonging to the same provider. For wide-scale commercial deployment of vehicular networks, the symmetric group key authentication has two main pitfalls. First, an attacker only needs to compromise one node (vehicle) to break the security of the system and paralyze the entire network. Second, mobile nodes (vehicles) can impersonate each other and can access each other messages breaking the nonrepudiation and the confidentiality security requirements. As for the symmetric pairwise key authentication, the main problem is in its inherent key establishment, nonscalability, as the number of keys grows linearly with the number of vehicles. Hence, public key cryptography with few performance enhancements seems to be the way to go.

- Nonrepudiation: Vehicles causing accidents or injecting malicious data must be reliably identified. Hence, a vehicle should not be able to deny the transmission of a message. If used carefully, digital signatures can provide the nonrepudiation property. However, the main reason for using signatures is not to provide nonrepudiation, but to allow authentication between two entities who have not previously encountered each other, without having to make an online query to a third party. Although nonrepudiation security requirement is mostly critical for ITS services, it may be desirable also to have it for some sensitive non-ITS services especially those involving online payments.

■ Privacy protection: As people increasingly worry about the big brother enabling technologies, private individuals' anonymity and nontraceability should be guaranteed. It should be noted, however, that anonymity and nontraceability are strictly conditional as nonrepudiation must be enforced. In fact, the privacy security requirement applies to both ITS and non-ITS services. Although traceability is a legitimate process for governmental authorities and networks operators, the nontraceability is an important security requirement to assure peoples' privacy. Thus a complex problem arises in this issue. Traceability in vehicular networks can include (i) who is talking to who, (ii) what one is sending, (iii) which site one is accessing or which application one is using, and (iv) where is the mobile client now (his location) and where is he going to be after a while [19]. In fact, a tough requirement in vehicular network environments is to manage traceability in terms of allowing this process for the concerned authorities and at the same time assuring the nontraceability between mobile clients (vehicles) themselves. Nevertheless, the latter is difficult to achieve and so far no promising solutions exist to resolve this issue in the vehicular network's dynamic and open environment.

From our investigation of the vehicular networks various security challenges and the different security requirements that should be achieved, we conclude that ITS service security requirements are a little bit different from non-ITS service security requirements. Figure 12.7 highlights the main security requirements for both services, showing the commonality and the differences between them.

12.5   Related Security Contributions

12.5.1   *Standardization and Deployment Efforts*

The most prominent industrial efforts in the domain of vehicular network security is carried out by the Car-2-Car Communication Consortium [20] in Europe and the DSRC [21] consortium in the United States, especially the

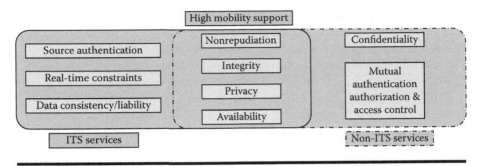

Figure 12.7 Main security requirements and security challenges for ITS and non-ITS services.

IEEE 1609.2 Working Group [22]. Also, some commercial products already make use of vehicular communication while considering some security aspects, where a related application in this context is the GPS car tracking [23].

Other efforts are also shown in a number of research projects focusing on vehicular network security and trying to develop a number of prototypes for deploying security-related solutions. It is noticed that the majority of these projects are concerned with road safety and traffic applications (e.g., SEVECOM [24]). The SEVECOM (SEcure VEhicular COMmunications) project [24] addresses the vehicular networks security and privacy with the objective to define the security architecture of such networks, as well as to propose a roadmap for integration of security functions, with the main focus on communication specific to road traffic. Aiming to enlarge the scope of road safety and traffic applications, the NOW (Network on Wheels) project [25] studied the data security in vehicular networks considering both safety applications and infotainment applications with infrastructure and between vehicles. This project tends to provide an open communication platform for a broad spectrum of applications.

12.5.2 Literature Review

Vehicular communication security is a young research domain, showing few research contributions and lacking real development solutions. The existing contributions mainly focus on securing ITS and safety-related services or pure ad hoc vehicular communications, neglecting the tremendous potential of infotainment services and infrastructure-based or hybrid ad hoc vehicular communication in driving the development and deployment of vehicular networks. Although extensively tackling security of ITS services, some of these contributions fail in considering some peculiarities of such services like broadcast transmission and its resulting effect on security requirements. Such limitations fuel the motivation in providing a broader and a more complete view of vehicular communications security challenges.

12.5.2.1 ITS-Services Security

1. Validating ITS data: For evaluating the validity of ITS data in vehicular networks, a general approach is proposed by Golle, Greene, and Staddon [5], in which a node searches for possible explanations for the data it has collected based on the fact that malicious nodes may be present. Explanations that are consistent with the node's model for the vehicular network are scored, where the node orders the explanations in accordance with a developed heuristic, named parsimony heuristic. This heuristic assumes that an attack involving a few malicious nodes is more likely than an attack that requires collusion between a large number of nodes. The data that is accepted by the node at the end is consistent with the highest ranking explanation or explanations. Although this approach is interesting in validating the critical ITS data that are transmitted among vehicles, the heuristic-based approach might necessitate large delays that can impact the real-time constraint. Also, more complexity is expected when coming to the deployment of a such solution.

2. Securing ITS service access: ITS service security in vehicular environment is also addressed by Zarki et al. [26], which is among the primary works on this issue. The authors use a broadcasting communication model that is one way with the base station (BS), while considering that there is no routing and no handover in their scenario. It is shown that providing efficient PKI and digital signatures mechanisms minimizes the need for confidentiality and key distribution. However, these mechanisms are required to have acceptable delay with respect to the ITS services real-time constraints. Indeed, this contribution does not support privacy issues; neither considers the security requirements for non-ITS services, thus limiting the possibility of its wide-scale deployment.

3. Privacy issues for ITS services: Privacy issues for ITS services are more specifically tackled by Hubeaux, Capkun, and Luo [27] through introducing the concept of entropy anonymity metric and recommending the use of PKI and some location verification tools to implement ITS security. The main advantage of this work is that privacy mechanisms, which are introduced can also be considered for non-ITS services. A more network-oriented work introducing threat analysis and security architecture for ITS services is presented in [1,28]. The authors provide a set of security protocols making use of PKI, digital signatures, anonymous public keys implementing privacy, and secure positioning among other points. Nevertheless, the security architecture is strictly relevant for vehicular networks running ITS services.

12.5.2.2 ITS and Non-ITS Service Security

1. Virtual infrastructure approach for security: An architecture for intervehicles communication (IVC) security is studied by Blum and Eskandarian [29]. This architecture comprises a PKI, distributed IDS (intrusion detection system), and a virtual network infrastructure. Although this work is general enough for securing ITS and non-ITS services, it is extremely restrictive in considering a stand-alone vehicular ad hoc network. Indeed, this does not represent a great interest for the envisioned business model of network operators who are expected to play a major role in the provision of infrastructure. Besides, the fact that vehicular network penetration is still weak necessitates an existing infrastructure to accelerate the evolution and deployment of such networks to allow reliable service access.

2. A real deployment approach for security on highways: A novel solution considering a real deployment approach from a network operator's perspective is presented by Moustafa, Bourdon, and Gourhant [30,31]. This solution aims at securing both ITS and non-ITS services in vehicular highway environments. A service access architecture and security mechanisms for intervehicular communication on highways are proposed, enabling authentication, access control, and service access authorization for mobile clients (vehicles) in such environment. In this work, each vehicle first authenticates with respect to the network operator or service provider at the highway entry point. Based on this authentication, each vehicle that is authorized to access the required services is configured with an IP address and obtains security credentials for latter use during the communication with other vehicles during the trip. The Kerberos authentication model is employed for the first authentication with the network operator or service provider and for issuing security credentials. In this case, the extensible authentication protocol (EAP) [32], precisely EAP-Kerberos [33], is used for authentication messages exchange. Then, 802.11i [34] authentication is used for authentication and secure link setup between two communicating vehicles during the trip, making use of the security credentials obtained during the vehicle's authentication with the network operator or service provider. Since vehicle communications mostly take place in a multihop fashion, 802.11i authentication and secure link setup are extended to the ad hoc multihop environment. This work presents a realistic approach, considering the benefits of mobile clients, as well as network operators and service providers, and allowing both types of applications—safety and leisure applications on highways. From a real deployment perspective, an extension of this solution is needed to support mobility

management, so that vehicles would not reauthenticate whenever they change their access points (APs), and hence service continuity would be guaranteed.

3. Securing intervehicular communication in urban environments: To enhance vehicular networks ubiquitous secure access from a network operator's perspective, novel architecture and security mechanisms are proposed by Tchepnda et al. [35] taking advantages of (i) the multi-hop ad hoc authentication concept, (ii) the smart card–based authentication, and (iii) the grid paradigm for security resource aggregation. This work mainly targets urban vehicular environments, where the multihop ad hoc communication allows communication between two vehicles far from each other (vehicle-to-vehicle communication) and also allows communication between vehicles and the infrastructure (vehicle-to-road communication), thus increasing the services access range. Thanks to the smart card–based authentication approach, vehicle authentication with the network operator can take place in an offline manner, before vehicle participation in the network. Vehicle authentication with the network operator can also take place within the vehicular network, even if vehicles do not have a direct link (i.e., single hop) with the infrastructure, thanks to the multihop ad hoc communication. Indeed, reauthentication optimization is guaranteed through using the grid paradigm, which aggregates the authentication information of vehicles among some dedicated fixed equipments. The ad hoc multihop authentication concept in this work is further developed by Tchepnda et al. [36] through introducing a security solution allowing authentication, access control, and service authorization, considering the network operator perspective. In this context, three main technical issues are developed: (i) A trust and security infrastructure, (ii) An authentication and credential delivery protocol, and (iii) An authentication transport protocol for transporting EAP (extensible authentication protocol) messages [32] over layer-2 in a multihop fashion. The proposed authentication message transport protocol is based on three main features: the stateless broadcast based geographic transport of authentication messages, the opportunistic relay of these messages (i.e., any node can relay a packet if the relaying is relevant to reach the destination), and their multipath transport (i.e., a packet can reach the destination through more than one path). Moreover, the authors' approach tends to increase the security of IP and upper layer services since the access to those services is denied until the authentication with the network operator is successfully completed at layer-2. Although this solution proves to provide a robust and pervasive authentication between vehicles and the network operator, a security association

derivation with access points in such a multihop architecture is still to be designed.

4. Security contributions in NOW and SEVECOM projects: Within the context of the NOW (Network on Wheels) project, Gerlach et al. [37] introduce a security architecture integrating existing individual solutions for vehicle registration, data integrity, data assessment, authentication, pseudonyms, certification, revocation, and so on. The authors distinguish in their architecture the high-level views, that is, the functional layers and the organizational structure describing how the overall security system should look. The implementation-near view is also shown in this work, describing an implementation design for the security system in vehicles' on-board unit while presenting the information flow among the architecture components.

On the other hand, Harsch, Festag, and Papadimitratos [38] propose a scheme for securing geographic position-based routing in the context of both NOW and SEVECOM projects. The proposed security scheme combines digital signatures, plausibility checks, and rate limitation. Indeed, the digital signatures in this work are based on a combination of hop-by-hop signature and end-to-end signature, aiming to provide strong authentication, integrity, and non-repudiation. While, plausibility checks (e.g., on velocity, on timestamp, on transmission range, and the like) can reduce the impact of false positioning information on the routing protocol. Finally, the rate limitation aims at reducing the effect of massive packets injection from malicious nodes that aim to overload the network. Although the authors' objective to enhance the network robustness can be met when using their solution in a pure ad hoc vehicular network, their approach can hardly meet the network operators deployment requirements as there seems to be no hierarchy either in the network architecture or in the security roles among the network's nodes.

5. Trust architecture for securing intervehicular communication: A trust authority architecture is proposed in [39], relying on a governmental authority together with private authorities to offer reliable communications between vehicles on top of which services can be deployed in a secure manner. These authorities are in charge of providing pseudonyms, key distribution, and key management for on-demand requesters, who are namely users and services providers. Single-hop and multihop secure service provisioning scenarios are illustrated in this work, although not sufficiently detailed to be thoroughly analyzed.

12.5.3 Summary and Outlook on the Existing Security Contributions

It is noticed that the security-related contributions are still few in vehicular networks. Although the use of digital signatures is generally discussed, none of the existing contributions consider practical issues such as key sizes, certificates lifetime and revocations, and security protocol or mechanism delays. On the other hand, some contributions focus on particular subjects in vehicular network security without defining the big picture where the proposed solutions would fit. The main open issues, in this context lie in designing multipurpose security architectures complying with vehicular network characteristics and meeting all the respective requirements of ITS and non-ITS services in a single framework. This type of framework can be depicted as a context-aware or a service-aware security framework for vehicular communications.

To sum up, future vehicular network security mechanisms should comply with the special characteristics of these networks, that is, high mobility, constrained bandwidth, dynamic topologies constrained by road topologies, large number of nodes, heterogeneous administration, and real-time constraints. These security mechanisms should be transparent to the underlying applications to serve network operators, and service providers, business needs and should allow each vehicle to authenticate and acquire security credentials whether in V2V or V2R mode. Optimization of security mechanisms is also an important issue that should not be ignored, for instance, authenticated vehicles should not reauthenticate from scratch whenever they change their point of attachment with the network. Some trusted (dedicated) vehicles can be delegated to have some security roles enabling them to serve new vehicles joining the network to decrease the charge on the authentication server or the certification authority. Vehicle security states and information must be aggregated a number of clusters constituted by fixed entities (grid-like structure) to decrease the delay in contacting a centralized server.

12.6 Conclusions

Security of intervehicular communication (IVC) is a major challenge, having a great impact on the future deployment and applications of vehicular networks. There exist a number of challenging attacks in the vehicular network environment, making it difficult to have a general solution for all types of scenarios. It is noticed that the existing security contributions are not general for all vehicular network environments and are restricted mostly to ITS services or standalone ad hoc vehicular networks. Also, these contributions hardly comply with the main security requirements that are important to be considered in the design of security solutions for real deployments of vehicular networks, and hence they can restrict vehicular networks potential, ignoring its importance in providing ubiquitous communications. A point that complicates the problem is that securing vehicular communication is service-related. For instance, safety-related services should be granted to every vehicle on the road while assuring secure messages transfer. On the other hand, from a commercial deployment perspective, only authorized mobile clients (vehicles) should be granted network access and hence service access. Consequently, appropriate security mechanisms, for each scenario type, should be in place providing authentication, access control, and trust and secure communication between vehicles. These mechanisms should also take into consideration the dynamic and nonfully centralized nature of vehicular networks, as well as the different types of applications and their specific requirements. To achieve this, a number of security requirements should be satisfied (e.g., mutual authentication, message source authentication, high mobility support, and nonrepudiation).

When we mention vehicular communication security, one should separate between two terms: safety and security. Safety rather concerns peoples' safety on roads including drivers and passengers; while security concerns the secure data transfer. Consequently, securing intervehicular communication should take into account both peoples' safety and secure data transfer. Intelligent traffic can partially assure safety through minimizing accident possibility, warning people in case of dangers (e.g., when passing the speed limit or when approaching near foggy or snowy roads), and allowing collaborative driving. Nevertheless, the nonsecure data transfer has an impact on peoples' safety even when intelligent traffic is in place. One can conclude that peoples' safety and secure data transfer in vehicular of networks are two faces of the same coin, which is vehicular communication security.

Since vehicular networks can be managed by more than one network operator or service provider, authentication should be performed during mobile clients (vehicles) roaming not only across different base stations or access points but also across different administrative domains. This, in turn, necessitates trust relationships among the stakeholders for authentication, authorization, accounting, and billing of end users.

Finally, traceability versus privacy is an important point that needs efficient management. Efficient mechanisms and systems should be in place to allow moving vehicles traceability only by legitimate authorities, while protecting their privacy with respect to other vehicles through mainly preventing privacy disclosure attacks.

References

[1] M. Raya and J-P Hubaux, "The security of vehicular ad hoc networks," In *Proc. of ACM Workshop on Security of Ad hoc and Sensor Networks (SASN'05)*, Alexandria, VA, November 2005.

[2] C. Tchepnda, H. Moustafa, H. Labiod, and G. Bourdon, "A panorama on vehicular networks security," In *Proc. of International Workshop on Interoperable Vehicles (IOV2008) Associated with Internet of Things 2008 Conference*, Zurich, Switzerland, March 2008.

[3] B. Parno and A. Perrig, "Challenges in securing vehicular networks," *In Proc. of HotNets-IV*, College Park, MD, November 2005.

[4] X. Yang, J. Liu, F. Zhao, and N. Vaidya, "A vehicle-to-vehicle communication protocol for cooperative collision warning," In *Proc. of International Conference on Mobile and Ubiquitous Systems (MobiQuitous)*, Boston, MA, August 2004.

[5] P. Golle, D. Greene, and J. Staddon, "Detecting and correcting malicious data in VANETs," In *Proc. of ACM International Workshop on Vehicular Ad Hoc Networks (VANET)*, Philadelphia, PA, October 2004.

[6] F. Picconi, N. Ravi, M. Gruteser, and L. Iftode, "Probabilistic validation of aggregated data in vehicular ad-hoc networks," In *Proc. of ACM International Workshop on Vehicular Ad Hoc Networks (VANET)*, Los Angeles, CA, September 2006.

[7] H. Guo, J. P. Baugh, and S. Wang, "A group based secure and privacy preserving vehicular communication framework," In *Proc. of IEEE Conference on Computer Communications (Infocom'07)*, Anchorage, AK, May 2007.

[8] A. Frier, P. Karlton, and P. Kocher, "The SSL 3.0 Protocol," Netscape Communications Corp., 18 November, 1996.

[9] T. Dierks and C. Allen, "The TLS Protocol Version 1.0," RFC 2246, January 1999.

[10] E. Rescoria and N. Modadugu, "Datagram Transport Layer Security," RFC 4347, April 2006.

[11] Wireless Application Protocol (WAP) Forum, "Wireless Transport Layer Security," November 1999.

[12] R. Rivest, A. Shamir, and L. M. Adleman, "A method for obtaining digital signatures and public-key cryptosystems," *Communications of the ACM* 21(2) (February 1978):120–126.

[13] N. Koblitz, *A Course in Number Theory and Cryptography*, Springer-Verlag, 1987.

[14] NTRU Communications and Content Security, http://www.ntru.com.

[15] AES Algorithm (Rijndael) Information, http://csrc.nist.gov/CryptoToolkit/aes/rijndael.

[16] American National Standards Institute, "American National Standard for Information Systems-Data Link Encryption," ANSI X3.106, 1983.

[17] W. Tuchman, "Hellman presents no shortcut solutions to DES," *IEEE Spectrum* 16(7) (July 1979):40–41.

[18] R. Atkinson, "Security Architecture for the Internet Protocol," RFC 1825, August 1995.

[19] H. Moustafa and G. Bourdon, "Vehicular network deployment view: Applications, deployment architectures and security means," In *Proc. of International Workshop on ITS for Ubiquitous ROADS (UBIROADS'07) Associated with IEEE GIIS Conference*, Morocco, June 2007.

[20] Car-to-Car Communication Consortium (C2C-CC), http://www.car-to-car. org.

[21] DSRC Consortium, http://www.leearmstrong.com/DSRC/DSRCHomeset. htm.

[22] IEEE Std 1609.2-2006, IEEE Trial-Use Standard for Wireless Access in Vehicular Environments Security Services for Applications and Management Messages, July 2006.

[23] ATTI GPS Vehicle Tracking System, http://www.advantrack.com/GPS-Datalogger.htm.

[24] SEVECOM (Secure Vehicular Communication), http://www.sevecom.org.

[25] NoW (Network-on-Wheels), http://www.network-on-wheels.de.

[26] M. Zarki, S. Mehrotra, G. Tsudik, and N. Venkatasubramanian, "Security issues in a future vehicular network," In *Proc. of European Wireless*, Florence, Italy, February 2002.

[27] J.-P. Hubeaux, S. Capkun, and J. Luo, "The security and privacy of smart vehicles," *IEEE Security & Privacy Magazine*, 2004.

[28] M. Raya and J.-P. Hubaux, "Securing vehicular ad hoc networks," *Journal of Computer Security (JCS)—Special Issue on Security on Ad Hoc and Sensor Networks*, 2007.

[29] J. Blum and A. Eskandarian, "The threat of intelligent collisions," *IT Professional—IEEE Computer Society Periodical*, 2004.

[30] H. Moustafa, G. Bourdon, and Y. Gourhant, "Providing authentication and access control in vehicular network environment," In *Proc. of International Information Security Conference (IFIP SEC)*, Kadstad, Sweden, May 2006.

[31] H. Moustafa, G. Bourdon, and Y. Gourhant, "AAA in vehicular communication on highways using ad hoc networking support: a proposed architecture," In *Proc. of the Second ACM Workshop on Vehicular Ad Hoc Networks (VANET 05) Associated with MobiCom 05*, Cologne, Germany, September 2005.

[32] B. Aboba, L. Blunk, J. Vollbrecht, J. Carlson, and H. Levkowetz, "Extensible Authentication Protocol (EAP)," RFC 3748, June 2004.

[33] S. Zrelli and Y. Shinoda, "Specifying Kerberos over EAP: Towards an integrated network access and Kerberos single sign-on process," In *Proc. of IEEE 21st International Conference on Advanced Networking and Applications (AINA'07)*, Niagra Falls, Canada, May 2007.

[34] IEEE Std 802.11i, "Amendment 6: Medium Access Control (MAC) Security Enhancements," June 2004.

[35] C. Tchepnda, H. Moustafa, H. Labiod, and G. Bourdon, "Securing vehicular communications: An architectural solution providing a trust infrastructure, authentication, access control and secure data transfer," In *Proc. of IEEE AutoNet—Global Communications Conference (Globecom)*, San Francisco, CA, November 27–December 2 2006.

[36] C. Tchepnda, H. Moustafa, H. Labiod, and G. Bourdon, "Performance analysis of a layer-2 multi-hop authentication and credential delivery scheme for vehicular networks," In *Proc. of IEEE Vehicular Technology Conference (VTC-Spring)*, Singapore, May 2008.

[37] M. Gerlach, A. Festag, T. Leinmller, G. Goldacker, and C. Harsch, "Security architecture for vehicular communication," In *Proc. of International Workshop on Intelligent Transportation (WIT)*, Hamburg, Germany, May 2007.

[38] C. Harsch, A. Festag, and P. Papadimitratos, "Secure position-based routing for VANETs," In *Proc. of IEEE Vehicular Technology Conference (VTC-Fall)*, Baltimore, MD, October 2007.

[39] E. Coronado and S. Cherkaoui, "Secure service provisioning for vehicular networks," In *Proc. of International Workshop on ITS for Ubiquitous ROADS (UBIROADS'07)*, Morocco, June 2007.

Chapter 13

Confidence Management in Vehicular Networks

Véronique Cherfaoui, Thierry Denoeux,
and Zohra-Leïla Cherfi

Contents

Due to the particularities of the vehicular network (ad hoc network, dynamical nodes, and broadcast messages), the data dissemination could be based on the confidence of redundant and distributed information. A model of confidence management based on the belief function framework is described considering spatial dispersion of data sources, delays due to the multihop transmission, and dependency between sources. Preliminary results are presented based on simulated messages referenced on a real map data. This chapter is a revised version of the paper by Cherfaoui, Denoeux, and Cherfi [2].

13.1 Introduction

13.1.1 Principle of Confidence for Redundant and Distributed Data

In the last years, more and more communication devices have been embedded in vehicles. Many applications based on wireless communication have been developed in which the vehicles are the nodes of an ad hoc network called VANET. In the VANET, all vehicles broadcast messages and each vehicle has knowledge about its neighborhood only through the messages it receives. Most research papers deal with communication protocols and routing and congestion problems. Due to the nature of applications (driving assistance systems or emergency braking alert), recent works have been dedicated to the security mechanism to avoid malicious node intercepting or modifying or sending erroneous data [7,8,9]. Supposing these problems to be (partially) solved, we propose a method to manage and exploit message information from the receiver node point of view.

In this chapter, we consider messages regarding safety such as accident, reduced visibility, traffic jam, and so forth, and we consider car-to-car (C2C) communications. It is the context of the SAFESPOT project [19,20] and the cognitive car [10]. Each car (node) is able to detect, localize, date, and characterize an event, and, if necessary, broadcast it in a message. Due to the multihop transmission protocol, the distance between sender and receiver nodes is not limited by the transmission power of antennas.

To localize and date the content of a message, we assume that each node is equipped with a global positioning system (GPS) receiver. When a node receives a message, it updates its database, and if necessary it broadcasts its updated information.

13.1.2 Confidence Management as a Solution for Information Dissemination in Vehicular Networks

The problem of information dissemination, that is, proposing a strategy to broadcast information, is one of the key issues of the C2C communications [13,18]. Indeed, the road traffic can be high, the bandwidth is limited, and the number of exchanging messages would have to be reduced. Different strategies have been proposed in the literature [6,1,17] concerning this problem. It should also be noted that algorithms for combining and fusing data are very different from algorithms developed for infrastructure vehicle (V2I) communication applications. In the latter case, a centralized module combines collected data and disseminates global information.

The objective of the work reported in this chapter is to develop a methodology for combining data included in messages arriving from other nodes. Since data are uncertain and could be the result of processing disseminated data, we focus on confidence management in a distributed and dynamical context. The confidence could be exploited to provide the driver with relevant information or to decide about the transmission of the result in the network. This contribution is intended to be a part of the information dissemination strategy to be developed in a decision process.

This work is based on the use of belief functions to combine degrees of confidence about events reported in exchanged messages. We first define the attributes of each message and then describe the methodology to combine data coming from distributed, dynamical, and asynchronous sources.

13.2 Knowledge Representation and Belief Functions

13.2.1 Knowledge Representation

The transferable belief model (TBM) [15] is a model to represent quantified believes based on belief functions [14]. It has the advantage of being able to explicitly represent uncertainty about an event. It takes into account what remains unknown and represents what is already known.

Let Ω be a finite set of all possible solutions of a problem. Ω is called the frame of discernment (also called state space). It is composed of mutually exclusive elements. The knowledge held by a source of information can be quantified by a belief function defined from the power set 2^Ω to $[0,1]$. Belief functions can be expressed in several forms. The basic belief assignment (bba) denoted m, credibility function bel, the plausibility function pl, and the commonality function q, which are in one-to-one correspondence. We recall that $m(A)$ quantifies the part of belief that is restricted to the proposition "the solution is in $A \subseteq \Omega$" and satisfies $\sum_{A \subseteq \Omega} m(A) = 1$.

Thus, a bba can support a set $A \subseteq \Omega$ without supporting any subproposition of A, which allows accounting for partial knowledge. The complete notation of a belief function is $m_{S,t}^{\Omega}\{X\}[BC_{S,t}](A)$ $A \subseteq \Omega$, where S is the information source, t the time of the event, Ω the frame of discernment, X a parameter which takes value in Ω, and BC the evidential corpus or knowledge base. This formulation represents the degree of belief allocated by the source S at time t to the hypothesis that X belongs to A. The notation is simplified in the following paragraph to clarify the combination formulas. In section 13.3, the complete notation including time, source, and parameter will be used.

Smets [15] introduced the notion of open world where Ω is not exhaustive; this is quantified by a nonzero value of $m(\emptyset)$. The other functions can be calculated from the bba m using the following formulas:

Credibility function: $bel^{\Omega}(A) = \sum_{\emptyset \neq B \subseteq A} m^{\Omega}(B)$
Plausibility function: $pl^{\Omega}(A) = \sum_{A \cap B \neq \emptyset} m^{\Omega}(B)$
Commonality function: $q^{\Omega}(A) = \sum_{B \supseteq A} m^{\Omega}(B)$

Another function that can be computed from q is the conjunctive weight function [16] defined by

$$w(A) = \prod_{B \supseteq A} q(B)^{(-1)^{|B| - |A| + 1}}$$

The w function is well-defined if m is nondogmatic, that is, if $m(\Omega) > 0$. Functions bel, pl, w and m are in one-to-one correspondence. In particular, formula to recover m from w are given by Denoeux [5] and Smets [16].

13.2.2 Information Fusion

Let n distinct pieces of evidence be defined over a common frame of discernment and quantified by bbas $m_1^{\Omega}...m_n^{\Omega}$. They may be combined using a suitable operator. The most common ones are the conjunctive and disjunctive rules of combination defined, respectively, as

$$m_{\cap}^{\Omega}(A) = \sum_{A_1 \cap ... \cap A_n} m_1^{\Omega}(A_1)...m_n^{\Omega}(A_n)$$

$$m_{\cup}^{\Omega}(A) = \sum_{A_1 \cup ... \cup A_n} m_1^{\Omega}(A_1)...m_n^{\Omega}(A_n)$$

The resulting bbas should be normalized under the closed world assumption. Dempster's rule [3] denoted by $\oplus$ normalizes the result of the conjunctive rule with $K = \frac{1}{1-m_\cap^\Omega(\emptyset)}$ and sets the mass on the empty set to 0. The conjunctive and disjunctive rules of combination assume the independence of the data sources. Denoeux [4,5] introduced the cautious rule of combination (denoted by $\bigwedge$) to combine dependent data. This rule has the advantage of avoiding double-counting of common evidence when combining nondistinct bbas. In particular, the combination of a bba with itself yields the same bba: $m = m \bigwedge m$ (idempotence property). The cautious rule of combination can be easily computed by taking the minimum of conjunctive weights, with obvious notations, $w_{1 \bigwedge 2} = w_1 \wedge w_2$, where $\wedge$ denotes the minimum operator.

13.2.3 Reliability and Discounting Factor

The belief function framework makes it possible to model the user's opinion about the reliability of a source [8]. The idea is to weight more heavily the opinions of the best source and conversely for the less reliable ones. The result is a discounting of the bba m^Ω produced by the source, resulting in a new bba $m^{\Omega,\alpha}$ defined by

$$m^{\Omega,\alpha}(A) = \alpha \cdot m^\Omega(A), \forall A \subseteq \Omega, A \neq \Omega$$

$$m^{\Omega,\alpha}(\Omega) = 1 - \alpha + \alpha \cdot m^\Omega(\Omega)$$

The discounting factor α can be regarded as the degree of trust assigned to the *source*.

13.3 Content of Exchanged Messages

13.3.1 Level of Information

Safety applications in VANET are being investigated to increase the vehicle visibility area and produce useful information in view of developing (advanced driver assistance system) (ADAS) functions. The level of exchanged data depends on the applications. The concept of cognitive car [10] assumes that vehicles communicate in crossing roads to avoid collision. Traffic information applications are based on the vehicle positions and speed exchange [12]. We propose to exchange data concerning a set of events to increase the "visibility" of the driver and to allow the anticipation of dangerous situations. These events are classified into three categories:

■ Static (or slowly evolving) and localized events such as accident (AC), working area (WA), or dangerous object (DO).

- Dynamical (quickly evolving) and localized events such as an animal on the road (AN), a counter sense vehicle (CV), or a dangerous vehicle (DV).
- Slowly evolving and diffuse events such as low visibility (LV), traffic jam or congestion zone (CZ), or low adherence area (LA). These events concern a whole geographical area.

In this work, it is assumed that vehicles are equipped with systems able to detect these events.

13.3.2 Spatial and Temporal References

When a node receives a message, it has to decide whether it is relevant according to the node location and the node itinerary. Knowing that an event is geolocalized with GPS, and assuming the node has a numerical map, it is possible to associate an event with a road segment. A road segment is an entity in a geographical information system (GIS) database. Each road segment is determined by a unique *Road ID*. Figure 13.1 describes the geometrical definition of a road segment. It is connected with other segments at the origin and end extremities.

This approach has three main advantages:

- The spatial data association between two events is made easier.
- The space representation by roads is discrete. Consequently, two messages geolocalized on the same *Road ID* concern the same event.
- It is possible to assign prior knowledge to each road segment. For example, frequent fog reported in an area can be associated to segments in this area.

As mentioned in the introduction, the time between the creation and reception of a message can be higher than transmission delay. This is due to the multihop and retransmission capabilities of communicating cars. Consequently, two attributes should be defined for dating an event. One for

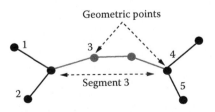

Figure 13.1   Definition on a road segment in GIS.

time stamping the event when it was detected and another for specifying the date of updating if the message was created from the combination of other messages.

13.3.3 Confidence Attributes

To analyze road situation according to set of events, we study the confidence of these events according to the confidence assigned by the vehicle that has detected the event, the redundancy of messages concerning this event, and the date and location attributes. We propose to model the confidence with belief functions to take into account the uncertainty of initial data. The operators defined in the belief function framework can be applied to compute a unique belief function by combining belief masses coming from many sources. A message describing the event *ev* is assumed to contain a mass function on $\Omega = \{0, 1\}$. The value 0 represents $\neg ev$ and 1 represents *ev*. The mass function *m* can then be represented as a quadruplet $[m(\emptyset), m(0), m(1), m(\Omega)]$. The value $m(\Omega)$ is interpreted as the degree of doubt and the value $m(\emptyset)$ represent the degree of conflict between sources.

There are two ways to initialize the mass function. First, when a vehicle *vehI D* detects an event ev at time *t* on road *Road ID*, it computes a degree of certainty *d*. This value is used to define the bba $m^{\Omega}_{vehID, roadID, t}\{ev\}$ as

$$m^{\Omega}_{vehID, roadID, t}\{ev\}(\emptyset) = 0$$

$$m^{\Omega}_{vehID, roadID, t}\{ev\}(0) = 0$$

$$m^{\Omega}_{vehID, roadID, t}\{ev\}(1) = d$$

$$m^{\Omega}_{vehID, roadID, t}\{ev\}(\Omega) = 1 - d$$

When a vehicle *vehID* predicts an event *ev* on a road segment *Road ID* and it does not detect it, it assigns a confidence value d' to *ev* and builds the following mass function for *ev*:

$$m^{\Omega}_{vehID, roadID, t}\{ev\}(\emptyset) = 0$$

$$m^{\Omega}_{vehID, roadID, t}\{ev\}(0) = d'$$

$$m^{\Omega}_{vehID, roadID, t}\{ev\}(1) = d$$

$$m^{\Omega}_{vehID, roadID, t}\{ev\}(\Omega) = 1 - d'$$

The confidence values d and d' are set based on the reliability of the detection system (driver observation, sensor processing).

Table 13.1   Message Attributes

Attribute	Description
ev	Type of event
roadID	Road ID for localization of event
subSeg	# of subsegments (for fineing localization)
coordgps	GPS coordinate
vehID	Vehicle ID having detected the event
src	Binary value indicating if the message content is the result of detection or the result of processing disseminated data
tacquisition	Time of event perception
tlastUpdate	Last updating date
m	Mass function (bba)

Our aim is to improve the level of knowledge from exchanged messages. The belief of the distributed sources is expressed by their mass functions. Distributed data fusion then consists of combining these masses with the appropriate operator. The conjunctive rule will be used when assuming the independence of messages, whereas the cautious rule will be applied in case of dependent messages. To determine which operator should be applied when combining information contained in a message with other information, an attribute *src* is set to 1 when the message is original and set to 0 when it results from the combination of other messages.

The contents of message are summarized in Table 13.1.

13.4   Message Information Combination

13.4.1   Temporal Persistence

The observed system is composed of events evolving in time and space. The delay between emission and reception can be small or large according to the routing and propagation algorithms. The confidence in a message also depends on the age of the received event.

Belief masses should thus be modified according to the delay between the date of data processing and the date of data emission. The objective is to maintain data consistency with or without new messages. Indeed, to maintain a high level of confidence about an event, new messages confirming this event are needed. Without such confirmation, the confidence should decrease. We propose to define a discounting factor γ according to the time difference $\Delta t = t_{\text{current}} - t_{\text{lastUpdate}}$ and a value $\rho(ev)$ depending on the event *ev*:

$$\gamma = \exp[-\Delta t / \rho(ev)]$$

The value $\rho(ev)$ characterizes the persistence of event *ev*. For example, $\rho(LV)$ is high while $\rho(AN)$ is shorter. Temporal extension is thus performed by a discounting operation:

$$m_{t_{\text{current}}}(A) = \gamma \cdot m_{t_{\text{lastUpdate}}}(A), \ A \neq \Omega$$

$$m_{t_{\text{current}}}(\Omega) = 1 - \gamma + \gamma \cdot m_{t_{\text{lastUpdate}}}(\Omega)$$

Notice that we have simplified the notation because the referential, event, road, and source are constant in this part of the algorithm.

13.4.2 Spatial Propagation

Message combination in ad hoc networks should also take into account spatial properties of the observed events. We can assume that some events observed on road segment S_i hold as well at positions close enough to S_i. It is true, for example, for weather observations. The size of the neighborhood depends on the type of the diffuse event. A previous approach to plausible reasoning from spatial observations was proposed by Lang and Muller [12]. These authors consider an observation point *o* (e.g., the current position of the vehicle) and try to infer beliefs about what holds in *o* from the properties of the other road segments $x_i(i = 1, \ldots, n)$. Their model of spatial persistence is based on an extrapolation of observation $m_{x_i \to o}$ calculated by discounting the mass function m_{x_i}, with a factor β:

$$\beta = \exp\left[-\frac{d(x_i, o)}{\lambda(ev)}\right]$$

$$m_{x_i \to o}(A) = \beta \cdot m_{x_i}(A) \ A \neq \Omega$$

$$m_{x_i \to o}(\Omega) = 1 - \beta + \beta \cdot m_{x_i}(\Omega)$$

where $d(x_i, o)$ is the distance between the focus point *o* and the road segment x_i and $\lambda(ev)$ represents the degree of spatial persistence. The belief at the focus point *o* is then the Dempster combination of the $m_{x_i \to o}$:

$$m_o = \oplus_{i:1..n} m_{x_i \to o}$$

However, the problem of the dispersion in space of points $x_i(i = 1, \ldots, n)$ has to be considered. Figure 13.2 shows two different situations. Since the

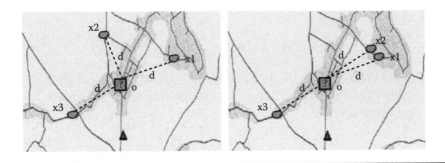

Figure 13.2 Dispersion configurations.

points x_1 and x_2 are close in the second configuration, their influence on point o should be reduced.

To remedy this problem, Lang and Muller propose to introduce a discounting factor when combining mass assignments. The discount rate grows with the proximity and thus the dependency between the points where observations have been made. The discounting factors are calculated based on geometrical criteria. We propose to make use of the numerical map to define the dependency between points x_i (defined, for example, by GPS coordinates). As previously mentioned, an event is localized on the numerical map and a road identifier is associated to it. The *Road ID* attribute can thus be used to define if events are located on the same portion of road. In this case, the masses are considered as nonindependent and combined using the cautious rule proposed by Denoeux [5]. If the messages containing the masses come from roads with different *Road IDs*, they are combined with the normalized conjunctive rule proposed by Dempster's rule [3]. These two rules are associative and commutative. The algorithm is then:

$$m_o = m_{x_i \to o}$$

`tabRoadID` $\leftarrow$ `roadID`(x_i)

`for` $i : 2..n$

    `if`   `roadID`$(x_i) \in$ `tabRoadID`

        $m_o = m_o \oslash m_{x_i \to o}$

    `else`

        `tabRoadID` $\leftarrow$ `roadID`(x_i)

        $m_o = m_o \oplus m_{x_i \to o}$

Here again the notation has been simplified because the referential, parameter, time, and source are constant.

13.4.3 Prior Knowledge

Using the numerical map to support the database makes it easy to attach prior information to road segments. For example, a congestion zone can be identified in some urban area, frequent fog condition can be observed near the wet zone, and so on. This information can be fused with extrapolated and combined data before a decision is made.

13.4.4 Global Algorithm

As the vehicle is moving continuously, two approaches can be considered for processing a received message. The first one is a message-triggered approach in which each message is processed when it arrives at the node. Since the frequency and number of messages are unknown, it is difficult to guarantee that all messages will be processed. We prefer the second approach, referred to as the road segment triggered approach, in which each received message is kept in memory. When the node moves on a road segment S_v, all messages in a specified neighborhood are processed to compute the belief in the situation on S_v. The neighborhood can be defined according to an area around the current position or according to the itinerary of the vehicle.

In the context of VANET, it is unrealistic to assume that messages are independent. Information could be relayed and completed by the nodes in the network. As mentioned previously, some message attributes can determine if two messages are independent or not. We can consider that independence can be assumed in the following cases:

1. Two messages sent by two different nodes with attribute *src* equal to 1. (*src* = 1 means that information is acquired by the node.)
2. Two messages sent by the same node on two different dates are regarded as independent if the node has made two distinct acquisitions. *src* equal to 1 and $t_{Acquisition}$ values are different.

In the other cases, the messages are processed as coming from dependent sources. The global algorithm is described in Figure 13.3. The gray box has not been implemented yet and is left for future work. The strategy to elaborate a decision concerning the segment and send the corresponding message is beyond the scope of this paper.

13.5 Preliminary Results

To test this approach, the above algorithm has been implemented in Matlab. The messages are simulated on the basis of a real numerical map (NavTeQ). A module extracts roads from the map in a specified area [11]. It uses the

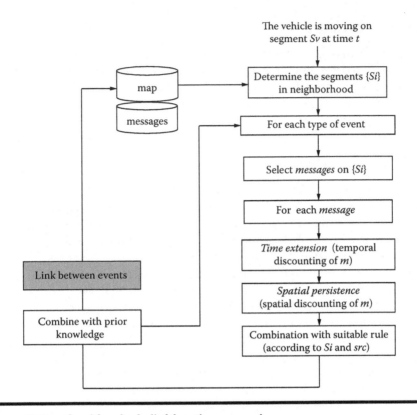

Figure 13.3   Algorithm for belief-function processing.

Benomad kit. Examples in this section are made with a map cache extracted from the GIS with a radius equal to 4 km.

13.5.1   Spatial Extension

To illustrate the spatial propagation we propose to simulate seven messages concerning the event "low visibility" (LV). Indeed, low visibility corresponds to weather conditions like fog or hard rain. These phenomena are spatially diffuse and can be spatially propagated. The first example shows the results of the combination of messages coming from seven distinct road segments situated in two distinct areas. Figure 13.4 shows the map for low visibility messages. The vehicle is on segment *V*. The fog area is localized in the gray area. *Road ID* of messages are shown. Figure 13.5 shows the bba contained in each message, and Figure 13.6 shows the spatial discounting on segment *V*: remains that 0 represents -LV and 1 represents LVevent.

The resulting belief function for LV event on segment *V* is given in Figure 13.7 and was calculated with normalized operators (mass on empty

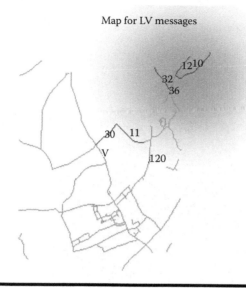

Map for LV messages

Figure 13.4   Example of low visibility (LV) messages distributed on real map.

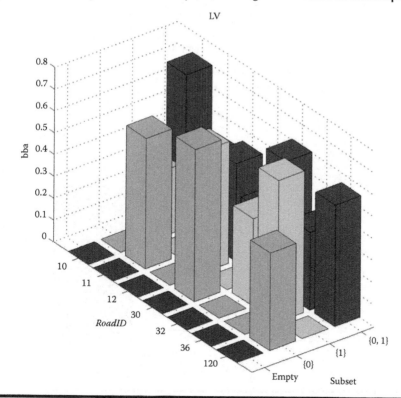

Figure 13.5   Initial bbas of message localized on roadId segment.

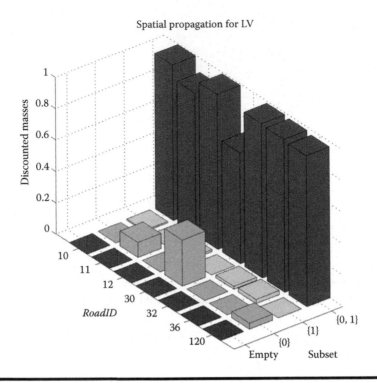

Figure 13.6   Discounted bbas by spatial propagation.

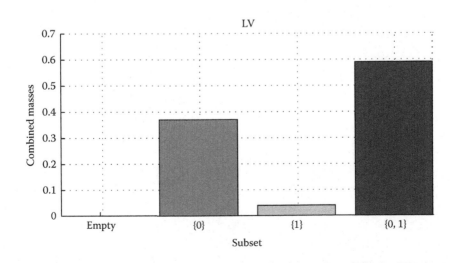

Figure 13.7   Resulting combination of bbas concerning LV messages.

Figure 13.8 **Example of spatial dispersion for seven messages localized on only three road segments.**

set was used to normalize). The attenuation factor computed from the distance between the current position of the vehicle and GPS data in messages discount efficiently the $m(1)$ values.

The next scenario concerns the problem of spatial dependency. To illustrate it, the mixed cautious or conjunctive rule was compared with conjunctive-only combination in the case where messages are localized on the same segments. Figure 13.8 shows the *Road ID* of segments with messages. The vehicle is on segment *V*. Figure 13.9 shows the bba contained in each message, and the spatial discounting on segment *V* according to the *Road ID* localization is shown in Figure 13.10. The result of the mixed cautious or conjunctive rule approach is compared with the use of conjunctive rule in Figure 13.11.

The use of cautious rule when messages are colocalized limits the reinforcement of confidence values as compared to well-distributed sources. The well-distributed sources are represented by the belief function resulting from the conjunctive rule only. This approach has a lower computational complexity than the spatial dispersion method [12].

The main difficulty for the implementation of this method is the definition of parameter λ. The spatial diffusion of real events is never constant and depends on a lot of context-dependent conditions.

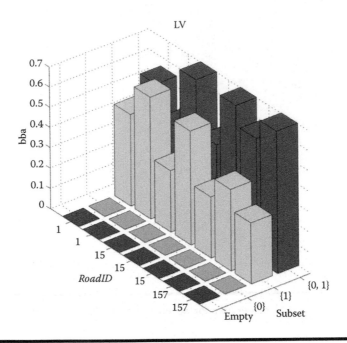

Figure 13.9 **Initial bbas of message localized on *Road ID* Segment.**

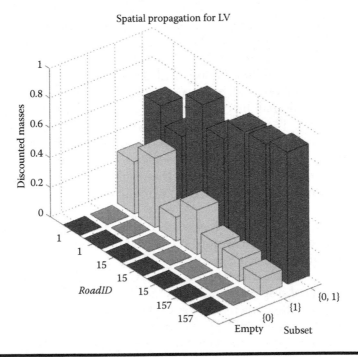

Figure 13.10 **Discounted bbas by spatial propagation.**

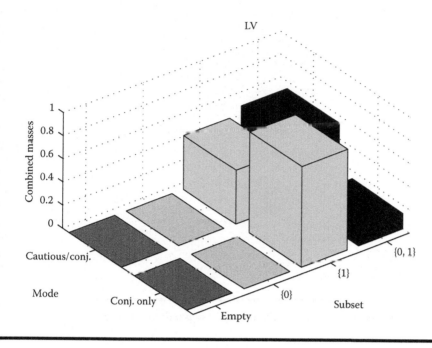

Figure 13.11   Resulting combination on segment V: comparison between two different strategies.

13.5.2   Source Dependency

The next scenario illustrates the global algorithm (Figure 13.12). It simulates messages concerning an event LV coming from a group of vehicles exchanging data in an area. Messages are relayed by other vehicles (thanks to multihop protocol). A subset of messages (localized at 11, 120, and 30 *Road ID*) are labeled with *src* = 0 (not original perception of event). Figure 13.13 gives the initial bbas. The Figure 13.14 gives the bbas after spatial discounting on segment *V* according the *Road ID* localization and *m*.

Figure 13.15 shows the belief functions on segment *V*. The result of our approach is compared with the use of the conjunctive rule only without taking into consideration the dependence between messages. Thanks to the cautious rule, all messages can be processed according to their distinctness. The unnormalized result (openworld) gives the degree of conflict (between LV and not LV) with $m(\emptyset)$. This value could be used in the decision process. The behavior of the cautious rule is highlighted in this example.

13.5.3   Temporal Discounting

Figure 13.16 shows an example of global combination with a message localized on segment *V* (current road segment). The message were dated with

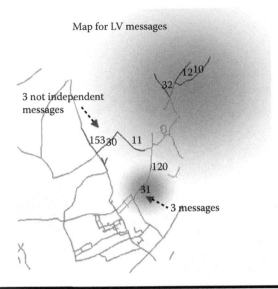

Figure 13.12 **Example of scenario of redundant and dependent messages.**

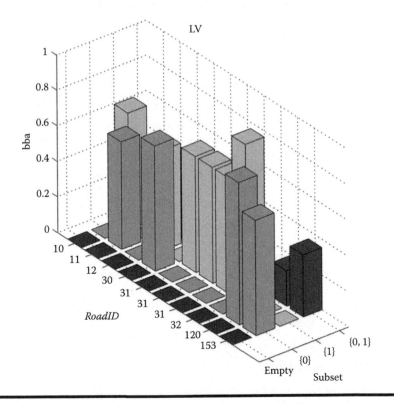

Figure 13.13 **Initial bbas of message localized on *Road ID* Segment.**

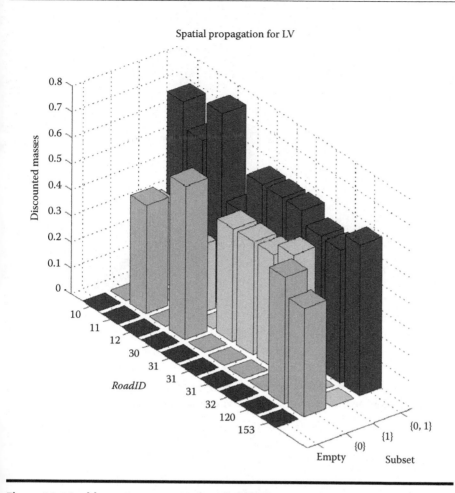

Figure 13.14 bbas on segment *V* after discounting according to the *Road ID* localization.

tAcquisition $= t$Current $- 3600s$. The *roadID* is the current segment V and $src = 1$. Temporal discounting was performed on the bba of this message, before combining it with the result reported in Figure 13.15.

The implementation of this algorithm in real conditions can be envisaged provided GPS data (position and global clock) are available. The discounting factor based on the decay function can be roughly estimated for different kinds of dynamical events. However, like the spatial parameter λ, we can already assume that an implementation in real conditions will require fine tuning of the γ parameter, as the life duration of an event is context-dependent.

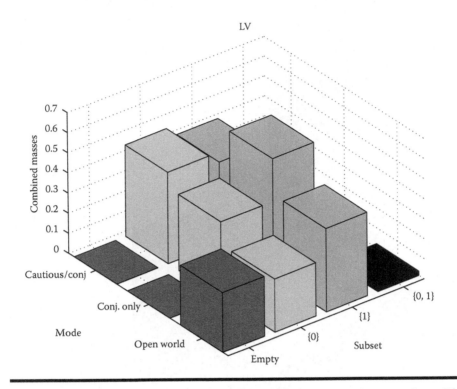

Figure 13.15   Resulting combination on segment *V*: comparison between three different strategies.

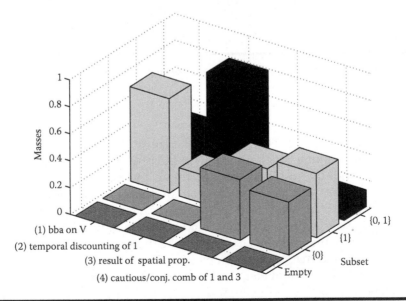

Figure 13.16   Resulting combination on segment V with temporal extension.

13.6 Conclusions

A distributed data fusion method for combining the confidence in ad hoc and dynamical networks has been presented. The method is based on belief functions and implements a strategy to combine confidence in messages. The preliminary results are promising, and this approach is still in the process of being developed to validate the principle. A decision rule and a sending message strategy have to be defined before considering more complex simulations, for example, the relation between events (A congestion event is often the consequence of an accident event). Coupling this program with an ad hoc network simulator will be a real added value for validation. This method presented in this chapter is receiving node point of view and it is well adapted for coupling this information with sending node confidence and contribute to be more robust to Sybil node attacks.

References

[1] C. Adler, R. Eigner, C. Schroth, and M. Strassberger, "Context-adaptive information dissemination in VANETs—Maximizing the global benefit," In *Proc. of Communications Systems and Networks*, 2006.

[2] V. Cherfaoui, T. Denoeux, and Z.-L. Cherfi, "Distributed data fusion: Application to confidence management in vehicular networks," In *Proc. of the 11th International Conference on Information Fusion*, Cologne, Germany, 2008.

[3] A. Dempster, "Upper and lower probabilities induced by multivalued mapping," *In Annals of Mathematical Statistics*, 38:325–340. American Statistical Association, 1967.

[4] T. Denoeux, "The cautious rule of combination for belief functions and some extensions," In *Proc. of the 9th International Conference on Information Fusion* 33(114) (2006):1–8.

[5] T. Denoeux, "Conjunctive and disjunctive combination of belief functions induced by non distinct bodies of evidence," *Artificial Intelligence* 172 (2008): 234–264.

[6] B. Ducourthial, Y. Khaled, and M. Shawky, "Conditional transmissions: A communication strategy for highly dynamic vehicular ad hoc networks," *IEEE Transactions on Vehicular Technology* (special issue on vehicular communication networks) 56(6) (November 2007):3348–3357.

[7] J. Douceur, "The Sybil attack," In *Proc. of First International Workshop on Peer-to-Peer Systems*, March 2002.

[8] Z. Elouedi, K. Mellouli, and P. Smets, "Assessing sensor reliability for multisensor data fusion within the transferable belief model," *IEEE Transactions on Systems, Man and Cybernetics, Part B* 34(1) (February 2004): 782–787.

[9] G. Guette and B. Ducourthial, "On the Sybil attack detection," In *Proc. of VANET International Workshop on Mobile Vehicular Networks (MoveNet 2007), co-located with IEEE MASS 2007*, Pisa, October 2007.

[10] A. Heide and K. Henning, "The 'cognitive car': A roadmap for research issues in the automotive sector," *Annual Review in Control* 30(2006): 197–203.

[11] M. Jabbour, P. Bonnifait, and G. Dherbomez, "Real-time implementation of a GIS-based localization system for intelligent vehicles," *EURASIP Journal on Embedded Systems* (ID 39350), 2007

[12] J. Lang and P. Muller, "Plausible reasoning from spatial observations," In *Proc. of the 17th Conference in Uncertainty in Artificial Intelligence* Scattle, WA, (August 2001):285–292.

[13] T. Nadeem, S. Daschtinezhad, C. Liao, and L. Iftode, "Trafficview: Traffic data dissemination in car-to-car communications," *ACM SIGMOBILE Mobile Computing and Communication Review Special Issue Mobile Data Management* 8(3) (2004):6–19.

[14] G. Shafer, *A Mathematical Theory of Evidence*. Princeton, NJ:Princeton University Press, 1976.

[15] P. Smets, "The combination of evidence in the transferable belief model," *IEEE Transactions on Pattern Analysis and Machine Intelligence* 12(May 1990):447–458.

[16] P. Smets, "The canonical decomposition of a weighted belief," In *Proc. of Int. Joint Conf. on Artificial Intelligence*: 1896–1901, San Mateo, CA, 1995.

[17] L. Wischof, A. Ebner, and H. Rohling, "Information dissemination in self-organizing intervehicle networks," In *Proc. of IEEE Transportation on Intelligent Transportation Systems* 6(1) (March 2005).

[18] H. Wu, R. Fujimoto, R. Guensler, and M. Hunter, "MDDV: A mobility-centric data dissemination algorithm for vehicular networks," In *Proc. of 1st ACM VANET*, New York: ACM Press, 2004.

[19] The Cooperative Intersection Collision Avoidance System (CICAS) initiative, http://www.its.dot.gov/cicas.

[20] The SAFESPOT project, http://www.safespot-eu.org.

Chapter 14

Geocast in Vehicular Networks

*Andreas Festag, Wenhui Zhang, Long Le,
and Roberto Baldessari*

Contents

The use of geographical positions for addressing and routing of data packets, commonly referred to as Geocast, has received attention from academia and industry. Originally proposed for mobile ad hoc networks, the concept was further refined. Feasibility studies indicate that Geocast is well suited as a candidate network protocol for vehicular ad hoc networks based on IEEE 802.11. This chapter presents a basic variant of Geocast and explains extensions of Geocast for various aspects, including reliability and efficiency of data transport, security and privacy, and Internet integration. These extensions represent state-of-the-art enhancements in the design of the Geocast protocol. The extended Geocast is also a comprehensive solution for a VANET network protocol supporting safety and infotainment applications in realistic environments. The overall solution can be regarded as a technical basis for future field operational tests of vehicular communication in Europe and ongoing standardization efforts.

14.1   Introduction

Geocast is a network protocol for ad hoc networks based on short-range wireless technology, such as WLAN IEEE 802.11. It provides wireless multihop communication in mobile environments without the need of a coordinating infrastructure as in cellular networks. Geocast utilizes geographical positions for data dissemination and packet transport. It offers communication over multiple wireless hops, where nodes in the network forward data packets on behalf of each other and extend the limited communication range of the short-range wireless technology. Originally proposed for general mobile ad hoc networks (MANETs), variants of Geocast have been proposed for other network types, such as vehicular ad hoc networks (VANETs), mesh networks, and wireless sensor networks (WSN). Therefore, Geocast can also be regarded as a family of network protocols based on usage of geographical positions for addressing and transport of data packets in different types of networks.

In VANETs, Geocast provides wireless communication among on-board units for vehicle-to-vehicle communication and among on-board and roadside units for vehicle-to-infrastructure communication. Geocast works connectionless and fully distributed based on ad hoc network principles, with intermittent or even without infrastructure access. It provides wireless multihop communication over multiple OBUs and RSUs to extend the coverage of the ad hoc network. The principles of Geocast meet the specific requirements of vehicular environments. First, it is well suited for highly mobile network nodes and frequent changes in the network topology. Second, Geocast flexibly supports heterogeneous application requirements, including applications for road safety, traffic efficiency, and infotainment. More specifically, it enables periodic transmission of safety status messages at high rate, rapid multihop dissemination of packets in geographical regions for emergency warnings, and unicast packet transport for Internet applications.

The principles of Geocast are simple and can be intuitively explained. Geocast basically provides two, strongly coupled functions: *geographical addressing* and *geographical forwarding*. Unlike addressing in a conventional network in which a node has a communication name linked to its identity (e.g., a node's IP address), Geocast can address a node by a position or address multiple nodes in a geographical region (geo-address). For forwarding, Geocast assumes that every node has a partial view of the network topology in its vicinity and that every packet carries a geo-address, such as the geographical position or geographical area as the destination. When a node receives a data packet, it compares the geo-address in the data packet and the node's view on the network topology, and makes an autonomous forwarding decision. As a result, packets are forwarded "on the fly," without need for setup and maintenance of routing tables in the nodes.

The origins of Geocast can be traced back to the 1980s, where geographical routing concepts were introduced as an alternative to topological routing in infrastructure networks [1,2]. *Greedy perimeter stateless routing* (GPSR) [3] can be regarded as one of the first Geocast approaches for routing of unicast packets in mobile ad hoc networks. Since then, many extensions of Geocast for general mobile networks have been published [4,5] and specific enhancements for Geocast in VANETs proposed [6–9]. Due to the availability of positioning devices and the widespread use of WLAN technology at low costs, Geocast received interest from academia and industry. It was studied in several R&D projects, such as *FleetNet—Internet on the Road* [10,11], *NoW—Network on Wheels* [12], and is further investigated in *GeoNet* [13], *SAFESPOT* [14], and the *PRE-DRIVE C2X* for preparation of field tests of vehicular safety systems. Geocast is also promoted by industry consortia, such as the *Car-2-Car Communication Consortium)* [15,16], and considered in standardization, as in *ETSI TC ITS* [17] and *ISO TC 204* [18].

This chapter presents a variant of Geocast, which is specifically designed and tailored for vehicular ad hoc networks. Starting from the basic Geocast protocol, we introduce advanced mechanisms and algorithms to enhance the basic Geocast, including per packet radio control, fair resource usage for periodic messages, reliable and efficient data dissemination in geographical areas, security, privacy, and Internet integration. Many of the advanced mechanisms have already been introduced in previous chapters of the book and are incorporated into the proposed Geocast variant as a comprehensive and well-balanced solution.

The chapter is structured as follows. We first identify requirements and scenarios for Geocast in VANETs in Section 14.2. After the introduction of the fundamental principles of Geocast in Section 14.3, we describe an architecture framework and present advanced mechanisms to enhance the basic Geocast for the specific VANET requirements in Section 14.4. Section 14.5 gives a summary.

14.2   Scenarios and Requirements

We categorize the basic communication scenarios in VANETs according to the following criteria:

> **Type of communication endpoints**[1]
> ■ Vehicle-to-vehicle communication (V2V)
> ■ Infrastructure-to-vehicle communication (I2V)
> ■ Vehicle-to-infrastructure communication (V2I)
>
> **Connection types**
> ■ Point-to-point
> ■ Point-to-multipoint
> ■ GeoBroadcast: Distribution of packets to all nodes in a geographical area
> ■ GeoAnycast: Distribution of packets to any node in a geographical area
>
> **How applications access the Geocast layer?**
> ■ Direct: Applications directly over Geocast, for example, safety and traffic-efficiency applications
> ■ Indirect: Applications on top of IP over Geocast

The combination of different criteria and modes result in requirements to be met by the protocol design for Geocast. Direct and indirect access of applications needs to be supported for all types of communication end points V2V, V2I, and I2V. The combination of connection types and types of communication end-points is illustrated in Figure 14.1 and shows that V2V and I2V should be possible for P2P, P2MP, GeoBroadcast, and GeoAnycast, but for V2I only P2P and GeoAnycast are reasonable. Based on these scenarios, we summarize the functional requirements for Geocast as follows:

> ■ Efficient dissemination of safety and traffic efficiency messages
> ■ Transparently transport of IP packets with minimum changes to IP
> ■ Support of privacy and security functions
> ■ Support of different communication media and interfaces

---

[1] The combination of different end-point scenarios is possible, for example V2I + I2V.

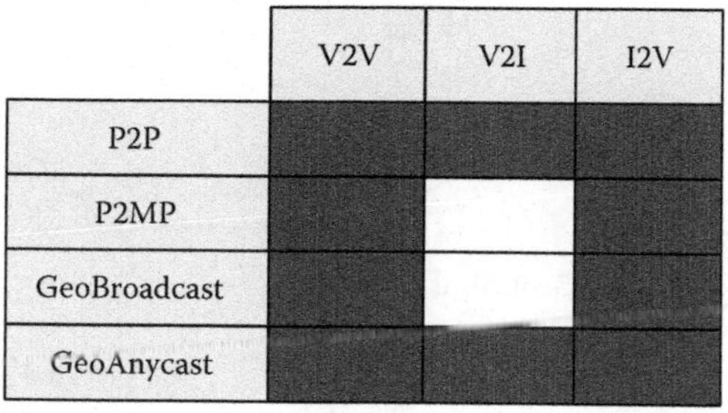

	V2V	V2I	I2V
P2P			
P2MP			
GeoBroadcast			
GeoAnycast			

Figure 14.1 Combination of scenarios.

The dissemination of safety messages in VANETs typically have performance-related requirements on latency, reliability, and dissemination area. For safety messages, Geocast needs to

- Provide low-latency communications
- Provide reliable communications with the highest reliability for safety messages
- Keep signaling, routing, and packet forwarding overhead low
- Be fair among different nodes with respect to bandwidth usage considering the type of messages
- Be robust against security attack and malfunction in communication nodes
- Be able to work in scenarios with low and high density of Geocast-enabled vehicles

14.3   Basic Geocast

In general, the basic Geocast offers a geographical addressing and best-effort packet transport between source and destination based on packet switching concepts. The packet transport does not assure any service quality in terms of packet delivery delay, sequence of packet delivery, or packet loss. Communication is purely sender-oriented, where the sender (more precisely the source) determines both the destination address (individual node or geographical area) and the type of packet transport.

In principle, Geocast assumes that every network node knows its geographical position, for example, by GPS or another positioning systems, and maintains a location table of geographical positions of other nodes as soft state. It supports point-to-point and point-to-multipoint communication. The latter case can be regarded as group communication, where the end points are inside a geographical region.

Core protocol components of basic Geocast are beaconing, location service, and forwarding. With *beacons*, also referred to as *heartbeats*, all nodes

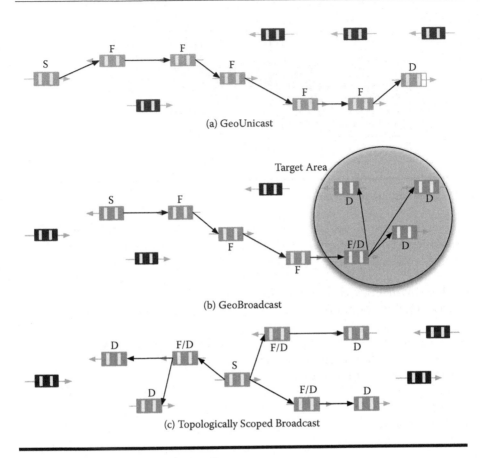

(a) GeoUnicast

(b) GeoBroadcast

(c) Topologically Scoped Broadcast

Figure 14.2   Basic Geocast forwarding types.

periodically broadcast short packets with their IDs, current geographical positions, speeds, and heading. On reception of a beacon, a node stores the information in its location table. The *location service* resolves a node's ID to its current position. When a node needs to know the position of another node, which is currently not available in its location table, it issues a *location query* packet with the searched node ID, sequence number, and hop limit. Upon receiving this packet, neighboring nodes rebroadcast it until it reaches the searched node (or the hop limit). If the received *location query* packet is not a duplicate, the searched node will reply with a *location reply* packet carrying its current position and timestamp. On reception of the *location reply*, the originating node updates its location table. *Forwarding* basically means relaying a packet toward the destination, and we distinguish

- *Geographical unicast (GeoUnicast)* (Figure 14.2a) provides packet transport between two nodes via single or multiple wireless hops. When a node wishes to send a unicast packet, it first determines the destination position (by means of location table look-up or the location service) and forwards the data packet to a node in the direction of the destination, which in turn reforwards the packet along the path until the packet reaches the destination.

■ *Geographical broadcast* (*GeoBroadcast*) (Figure 14.2b) distributes data packets by flooding, where nodes rebroadcast a packet if they are located in the geographical region determined by the packet. Packet is forwarded only once. If a node receives a data packet, which it had received previously, the packet is dropped. *GeoAnycast* is similar to *GeoBroadcast* but addresses a single (i.e., any) node in a geographical area.

■ *Topologically scoped broadcast* (TSB) (Figure 14.2c) provides re-broadcasting of a data packet from a source to all nodes in n-hop neighborhood. Single-hop broadcast is a specific case of TSB, which is used to send *heartbeats* including application data payload.

For optimization, a node processes all data packets on the wireless links, regardless of being forwarder or destination,[2] to keep track of the nodes in its surrounding. Also, each data packet carries the source's and previous forwarder's positions at the expense of a moderate packet overhead and updates its location table accordingly. Consequently, Geocast defines packet headers with fields for node identifier, position, and timestamp for source, sender, and destination, and more.[3] In the header we distinguish between immutable and mutable fields. *Immutable* fields are not altered in the forwarding process, while *mutable fields* can be updated by forwarders. This allows a forwarder to alter header fields on the fly, for example, if it has more recent information in its location table about a given destination.

Another optimization is the use of feedback from the MAC protocol layer beneath the Geocast protocol to provide the link status to individual neighbor nodes. For example, a MAC "excessive retry" indicates that the connectivity to a node is lost, and the corresponding entry can immediately be removed from the location table.

For forwarding, basically two algorithms are applied: *greedy forwarding* for GeoUnicast and simple GeoBroadcast.

Greedy forwarding: Greedy forwarding assumes that (i) every node is aware of its own geographical position, and (ii) every node knows the position of its direct neighbors, which is obtained by means of the periodic *beacons*. Before two nodes can communicate, the source node determines the current position of the destination

---

[2] It requires to run a network interface card in promiscuous mode.
[3] The originator of a packet is referred to as source, and the last forwarder as sender.

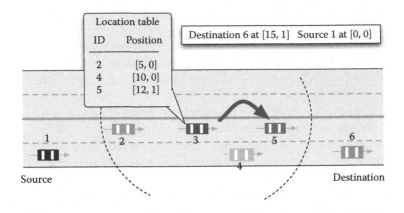

Figure 14.3   Principle of greedy forwarding: Node 3 selects node 5 as the next hop.

node. This is accomplished by a location service, which maps an arbitrary node identifier to its current position if multihop connectivity to the node exists. The main principle of greedy forwarding is as follows (Figure 14.3). On reception of a data packet that needs to be forwarded, the forwarding node calculates the distance between all direct neighbors and the destination and determines the neighbor that is closer to the destination. The *most-forward within radius* (MFR) policy [19] selects the node with the minimum remaining distance to the destination's position.

Simple GeoBroadcast: With simple GeoBroadcast, the source of a packet defines a geographical area, approximated by a geometrical shape (e.g., circle or rectangle), appends the information to the packet header and broadcasts the packet to all neighbors. The receiving node executes two main steps. First, it checks whether it has already received the packet based on the source ID and sequence number carried in the packet header and drops the packet if it is a duplicate. Second, the node rebroadcasts the packet if geographic position is inside of geographical target area. If the node is outside, the packet is forwarded toward the center position of the target area by means of greedy forwarding. It is worth noting that simple GeoBroadcast faces some problems, such as redundant retransmissions [20] and network fragmentation if nodes lack connectivity inside of the geographical target area.

14.4 Advanced Mechanisms and Algorithms for Geocast in VANETs

The preceding section has introduced the basic principles and concepts of Geocast in VANETs. Next, we describe advanced mechanisms and algorithms that enhance the basic Geocast and adopt it to the requirements and scenarios from Section 14.2. In detail, the mechanisms and algorithms include

- Radio control on a per packet basis (Section 14.4.2)
- Efficient bandwidth utilization and fairness for periodic messages (Section 14.4.3)
- Efficient and reliable dissemination of safety information in geographical areas (Section 14.4.4)
- Data dissemination in sparse networks (Section 14.4.5)
- Data security (Section 14.4.6)
- Privacy by means of changing and revocable pseudonyms (Section 14.4.7)
- Integration of Geocast with the Internet protocol IP version 6 (Section 14.4.8)

Before we explain the mechanisms and algorithms, we present an architecture framework for Geocast in VANETs.

14.4.1 Architecture Framework for Geocast

The assumed system architecture involves several entities and network domains, as depicted in Figure 14.4. A car is equipped with an *on-board unit* (OBU) that implements the communication protocol stack. OBUs in different cars can communicate with each other or with fixed network nodes installed along roads termed *roadside units* (RSUs). OBUs and RSUs implement the same protocol functionality and form a self-organizing network, here referred to as the *ad hoc domain*. OBUs and RSUs differ from each other with respect to the networks they are attached to. OBUs offer an interface to driver and passenger devices—called *application units* (AUs)—present in a car. The mobile network, composed of AUs, defines a domain termed *in-vehicle domain*. An RSU can either be standalone or attached to a larger structured network. In the first, *standalone* case, an RSU distributes information (safety or other) and simply extends the VANET's coverage by acting as forwarding entity. In the latter, *attached* case, an RSU distributes information toward or from a remote entity (e.g., road traffic management center). An RSU can also interconnect the vehicular network to an infrastructure network and the Internet, which is generally referred to as *infrastructure domain*.

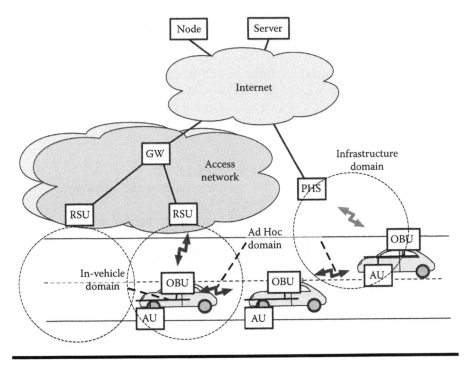

Figure 14.4   System architecture.

For the protocol stack, we assume that applications for road safety and infotainment have fundamentally different communication requirements. Safety applications typically disseminate information about events or other vehicles in the local vicinity or a geographical region. Vehicles either broadcast short status messages periodically and with high frequency (so called *awareness messages*), or they generate messages when they detect a safety event and distribute them by multihop communication in a certain geographical area (*event-driven messages*). In contrast, infotainment applications typically establish sessions and exchange unicast data packets in greater numbers, bidirectionally, and over multihop. Consequently, the OBU protocol stack in Figure 14.5 depicts basically a two-column protocol stack, dedicated to road safety and traffic efficiency (left) and infotainment applications (right). For safety applications, Geocast provides ad hoc communication among OBUs, as well as among OBUs and RSUs over IEEE 802.11p radio. Infotainment applications access the traditional IP protocol stack and can use the ad hoc and multihop capabilities of Geocast as a sub-IP layer. The information connector, drawn vertically in Figure 14.5, offers efficient and structured information exchange among the protocol layers.

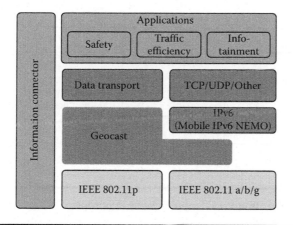

Figure 14.5   Protocol architecture.

The protocol architecture assumes that the two stacks are not fully integrated. The loose coupling of the two protocol columns allows safety applications to transceive safety information via the stack for infotainment applications, and infotainment applications to access data structures (e.g., a location table) of the safety stack. Moreover, it facilitates the definition of a basic system (where only the safety stack is present) and an extended system (for both, safety and infotainment).

14.4.2   Radio Control

An important goal of vehicular networks is to disseminate safety messages with high reliability and low latency. These requirements, the time-varying characteristics of the wireless medium, the limited available bandwidth allocated to safety application, and the highly dynamic nature of vehicular networks pose a great challenge to the design and implementation of vehicular networks. For this reason, effective and flexible control of hardware radio is necessary. Based on the allocation of a dedicated frequency band for safety and traffic efficiency in Europe, the concept of radio parameter control for IEEE 802.11 on a per packet basis is presented.

In Europe, the frequency band 5.875–5.925 GHz will be allocated for road safety and traffic efficiency [21]. This frequency band will be divided into two parts 5.875–5.905 GHz and 5.905–5.925 GHz for an initial and a later deployment phase. Further, the frequency band 5.855–5.875 GHz will be allocated to nonsafety ITS applications. An overview of the spectrum allocation for ITS applications in Europe is illustrated in Figure 14.6.

For the initial deployment phase of vehicular networks in Europe, attention has been focused on the frequency band 5.875–5.905 GHz allocated

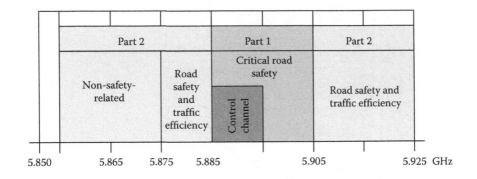

Figure 14.6   Spectrum allocation for ITS applications in Europe.

for road safety and traffic efficiency. A recent paper provided a comprehensive overview of existing approaches for the usage of this frequency band [22]. Advantages and disadvantages of these approaches were analyzed using an extensive set of evaluation criteria: usability, robustness, cost, efficiency, scalability, and development effort. The analysis recommended dividing the allocated frequency band into three 10-MHz channels: one control channel (CCH) and two service channels (SCH1 and SCH2). Further, it considered two usage schemes for operating these three channels: CCH + SCH1 + SCH2 with two transceivers (Scheme A) and CCH + 2 × SCH with a single transceiver operating channel switching (Scheme B). The analysis compared the two schemes based on the aforementioned evaluation criteria and recommended the scheme with two transceivers and low transmit power on SCH2 (Scheme A). This usage scheme is depicted in Figure 14.7.

Various effects such as fading, collisions, interference, and multipath signal propagation can cause packet losses on a wireless channel. The IEEE 802.11 protocol counters unreliable wireless channels by implementing a semireliable mechanism for packet transmissions. After transmitting a data frame, the sender expects an ACK frame from the recipient. If the sender does not receive the ACK frame within a short time interval, it retransmits the data frame. The sender continues to retransmit the data frame until the number of retransmissions reaches a configurable parameter. We note that

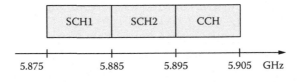

Figure 14.7   Proposed channel allocation for safety-related frequency band in Europe.

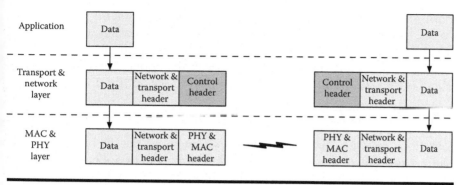

Figure 14.8   Control header for per-packet control.

this semireliable mechanism only works for unicast transmissions because a receiver does not send an ACK frame for broadcast data frames.

In summary, three mechanisms for radio control are required to provide efficient support of Geocast: multichannel operation, semireliable packet transmission, and five-grained transmit power control. These three mechanisms require fine-tuned control of radio interface on a per-packet basis the radio control allows an upper protocol layer, that is, Geocast, to append a control header to a data packet when it is passed to the IEEE 802.11 MAC and vice-versa. The control header specifies various radio parameters for both sending and receiving side.

As illustrated in Figure 14.8, on the sender's side, the network layer passes data frames and a control header down to the MAC layer. The control header contains several parameters such as transmit power, data rate, and number of retransmissions. These parameters specify how data frames are to be sent on the wireless channel. The MAC layer removes the control header before transmitting the data frames on the wireless channel.

On the receiver's side, the MAC layer prepends a control header to the received data frames and passes them to the network layer. The control header contains the data rate and received signal strength of the received data frames. These parameters provide information for the network layer to adjust their mechanisms, for example, power-aware greedy forwarding. Figure 14.8 illustrates how the MAC and network layer handle the control header.

The per packet control is regarded as a technical means to implement advanced networking algorithms. One example algorithm is *power-aware greedy forwarding* [25], which extends greedy forwarding to limit interference among nodes and to achieve spatial reuse of wireless channels. The algorithms adjust the transmit power for each packet in a multihop forwarding chain, such that every forwarder uses the minimum transmit power needed to reach the next hop. The algorithm requires every node to maintain and update the channel gain and interference level to each of its neighbors. After a forwarder selects the next hop among its neighbors on a greedy basis, the transmit power necessary for a successful packet reception at the next hop is calculated. The packet is transmitted with the calculated transmit power. Another important example for the application of per packet control is for load control of periodic broadcasts (see Section 14.4.3).

14.4.3 Periodic Broadcasting

Many applications for road-safety rely on the exchange of periodic broadcast messages among all nodes in the surrounding area. The periodic messages carry position, speed, heading, and other information to inform other cars about their actual status. The periodic messages are broadcast as beacons or as single-hop broadcast packets.[4] They are typically sent at a high rate and, thus, can consume a considerable portion of the available wireless bandwidth in medium to high road traffic densities.

Due to the limited bandwidth resources allocated to safety applications and the broadcast nature of the wireless medium, it is crucial to avoid scenarios where the wireless channel is congested and safety messages cannot be delivered. Thus, effective mechanisms for load control in vehicular networks are necessary. In principle, for load control two concepts can be identified: (i) transmit power control and (ii) control of the packet generation rate. Both concepts require distributed algorithms and are based on per packet radio control introduced in Section 14.4.2.

> **Transmit power control (TPC):** In order to control the load from periodic broadcasts on the wireless channel, TPC can adjust the transmission power and hence control the radio coverage for the packet. One of the presented algorithms, *distributed fair power assignment in vehicular ad hoc networks* (D-FPAV) [23], defines a *maximum beacon load* (MBL), which must not be exceeded by the periodic broadcasts. In addition to the load threshold, D-FPAV considers fairness as an important safety aspect. It means that every vehicle needs to get assigned a fair share of the wireless bandwidth that can be consumed by periodic broadcasts by all nodes. Basically, D-FPAV achieves fair power assignment by piggybacking information about nodes in the carrier sense range[5] of each node in the periodic broadcast. Based on the information about interfering nodes and an estimation of the transmission range of the own broadcast, every node adjusts its transmit power such that the available bandwidth is assigned following the max-min fairness criterion (Figure 14.9).

---

[4] The latter packets carry additional application data as payload.

[5] The carrier sense range is the distance between two nodes in which a transmission of one of them can be sensed by the other one, that is, it prevents the second node from accessing the channel.

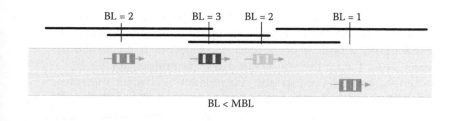

Figure 14.9 Use of transmit power control to adjust the beacon load to a maximum.

Control of the packet generation rate: Alternatively to the use of TPC, the load of periodic broadcasts can be controlled by adjusting their generation rate. With MHVB [24], every node monitors the number of neighbors in its communication range. If this number exceeds a threshold, the packet generation rate is reduced to a value that is inverse-proportional to the number of neighbors.

It is worth noting that load control of periodic broadcasts is part of an overall concept for congestion control that comprises different mechanisms operating at different protocol layers.

14.4.4 Multihop Communication

Basic Geocast provides two primary forwarding types (Section 14.3): The first is forwarding based on flooding, that is, to forward a packet by means of broadcasting the packet toward the destination. The second is forwarding a packet along a selected path, for example, using greedy forwarding to forward a packet along a route to the destination. For both types, several improvements and enhancements have been proposed, from which we present relevant ones.

The simple GeoBroadcast scheme presented in Section 14.3 leads to the well-known *broadcast storm* problem, which features a large number of redundant rebroadcasts, increased contention and collision, and increased channel load [20]. To cope with this problem, flooding could be limited to certain areas. For example, with the *restricted directional flooding*, the sender of a packet restricts flooding by forwarding the packet only to one-hop neighbors in the direction of the destination [8]. A similar approach is also used by *location-aided routing* [9]. In addition to limiting the flooding area, various mechanisms have been proposed [20]. For example, with the *probabilistic scheme*, a communication node will rebroadcast a received packet only with a certain probability. With the *counter-based scheme*, a node will rebroadcast a packet only if the number of receptions of the same packet is less than a threshold value. Using the *distance-based scheme*, a node will rebroadcast a packet if it does not receive any packet from a node at a distance less than a threshold value after a random delay. Applying the *location-based scheme*, a node will rebroadcast a packet if its additional coverage area is larger than a threshold value after a random delay.

These flooding algorithms reduce the redundancy of packet transmissions to a certain degree, but cannot eliminate them. With the IEEE 802.11 MAC, a broadcast frame may be received by many nodes. These nodes may forward the received packets simultaneously if they are not well coordinated. Simulations show that the random delay has to be properly tuned to maximize packet delivery ratio and reduce end-to-end delay using these flooding algorithms [26].

To disseminate messages in VANETs quickly, efficiently, and reliably, the *contention-based forwarding* (CBF) scheme was proposed [27]. The CBF scheme assumes that each node knows its own position, and a node sends a packet as broadcast to all its neighbors. Upon reception of the packet, all neighboring nodes will contend for forwarding the packet by waiting for a locally calculated contention time. The node with the maximum progress to the destination will have the shortest contention time and rebroadcast the packet first. Other nodes will receive the packet and hence cancel the rebroadcast. Without losing generality, the waiting time WT may be formulated as

$$\mathrm{WT}(d) = \mathrm{MaxWT}(1 - d/\mathrm{range}) \qquad (14.1)$$

where d is the distance to the sender of the packet, MaxWT is the predefined maximum contention time and *range* is the communication range or the maximum forwarding range. CBF is efficient since it allows each node to set its waiting time in a distributed way, and nodes with more progress in the dissemination direction will have less waiting time thus will win the contention of forwarding.

The concept of CBF has also been used for unicast packet forwarding as a substitute of greedy forwarding scheme (Section 14.3). If CBF is used for unicast forwarding, the sender of a packet does not make the decision to select the next hop, instead, possible forwarding nodes themselves will contend to be the forwarder. Such CBF-based unicast algorithms have been reported by FüBler et al. and FüBler [28,29], and also in a similar approach called *beacon-less routing* [30]. Both approaches also propose to reduce duplicated rebroadcast by confining the forwarding area so that only nodes in limited areas in the forwarding direction will join the contention [28,30]. This is useful in areas with high node density, but may also have the consequence that no node will join the contention in areas with low node density.

In the ideal case, duplicated retransmissions with CBF are possible only if potential forwarding nodes have the same progress. However, with the

IEEE 802.11 MAC, retransmission is still possible if potential forwarding nodes do not have the same progress but are near to each other. This is because these nodes may have a very small difference in their contention time due to the small distance to each other, thus if one nodes sends out a packet, other nodes may already send the same packet to the MAC layer before the packet is correctly received.

In principle, the longer the maximum contention time, the less the number of duplicated packets sent on the wireless link. However, a longer maximum contention time will also lead to long end-to-end latency. Therefore, there is a trade-off between the maximum contention time and the possible duplicated retransmission.

Compared with simple GeoBroadcast, CBF avoids the signaling overhead due to beacons and thus saves bandwidth. It deals well with the uncertainty of packet reception since multiple nodes may join the contention process and failed reception of a packet at a node will only prevent itself from joining the contention. It does not suffer from the outdated neighboring entries and neighbor positions, thus avoiding excessive retransmissions at the MAC layer in comparison with greedy forwarding. With reduced packets transmitted over the air, CBF also encounters less collision. Therefore, CBF also provides a promising forwarding algorithm for GeoUnicast.

To apply CBF for dissemination of critical safety messages in VANETs, an improved CBF scheme called *emergency message dissemination for vehicular environments* (EMDV) has been proposed [23]. EMDV takes beacons for granted on a communication channel for safety applications because beacons are needed by safety applications for cooperative awareness.

To minimize the delay of packet delivery, CBF is complemented with an explicit selection of forwarder by the sender of a packet at transmission. The selection of the forwarder is done in the same way as greedy forwarding selects the next hop forwarder. If the packet is correctly received by the selected forwarder, it will immediately forward the packet without any delay, and the forwarded packet will suppress the forwarding of other nodes. If the selected forwarder fails to receive the packet, other nodes will use the CBF scheme to start the contention for forwarding. In this way greedy forwarding is combined with CBF, and CBF functions as a fall-back solution of greedy forwarding.

For increase of the dissemination reliability, two algorithms have been proposed. First, the maximum forwarding range is configured shorter than the theoretical communication range. The range is adjusted according to the reception probability of single-hop broadcast so that a relatively high reception probability near the boundary of the forwarding area can be achieved. Second, a packet retransmission scheme within the dissemination area is applied, which allows a controlled number of redundant retransmissions.

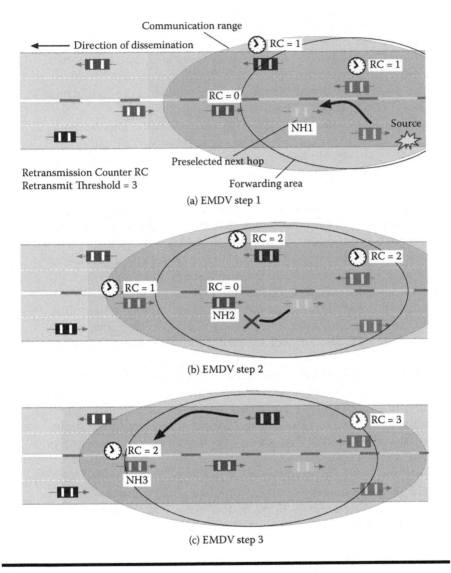

Figure 14.10 Explanation of EMDV in an example scenario.

A counter, referred to as retransmission counter RC, for a packet will be incremented each time the same packet is received during the contention time, and a node will exit the forwarding process if the counter reaches a predefined threshold. In this way, a very high reception probability is ensured.

Figure 14.10 explains EMDV in an example highway scenario, where a node detects a hazard and distributes a packet in the opposite driving direction. The source node selects NH1 as next hop within its forwarding range and broadcasts the packet. Nodes receiving the packet, cache it, set the retransmit counter RC (RC = 1) and start a timer (Figure 14.10a). RC = 0 in Figure 14.10a indicates that the node has not received the data packet correctly. As shown in the next Figure 14.10b, NH1 immediately rebroadcasts selecting NH2 as next hop, but the packet to NH2 is again lost (RC = 0 for NH2). The other nodes in the transmission range increment their RC. Next, the retransmit timer (Figure 14.10c) in the upper node expires and the node rebroadcasts the packet to improve the robustness of the protocol. RC is incremented each time a packet is overheard and when RC reaches the threshold (RC = 3), it is discarded.

14.4.5 Caching

Temporary caching of data packets can potentially improve the reliability of packet delivery, particularly in scenarios with a low density of vehicles. In certain situations, nodes may cache data packets and retransmit these packets when suitable new neighbors enter the transmission range of the caching node. Caching can apply to GeoUnicast and other Geocast packets as explained below.

The caching of GeoUnicast packets may prevent packet loss in case of forwarding failures: A forwarding node caches a packet when no further vehicle closer to the destination is available (no greedy route). In this case, the node caches the packet without transmitting it. When a new neighbor occurs, the caching node checks whether this new neighbor is closer to the destination. If so, the previously unroutable packet is forwarded by selecting the new neighbor as next hop.

For GeoBroadcast another type of caching can be applied. It attempts to keep information inside a geographical area alive for a certain validity duration. A GeoBroadcast packet may not reach all nodes inside the target area due to network partition in a sparse scenario or further vehicles may enter the target area during the lifetime of the event after the initial distribution. Consequently, the retransmission of packets upon detection of new neighbors inside the geographical area may inform all nodes inside a target area when connectivity resumes or inform nodes that enter the geographical area later.

14.4.6 Security

The security extensions of Geocast protect the protocol from misconfiguration, attacks, and misuse. In general, common security objectives such as integrity, authentication, nonrepudiation, and (for selected use cases) confidentiality are also applicable to Geocast operations. Specific security aspects of Geocast are related to the use of positions, where an attacker could potentially manipulate and broadcast its own position or forge

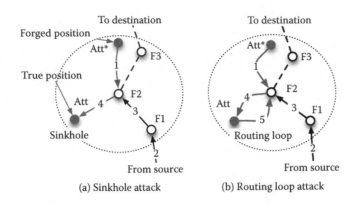

(a) Sinkhole attack   (b) Routing loop attack

Figure 14.11   Attack on Geocast.

positions carried in the Geocast packet headers. These attacks may disrupt communication or result in a routing loop as illustrated in Figure 14.11. A systematic attack analysis is presented by Bochow et al. [31].

The security extensions for Geocast rely on the combination of a set of measures, that is, cryptographic protection, plausibility checks, rate limitation, and trustworthiness assessment of nodes [32]. For the use of cryptographic primitives, we assume a public key infrastructure (PKI) with a certification authority (CA), which issues public or private key pairs and certificates to vehicles. A certificate contains a node's public key, attribute list (to distinguish between regular vehicles from, e.g., road-side units), the CA identifier, the certificate lifetime, and the CA signature.[6]

14.4.6.1   Cryptographic Protection

We use asymmetric cryptography and digital signatures for Geocast packets. In the case of *beacons* a single signature is applied, with the source node signing the whole Geocast packet. This is straightforward since there are no intermediate nodes that change Geocast header fields. In contrast, for multihop communication, additional protection is necessary for the mutable fields in the Geocast headers (see Section 14.3 for a definition of mutable and immutable fields).

An end-to-end signature by the packet's source can only cover the immutable fields. To enhance the protocol robustness, we propose a combination of hop-by-hop (neighbor-to-neighbor) and end-to-end (source-to-destination) security and to protect the packets with two signatures: the *source signature*, calculated by the source over the immutable fields (and

---

[6] For anonymity support we assume that a vehicle is equipped with a set of credentials (see Section 14.4.7.)

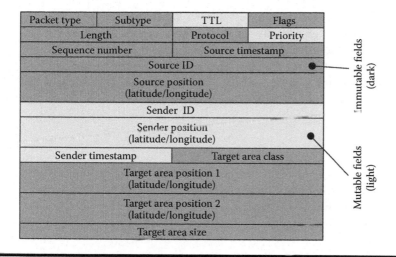

Figure 14.12   Example Geocast packet header with mutable and immutable fields.

the payload of the packet), and the *sender signature*, generated by each sending node over the mutable fields (see Figure 14.12 for an example packet).

As illustrated in Figure 14.13, on reception of a packet, a forwarding node verifies both the source and sender signature. Then, it updates the

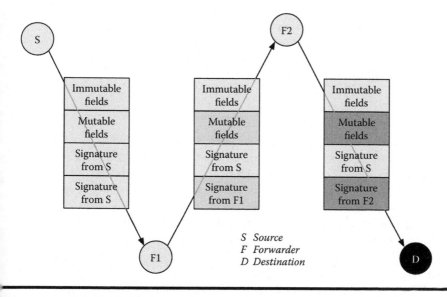

Figure 14.13   Hybrid signature scheme: a packet holds source and sender signature/certificate.

mutable field values and generates a new sender signature, replaces the old signature by the new one, and finally reforwards the packet. The destination node verifies both the sender and source signatures.

It is worth noting that the hybrid signature scheme can be applied to any Geocast packets that are sent over multiple hops, such as data packets (GeoUnicast, GeoBroadcast, and others), and also location service query and response.

14.4.6.2 Plausibility Checks

In principle, the plausibility checks compare received data values with some expected value by means of heuristics. On reception of a packet, a node executes the different checks in sequence. Below is a selection of checks, similar to those presented by Leinmüller et al. [33].

- Time: A timestamp is checked for being within a time window. It ensures that the timestamp is neither outdated (to prevent replay attacks), nor lies in the future. In fact, nodes update their location table only if the information of the *Geocast* header is newer. The check ensures that an attacker cannot alter the destination position of a multihop packet and imposes this information to all subsequent forwarders by setting the timestamp to a future value.
- Acceptance range: Assuming that communication devices have a maximum transmission range Δ_{max}, no neighbor can be further away than Δ_{max}.
- Velocity: The maximum velocity of vehicles is limited by natural law to v_{max}. Therefore, a claimed position update needs to be within a predicted space window, calculated around the node's previous position and a radius of $\Delta_{time} \times v_{max}$ (Δ_{time} is the time between two position updates).

14.4.6.3 Rate Limitation

As the injection of false multihop Geocast packets wastes resources of a large network part, mechanisms according to the attributes of the sending node limit their data rate. If the rate of such traffic originating from a node exceeds a protocol-specific threshold, its packets are not forwarded any further. Digital signatures and unique identification (source timestamps) of the sender and the transmissions allow this throughout the network. To exert even tighter control, yet maintain effectiveness, we define distinct thresholds for different types of nodes. Furthermore, the description of the transmission (e.g., the target geographical area) can correspond to different thresholds. For example, private vehicles can be disallowed to initiate Geocast packets beyond a given area size and allowed to do so at the lowest rate, while RSUs or emergency vehicles can do the same for larger areas and at higher rates.

14.4.6.4 Trustworthiness Assessment

The trustworthiness in a certain node is assessed by plausibility checks. Plausible data in a packet causes an increase of the trust value, inplausible data decrease it. In practice, trustworthiness is expressed as a variable assigned to a node identifier and maintained in a node's local database (location table). The trust value is between a minimum and maximum value defined to 0 and C_{max}. Starting from a predefined default value, the trust value is incremented or decremented for every executed plausibility check up to a maximum or minimum. The trust threshold determines the minimum value at which a node is considered to be trustworthy. When a node has a trust value smaller than the threshold, it is *untrusted*, and the following restrictions apply to those nodes: (i) Packets originated or forwarded from an untrusted node are not reforwarded and delivered to the application, but dropped. (ii) Packets with an untrusted node as destination are not forwarded, but dropped. (iii) Information about other nodes, carried in the Geocast header and signed by the node, are not accepted. Once a node is untrusted, it is isolated from the network. In order to reintegrate the node in the network, the node needs to send packets with plausible data. When the trust value assessed by the neighbor nodes exceeds the threshold, the neighbors start forwarding data packets from and to the node.

Figure 14.14 shows the sequence of security operations for a forwarding node. A source node applies only the last two steps, a destination node all steps, except the last two. In case any of the checks fail, the packet is discarded and subsequent steps are not executed.

14.4.7 Privacy

By use of Geocast, nodes publicly disclose data, such as node address, position, speed, heading, and time since these data are an integral part of the protocol. An attacker can potentially link the data to the user's identity and invade the privacy of the driver or other users, even against the user's will. Encryption of all those data is no option as the data represents a basic requirement for network operation and hence have to be available to all nodes.

The key mechanism to enforce privacy in Geocast is the use of pseudonyms. Basically, the pseudonym represents a node in the VANET. It is utilized as a routing address and carried by the Geocast packet sent over the wireless links. This implies that a pseudonym has removed all information which allows to link the pseudonym to the identity of the node or user. To prevent tracking, a node changes its pseudonym frequently.

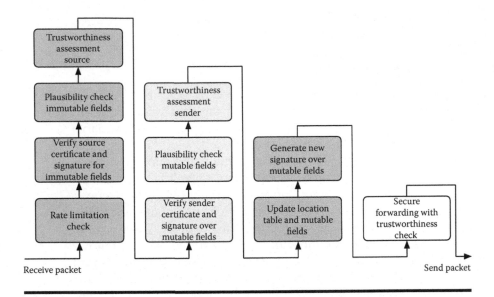

Figure 14.14   Sequence of security operations for a multi-hop event-driven message in a forwarder node.

When applying pseudonyms with Geocast, the following aspects need to be considered [34]

Cross-layer addressing and information exchange: In the layered protocol architecture (Figure 14.5), a node uses multiple addresses at different protocol layers simultaneously. If a pseudonym is changed only at one layer, an adversary can link two pseudonyms by the unchanged address at another protocol layer. Consequently, all addresses across a node's protocol stack should be changed at the same time. To generate Geocast addresses, we utilize the concept of cryptographically generated addresses (CGA) [35]. A CGA is an IPv6 address where the host part is defined by the hash value of the public key. This links the public key to the IPv6 address. This concept is adapted to other protocol layers, where the same hash value of the public key is also converted to a MAC address and an address for geographical routing (see Figure 14.15). In this way, all addresses are linked to the currently used pseudonym. The main benefit lies in the process of pseudonym change. By generating a single hash value of the actual pseudonyms, the addresses on all protocol layers can be changed simultaneously and privacy is provided on all layers.

Resolution service for pseudonyms: To establish a communication session among nodes, the communication peers need to identify each other. Therefore, Geocast offers a service that resolves

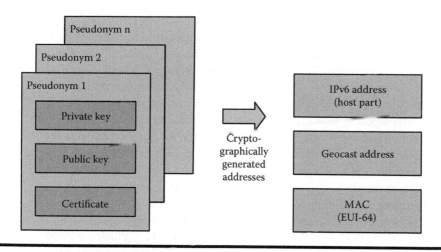

Figure 14.15   Cross-layer address assignment scheme.

the true identifier to the current pseudonym. To preserve the users' privacy, the resolution service discloses the link between ID and pseudonym only to trusted nodes and communication peers.

Pseudonymity-enhanced packet forwarding: Due to frequent topology changes in VANETS, it frequently happens that nodes have outdated entries in their location table. From the perspective of a certain node, a pseudonym change of another node cannot be distinguished from the case that a new neighbor leaves its transmission range. In both situations, the node would have an outdated entry for forwarding of unicast packets, which might result in a wrong forwarding decision and packet loss. Arguing that the change of addresses aggravates the problem of stale entries in the location tables, pseudonymity can significantly degrade system performance by increasing unicast packet loss. The problem can be alleviated by enhancement of the forwarding scheme [34]. A node keeps its previous pseudonyms for a small duration of time in addition to its current pseudonym. Then, a forwarder can still use the old pseudonym of its neighbor for packet forwarding. Eventually, the enhanced forwarding scheme eliminates stale entries and reduces packet loss, with reasonable compromises for the privacy.

14.4.8   Internet Integration

The Internet protocol suite was originally designed by following a packet-centric topological approach, which meets the needs of interconnecting networks with heterogeneous technologies and delivering information irrespective of its content. The Internet protocol suite was designed for generic purposes, whereas Geocast has to meet specific requirements. For this reason, while it is possible to extend IP to support Geocast, this approach is rather complex and inefficient. Therefore, Geocast is regarded as an independent network protocol, which offers more design freedom to fulfill VANET requirements. The support for IP-based communication, though, is fundamental for VANETs. The remarkable number of noncritical applications envisaged in automotive scenarios (traffic notification, remote diagnostic, point-of-interest notification, multimedia, etc.) and their potentially large-scale deployment justify the design of a system that tackles the challenging goal of attaching a short-range and highly mobile, spontaneous network to the Internet. In this section, we present the major goals, challenges, and currently available solutions that address this issue.

The goals for usage of the Internet protocol suite in Geocast VANETs can be roughly classified in two categories. The first one includes all the necessary functionality to connect an isolated VANET to an infrastructure network such as the Internet. With these functionalities, a vehicle can acquire a globally valid IP address, send and receive packets from arbitrary nodes in the Internet in both cases of direct connectivity to a point of attachment (RSU) and connectivity via multihop communication paths. Further, changes of attachment points should be supported efficiently and transparently for the whole *in-vehicle network*, that is, the network composed of drivers' and passengers' devices. This requires that protocols like *network mobility basic support* [36] are also supported. Finally, related to this first category is the capability of a node in the Internet to address all or certain vehicles located in a specific geographical area.

The second category of required functionality concerns the use of applications based on the Internet protocol suite within the VANET, which might be often disconnected from any network infrastructure. In particular, even when the VANET is disconnected, connectivity at IP layer should be possible as long as physically feasible, that is, as soon as a single- or multihop path connecting the communicating peers exists. This category includes OBU-to-OBU and AU-to-AU communication, where application units of different vehicles communicate relying on the communication facilities provided by the OBUs. In the last scenario, an important goal is that no additional requirements are put on the AU's protocol stack, which should be assumed to be a standard Internet protocol suite.

The above described goals pose important challenges on the design of IP integration with Geocast. With respect to IPv6, the main challenge in ad hoc scenarios[7] consists of providing IPv6 with a *link* as defined by IPv6

---

[7] The challenging nature of this aspect is also testified by the still ongoing debate withing the IETF AUTOCONF work group on the link model in MANET.

standard [37] that provides link-local multicast addressing as required by
IPv6 basic operations such as neighbor discovery [38] and stateless address
autoconfiguration [39]. It is important to note that *link* in IPv6 terminology does not necessarily refer to a physical medium but more generally
to a communication facility located in the layer immediately below IPv6.
Another important challenge is related to the in-vehicle networks, whose
network prefixes cannot change over time to not pose requirements on
AUs. Thus, these prefixes are not topologically correct with respect to the
VANET nor to the Internet, which requires the design of a flexible and
scalable route management. Finally, for an Internet node to be able to
address geographical areas, the topological addressing or forwarding of
packets in the Internet needs to be extended with geographical information, which has already been the target of many projects but has never
found enough justifications for wide-scale deployment.

To date, a limited number of solutions for integration of Geocast and
IPv6 for VANETs have been proposed. The solution described by Baldessari,
Bernardos, and Calderon [40] is based on Geocast as a separate, sub-IP
communication protocol. It provides an adaptation layer that is presented
to IP as a link-local multicast–capable link. In particular, the link is modeled as a geographical area assigned to and announced by RSUs. All nodes
located in the geographical area are considered by IP as belonging to the
same link. Packet delivery is achieved by tunneling IP packets into Geocast
packets, assuming a static mapping between identifiers of different layers.
The same approach is currently assumed by the Car-2-Car Communication
Consortium [16]. Further, the European research project *GeoNet* [13] is currently working on integration of IPv6 and Geocast by following a similar
approach.

The application of *network mobility basic support* protocol to generic
ad hoc networks, instead, has been explored for a longer time. As described and analyzed by Baldessari, Festag, and Abeillé [41], approaches
can be classified as NEMO- and MANET-centric. With respect to these categories, Geocast-based NEMO deployments fall into the MANET-centric
type, because the geographical routing and forwarding policies of Geocast
cannot be implemented directly as NEMO protocol extensions. Application
of NEMO basic support to Geocast-based VANET has been described and
investigated by Baldessari et al. [42]. The Car-2-Car Communication Consortium is considering NEMO basic support as default protocol to provide
IP session continuity and global reachability and is also currently adopting a MANET-centric approach. Further, the consortium is providing IETF
with automotive requirements for the design of NEMO route optimization
techniques, which are explained by Baldessari et al. [43] together with the
overall integration approach.

14.5 Summary

This chapter presented Geocast as a potential network protocol for vehicular ad hoc networks based on WLAN IEEE 802.11 and multihop communication. Starting from basic Geocast, the chapter presented optimizations and extensions for various aspects ranging from radio control to Internet integration. Two aspects are particularly considered: (i) enhancements for reliability and efficiency of data dissemination and (ii) data security and user privacy. Geocast with its optimizations and extensions represents a solid technical basis to bring ad hoc networking technology into the real world.

References

[1] G. Finn, "Routing and addressing problems in large metropolitan-scale Internetworks," ISI Research Report ISI/EE-87-180, University of Southern California, March 1987.

[2] J.C. Navas and T. Imielinski, "GeoCast—geographic addressing and routing," In *Proc. of 3rd ACM/IEEE International Conference on Mobile Computing and Networking (MobiCom)*:66–76, Budapest, Hungary, September 1997.

[3] B. N. Karp and H. T. Kung, "GPSR: Greedy perimeter stateless routing for wireless networks," In *Proc. of 6th ACM/IEEE International Conference on Mobile Computing and Networking (MobiCom)*:243–254, Boston, MA, August 2000.

[4] M. Mauve, J. Widmer, and H. Hartenstein, "A survey on position-based routing in mobile ad-hoc networks," *IEEE Network* 15(6)(November/December 2001):30–39.

[5] X. Jiang and T. Camp, "A Review of geocasting protocols for mobile ad hoc networks," Grace Hopper Celebration (GHC), 2002.

[6] H. Füßler, M. Mauve, H. Hartenstein, M. Käasemann, and D. Vollmer, "Poster: A comparison of routing strategies for vehicular ad hoc networks," In *Proc. of 8th ACM/IEEE International Conference on Mobile Computing and Networking (MobiCom)*, Atlanta, GA, September 2002.

[7] C. Maihöfer, "A survey of geocast routing protocols," *IEEE Communication Surveys & Tutorials* 6(2) (2nd Quarter 2004), http://www.comsoc.org/pubs/surveys (accessed November 25, 2008).

[8] S. Basagni, I. Chlamtac, V. R. Syrotiuk, and B. A. Woodward, "A distance routing effect algorithm for mobility (DREAM)," In *Proc. of 4th ACM/IEEE International Conference on Mobile Computing and Networking (MobiCom)*:76–84, Dallas, TX, 1998.

[9] Y.-B. Ko and N. H. Vaidya, "Location-aided routing (LAR) in mobile ad hoc networks," *ACM/Baltzer Wireless Networks* 6(4) (July 2000):307–321.

[10] W. Franz, H. Hartenstein, and M. Mauve (eds.), "Inter-vehicle-communications based on ad hoc networking principles—the fleetnet project," Universitätsverlag Karlsruhe, http://www.uvka.de/univerlag/volltexte/2005/89/, (accessed November 25, 2005).

[11] A. Festag, H. Füßler, H. Hartenstein, A. Sarma, and R. Schmitz, "Fleet-Net: Bringing Car-to-Car Communication into the Real World," ITS World Congress and Exhibition, Nagoya, Japan, November 2004.

[12] A. Festag, G. Noecker, M. Strassberger, B. Lübke, A. Bochow, M. Torrent-Moreno, S. Schnaufer, R. Eigner, C. Catrinescu, and J. Kunisch, "'NoW—network on wheels': Project objectives, technology and achievements," *In Proc. of 5th International Workshop on Intelligent Transportation (WIT)*:123–128, Hamburg, Germany, March 2008.

[13] GEONET Project, http://www.geonet-project.eu (accessed November 25, 2008).

[14] SAFESPOT Project, http://www.safespot-eu.org (accessed November 25, 2008).

[15] Car 2 Car Communication (C2CC) Consortium, http://www.car-to-car.org. (accessed November 25, 2008).

[16] Car-to-Car Communication Consortium, "C2C-CC Manifesto," Version 1.1, September 2007, http://www.car-to-car.org. (accessed November 25, 2008).

[17] ETSI Technical Committee ITS, http://www.etsi.org. (accessed November 25, 2008).

[18] ISO Technical Committee 204 Working Group 16, http://www.isotc204.com. (accessed November 25, 2008).

[19] H. Takagi and L. Kleinrock, "Optimal transmission ranges for randomly distributed packet radio terminals," *IEEE Transactions on Communications* 4(32) (1984):246–257.

[20] S.-Y. Ni, Y.-C. Tseng, Y.-S. Chen, and J.-P. Sheu, "The broadcast storm problem in a mobile ad hoc network," In *Proc. of 5th ACM/IEEE International Conference on Mobile Computing and Networking (MobiCom)*:151–162, Seattle, WA, August 1999.

[21] ETSI Technical Committee Electromagnetic Compatibility and Radio Spectrum Matters (ERM), "Technical characteristics for pan-european harmonized communication equipment operating in the 5 GHz frequency range and intended for critical road-safety applications," Technical Report ETSI TR 102 492-1/2, ETSI, 2005.

[22] L. Le, W. Zhang, A. Festag, and R. Baldessari, "Analysis of approaches for channel allocation in car-to-car communication," In *Proc. of 1st International Workshop on Interoperable Vehicles (IOV)*, Zurich, Switzerland, March 2008.

[23] M. Torrent-Moreno, "Inter-vehicle communication: achieving safety in a distributed wireless environment, challenges, systems and protocols," PhD thesis, University Karlsruhe, Karlsruhe, July 2007, http://www.uvka.de/univerlag/volltexte/2007/263. (accessed November 25, 2008)

[24] T. Osafune, L. Lin, and M. Lenardi, "Multi-hop vehicular broadcast (MHVB)," In *Proc. of 6th International Conference on ITS Telecommunications (ITST)*:757–760, Chengdu, China, June 2006.

[25] A. Festag, R. Baldessari, and H. Wang, "On power-aware greedy forwarding in highway scenarios," In *Proc. of 4th International Workshop on Intelligent Transportation (WIT)*:31–36, Hamburg, Germany, March 2007.

[26] B. Williams and T. Camp, "Comparison of broadcasting techniques for mobile ad hoc networks," In *Proc. of 3rd ACM International Symposium on*

Mobile and Ad Hoc Networking & Computing (MobiHoc):194–205, Lausanne, Switzerland, June 2002.

[27] L. Briesemeister, L. Schäfers, and G. Hommel, "Disseminating messages among highly mobile hosts based on inter-vehicle communication," In *Proc. of IEEE Intelligent Vehicles Symposium*:522–527, Dearborn, MI, October 2000.

[28] H. Füßler, J. Widmer, M. Käsemann, M. Mauve, and H. Hartenstein, "Contention-based forwarding for mobile ad-hoc networks," *Elsevier's Ad Hoc Networks* 1(4) (2003):351–369.

[29] H. Füßler, "Position-based packet forwarding for vehicular ad-hoc networks," PhD thesis, University Mannheim, Mannheim, April 2007. http://madoc.bib.uni_mannheim.de/madoc/volltexte/2007/1406 (accessed November 25, 2008).

[30] M. Heissenbüttel, T. Braun, T. Bernoulli, and M. Wälchli, "BLR: Beacon-less routing algorithm for mobile ad-hoc networks," *Elsevier's Computer Communications Journal* (Special Issue) 27(11) (July 2004):1076–1086.

[31] B. Bochow, Dötzer, A. Festag, G. Gerlach, T. Leinmüller, and M. Schäfer, "Attacks on inter vehicle communication systems—an analysis," In *Proc. of 3rd International Workshop on Intelligent Transportation (WIT)*:189–194, Hamburg, Germany, March 2006.

[32] C. Harsch, A. Festag, and P. Papadimitratos, "Secure position-based routing for VANETs," In *Proc. of 66th Vehicular Technology Cconference*, Baltimore, MD, October 2007.

[33] T. Leinmüller, C. Maihöfer, E. Schoch, and F. Kargl, "Improved security in geographic ad hoc routing through autonomous position verification," In *Proc. of 3rd ACM International Workshop on Vehicular Ad Hoc Networks (VANET)*:57–66, New York, 2006.

[34] E. Fonseca, A. Festag, R. Baldessari, and R. Aguiar, "Support of anonymity in VANETs—putting pseudonymity into practice," In *Proc. of IEEE Wireless Communications and Networking Conference (WCNC)*, Hong Kong, March 2007.

[35] T. Aura, "Cryptographically Generated Addresses (CGA)," Internet RFC 3972, IETF, 2005.

[36] V. Devarapalli, R.Wakikawa, A. Petrescu, and P. Thubert, "Network mobility (NEMO) basic support protocol," RFC 3963 (Proposed Standard), January 2005.

[37] S. Deering and R. Hinden, "Internet protocol, version 6 (IPv6) specification," RFC 2460 (Draft Standard), December 1998.

[38] T. Narten, E. Nordmark, and W. Simpson, "Neighbor discovery for IP version 6 (IPv6)," RFC 2461 (Draft Standard), December 1998.

[39] S. Thomson and T. Narten, "IPv6 Stateless Address Autoconfiguration," RFC 2462 (Draft Standard), December 1998.

[40] R. Baldessari, C. J. Bernardos, and M. Calderon, "GeoSac—scalable address autoconfiguration for VANET using geographic networking concepts," In *Proc. of IEEE International Symposium on Personal, Indoor and Mobile Radio Communications (PIMRC)*, Cannes, France, September 2008.

[41] R. Baldessari, A. Festag, and J. Abeillé, "NEMO meets VANET: A deployability analysis of network mobility in vehicular communication," In *Proc. of 7th International Conference on ITS Telecommunications (ITST)*:375–380, Sophia Antipolis, France, June 2007.

[42] R. Baldessari, A. Festag, W. Zhang, and L. Le, "A MANET-centric solution for the application of NEMO in VANET using geographic routing," TridentCom, Innsbruck, Austria, March 2008.

[43] R. Baldessari, T. Ernst, A. Festag, and M. Lenardi, "Automotive industry requirements for NEMO route optimization," Internet Draft, draft-ietf-mext-nemo-ro-automotivereq-01, work in progress, July 2008.

Chapter 15

Market Introduction and Deployment Strategies

Robert Lasowski and Markus Strassberger

Contents

Driver assistance systems are the key technology to improve traffic safety and lower the number of accidents, casualties, and fatalities. The direct co-operation of vehicles will further enhance this field of traffic safety and traffic efficiency. However, such cooperative systems will only become effective, if a reasonable number of vehicles are equipped with the respective communication technologies and services. This chapter gives an insight on the challenges of market introduction, delineates the key stakeholders, and describes the opportunities of different deployment strategies. In particular, a promising hybrid approach is presented that exploits the advantages of cellular and ad hoc communication and provides both cooperative and noncooperative services to the customers.

15.1   Introduction

Today's state-of-the-art Car2X communication technology has already met a high level of maturity and the essential evidence and technical proof of concepts has been carried out successfully. It can be assumed that market introduction of Car2X technology has already started in an initial stage. It can be seen that numerous car manufacturers, suppliers, and telecommunication service providers already offer telematics services like "BMW Assist" (BMW) [1], "HD Traffic" (TomTom/Vodafone) [2], and the "Go-Box" tall system (ASFINAG) [3]. Although these activities have been made independent from each other, they all seem to have the same objectives—namely "connected vehicles."

However, the idea of interconnected vehicles has been evolving over years. The first telematics systems were launched in the 1990s, unfortunately, with minor success [4]. This had several reasons, for example, high hardware costs and technological boundaries. Thus, customer acceptance was not established and a successful and profitable roll-out of the new technology has failed.

Since then new ideas regarding applications and technologies have been worked out with more promising prospects for customer benefits. The key focus have been shifted on cooperative safety and traffic efficiency applications. Based on technological innovations, an advanced Car2X system shall provide a variety of new value-added services. However, to assure the aspired functionality, such cooperative systems are dependent on minimum amount of nodes that are participating in the overall system, which in turn leads to the necessity of a certain level of market penetration. This leads to new challenges within the context of market introduction that longs for new innovative deployment strategies.

This chapter describes various ideas and concepts of deployment and marketing strategies for the Car2X technology. It specifies the necessary preconditions, the involved parties and their tasks, and concludes with a description of a successive model as key enabler for successful deployment of Car2X systems.

15.2 Market Development

The idea of connected vehicles was born more then 10 years ago when research projects like intervehicle hazard warning (IVHW) [17] were invented and systems like OnStar (GM) [5] were made available to customers. Connected vehicles seemed to be the new hype at the end of the mid-1990s [4], which itself obviously was the answer to the boom in the telecommunication industry. This hype period lasted until the ebiz crash in the year 2002. Thereupon, based on the knowledge of the mistakes that have been made in this first stage of connected vehicles, for example, isolated product development by each company itself, a new form of collaboration was conceived in Europe—the Car 2 Car Communication Consortium (C2C-CC). The objective of the C2C-CC is the enhancement of road traffic safety and efficiency by means of intervehicle communication where car manufacturers, suppliers, and research institutes are working together [6] .

Beside the new attitude toward intercompany cooperation other effects exists that are conducive for further development of intervehicle communication, which can be grouped into

- Technological drivers
- Global drivers

15.2.1 Technological Drivers

Technological development has made a big step forward in the last couple of years. Hence, problems that have existed in the past are solved now and new opportunities are opening. For example, programmable controllers became smaller by enhancement of processing power. This, of course, simplifies the decision to utilize more embedded electronic devices inside vehicles. Car manufacturers exploited the potential of electronics in vehicles and started to equip them with control units and communication buses like CAN and MOST. Thus, a great amount of diverse information is available in modern vehicles, which will drive innovative Car2X applications.

Furthermore, wireless communication hardware enjoys great popularity and became a mass product [7], which today is widely used in private and business sectors for mobile communication. Organized ad hoc networks made it possible to obtain a high level of knowledge about characteristic behavior in wireless networks and let us learn more about potential problems and improvements.

In addition, a powerful back-end infrastructure has been built up making almost any information globally available. Thus, we have access to a wide range of information and can use it in new interoperable services. Also new flexible pricing models like flat rates for Internet access and mobile communication has been elaborated by the telecommunication industry, making information technology affordable for everyone.

15.2.2 Global Drivers

In addition to the technological development also a global market driver for Car2X technology evolved. One of the issues within this context is the raising of oil prices which, amplified the demand on intelligent and efficient navigation systems, for example, for the calculation of shortest travel routes. Beyond that, the continuously growing congestion on rural roads, in cities, and on highways and the related stress, environmental pollution, and the rising rate of road accidents are calling for a new innovative traffic management solution [7] . Considering the traffic growth at is increasing each year, especially heavy traffic has an enormous impact on safety and traffic management. Just for the freight traffic a growth of 50% until the year 2020 is forecasted in Europe [8]. Due to these facts regular measures like road widening or similar constructional activities are not enough anymore to solve the current traffic challenges.

Altogether, the technological development in the recent past and the current global condition open completely new opportunities and possibilities for innovative interoperable services and use cases in the vehicular environment.

15.2.3 Future Market Development

Due to the highly cooperative character and the altruistic trait of many Car2X applications, it is important to reach an appropriate market penetration to achieve a benefit for the customers. However, obviously, every customer expects perceptible added value in return for his financial investment. Thus, the inherent system characteristics and the customer's expectation seem to be a big discrepancy and represent accordingly the biggest challenge regarding the market introduction strategy. Concerning the overall requirements of Car2X services, a lower limit of 10% system penetration is assumed [6].

Figure 15.1 shows an overview about the predicted market development regarding the system penetration concerning four potential deployment strategies [18]. These results are based on the assumption 8% admittance for new cars each year. First of all an optimal scenario is being examined where all new cars are equipped with the Car2X technology from a defined point in time. This would cause a penetration rate of 10% after the time period of 1.5 years [16]. In this context a penetration of 50 percent would be achieved after 7.5 years. However, it is rather unrealistic that all car manufacturers will offer all their vehicles equipped with the Car2X system and that the customers will agree and pay for the technology also.

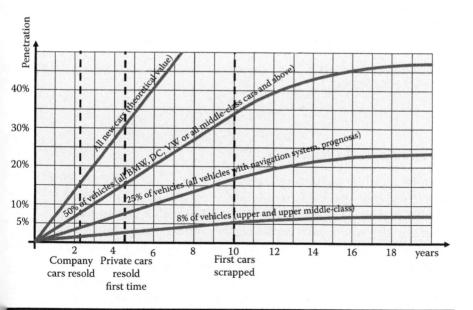

Figure 15.1 Evolution of Car2X system penetration with respect to different deployment strategies [18].

Another more probable scenario would be to equip only the upper-class and upper middle-class vehicles with Car2X systems hence the related customer base is most likely willing to pay for new technology. By this means, a penetration rate of 5% would be reached after 10 years where first cars already will be scrapped. Further, a rate of more than 8% will never be reached in this case. Thus, such a deployment strategy is not feasible because the quality and variety of the system could not be enhanced if there is no growth of penetration.

Other strategies are scenarios where every second or every fourth car is equipped, which means all new middle-class vehicles and higher (50%), and all cars with navigation systems (25%). In this case, the achievement of a 10% penetration value would take between three and six years. A higher penetration rate, for example, 25% will not be achieved before seven years.

Obviously it is necessary to rethink and optimize those static deployment strategies to generate customer value in a shorter time period or by using more attractive services to speed up the deployment process.

15.3   Challenges

The new market development situation comes along with new challenges for Car2X communication technology. To derive a viable market introduction strategy and prepare the roll-out of Car2X systems, obviously those specific challenges have to be tackled. In general, there are three domains that have to be considered, namely:

■ Technological challenges
■ Customer acceptance
■ Costs

15.3.1   Technological Challenges

Although wireless ad hoc technology already is part of our everyday life the intervehicle use cases have different requirements this kind of network. A vehicular ad hoc network, for example, has to deal with a big amount of communication nodes where each is potentially running a large number of different applications or services. This leads to high requirements from the respective communication channel. Therefore innovative architecture and intelligent algorithms have to be defined to deal with the limited channel capacity and high channel congestion. Concerning a competitive development of Car2X services and communication units, the compatibility of the deployed systems has to be guaranteed. Finally the issue concerning the long lifetime of vehicles compared to the fast moving technology domain has to be solved. In the following, the four mentioned aspects are described in detail.

15.3.1.1 Architecture

To ensure a future-proved overall solution, the overall architecture is one of the core long-term challenges of the Car2X work flow. The architecture design has to deal with different kinds of applications like active safety, traffic efficiency, and infotainment services and even more that could be defined in future activities. Therefore it is important to develop an architecture design that is able to deal with different service requirements like low latency, TCP/IP integration, mobility, and even roaming. In addition to the application requirements, the architecture has to be capable of integrating different communication protocols and standards like regular IEEE 802.11 a/b/g communication, IEEE 802.11p [12] as the most probable standard for vehicular ad hoc networks, and cellular communication technologies like GPRS and UMTS.

From a software engineering point of view, the architecture components and layers have to offer well-defined interfaces, which are easy to use and flexible for future enhancement. Furthermore it must be possible to exchange or add particular layers if they are required in the future. For that purpose, different projects like AKTIV [9], CALM [10], and NoW [11] concentrate on the conceptual definition of an appropriate architecture design. However, based on the fact that the actual focuses of the projects vary, also the architecture proposals vary in some details. In addition, intervehicle communication poses an international challenge in a global market. With respect to the necessary development and deployment efforts, the key goal is to define a common system that can be used in all markets. Unfortunately, the prerequisites and constraints imposed by different markets vary to a great extent. Beside the availability of dedicated frequency bands in different countries, varying customer expectations and different legal aspects and regulations have to be considered. For these reasons, global activities in the context of developing and standardizing Car2X systems vary, although there is a common goal and similar approaches. In particular, the U.S. proposals like the IEEE 1609 protocol stacks [12] differ from the European in its components and especially the defined system work flow. This could cause problems at consolidating the results of those two efforts to one global view. Nonetheless, to consolidate these individual developments on a pan-European basis, the European Commission initiated the COMeSafety project [13], which among other activities is chairing an architecture task force that is coordinating the European view on the Car2X architecture. In addition, the C2C-CC is collecting and consolidating requirements regarding the architectural framework. The consortium is responsible for discussing and reviewing existing

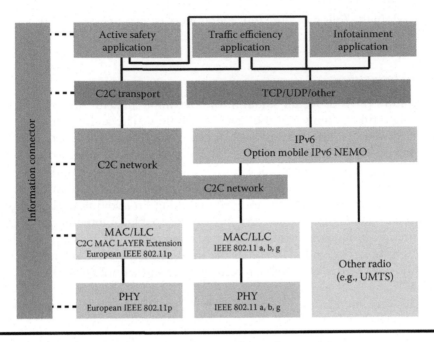

Figure 15.2   C2C-CC common architecture view [16].

architecture proposals from several international research projects and consolidating them to a first general admitted version of a European communication stack. Figure 15.2 delineates the common view on an overall communication architecture as it is currently discussed in the scope of the C2C-CC [16].

15.3.1.2   Algorithms

For an efficient usage of the limited resources, in particular, power and the overall available channel capacity, smart algorithms, for example, for intelligent message dissemination and message aggregation have to be developed. Therefore one of the core aspects of the communication unit is the scalability of the system regarding a high channel load where a high number of network nodes exist. Nevertheless, scalability does also mean a low penetrated network where information dissemination also has to be assured. In the current state the development of algorithms is rather complicated because no flexible communication architecture is available yet and practical evaluation can hardly be performed. Hence, results are so far based on theoretical assumptions and network simulations only, lacking further experience in large-scale real world scenarios.

15.3.1.3 Compatibility

Compatibility of the Car2X system is one of the most important issues for a successful introduction scenario. Obviously, compatibility always comes along with standardization. However, in the initial stage it is not necessary to standardize every system characteristic in detail. In a first step it is important to standardize just those parts of the overall system that are necessary to enable sustainable product development. In particular, this comprises communication frequency, network protocols, and the application protocol.

The first step toward compatibility is to negotiate on a radio spectrum for the physical layer. Taking in to account the radio spectrum allocation tables worldwide, exploiting resources at about 5.9 GHz have been considered to be most appropriate [6]. One of the reasons for this decision is the possibility of exclusive usage of this frequency for safety and traffic-related applications. Currently, similar frequency bands have already been protected in Europe, the United States, and Japan. The advantage of this global settlement is the related growth of market demand for 5.9-GHz communication hardware. Hence, mass production can be started and overall system costs are supposed to decrease. This is an important precondition for the desired customer acceptance.

The next requirement regarding compatibility is standardized protocols. Obviously, it has been negotiated that the IEEE 802.11p protocol will be used for basic communication at the physical layer, which has been especially designed for automotive requirements. Furthermore, a general and extensible application and network protocol have to be assumed, for example, the SAE Message Directory [14] and the NoW [11] network protocol. In this case also flexibility is essential, which enables the compatibility of different versions of applications and guarantees a future-proved approach.

15.3.1.4 Technology Life Cycle

Due to the long life cycles of vehicles, vehicles that are provided with a Car2X communication system have to be up-to-date for decades. Obviously, this will cause problems in comparison to the fast development in micro technology, embedded devices, and multimedia solutions. Concerning the technology development within the infotainment and multimedia sector it can be assumed that new opportunities and potentials for more powerful services and use cases will rise in a range of, approximately, only a few years. However, it stands to reason that customers expect a future-proved solution of the integrated Car2X systems. To guarantee customer satisfaction these two conditions have to be consolidated.

15.3.2 Customer Acceptance

It is obvious that a successful market introduction of a Car2X technology depends on the acceptance and the satisfaction of the customers. As long as the customers do not feel convinced about their benefit and the added value of the system, respectively, they will not invest in it. Hence, to achieve the necessary demand it is significant to pay attention to the following aspects.

15.3.2.1 Added Value

As mentioned, in particular applications traffic safety and efficiency are cooperative and altruistic. Hence, there is no immediate benefit for the customers. Obviously, the benefit of participation will rise with the number of vehicles that will be equipped with the Car2X unit. However, this obviously raises the question of who could be the first to buy this new communication technology. The answer strongly depends on the approach of the market introduction strategy. Thus, customers need a perceptible and sensible added value to be willing to pay for the existing and upcoming services and the follow-up costs in the future.

The challenge is to develop applications that provide a generally accepted benefit to a great number of potential customers. In this context, the idea of a "killer application" was born in the past. Such killer application has to be just one Car2X service that is desired by every customer. Unfortunately, this service is still not yet defined—and probably will never be found. The reason for that may be that this approach presumes a completely homogeneous customer base. Instead, the total opposite condition exists where it can be assumed that the customers' expectation of Car2X technology is rather heterogeneous due to, for example, different gender, age, interests, and cultures. Thus, it is unlikely to come up with just one initial static application. Instead, to satisfy customers' demands, a dynamic content and service architecture is necessary that enables individual and personalized service compositions. However, this dynamic architecture requires a much more complex market introduction strategy, which implies the coordination of several parties to guarantee service provisioning, content offering, and billing.

15.3.2.2 Pricing

The first perceptible experience for the customer will be the price he will have to pay for the Car2X technology and the bundles services, respectively. Concerning the different customer classes the readiness to pay for pursuant service depends on the individual financial possibility. Therefore a flexible pricing model is required. As we know from telecommunication and Internet providers there could be basically two models—the pay per use and the flat-rate model. Both of them have advantages and disadvantages for customers and providers. Nevertheless it should be possible, especially, in the first stage of market introduction to have the choice at least between two opportunities. As one possibility, initially there may also be free services launched to demonstrate the advantages of the overall system.

15.3.2.3 Usability

The issue of usability is certainly very important with respect to the acceptance of the overall Car2X technology. The challenge, therefore, is the design and the specification of a flexible and extensible human machine interface (HMI) that allows an intuitive and convenient usage of the available services. However, for a global strategy of market introduction it is of subsidiary relevance, because each supplier will come up with its own implementation and integration.

15.3.3 Costs

The introduction of a completely new technology is usually inherently related to high investment costs from the provider point of view where, for example, car manufacturers, suppliers, service providers, and also government and public authorities can be depicted. Hereby not only the hardware development costs but also software development and research costs have to be calculated. In particular, the following issues have to be taken into account.

> **System costs:** The overall Car2X system consists of several subsystems like the application unit, the communication unit, wireless communication devices, and further input or output devices. Each of those parts includes hardware and software modules that have to be developed before offering a final product, comprising years of research and standardization activities. In addition to the development process, the integration of single components like wiring, antennas, and possible changes of the design of a vehicle have to be kept in mind.
>
> **Infrastructure costs:** To enable Car2I communication, additional hardware and software components on the infrastructure side are necessary. This could be roadside units that are installed along the roads, as well as communication end points within the private and public sector, for example, hot spots. In this case besides the development process the installation and maintenance of those components mean high investments. Furthermore, an intelligent and scalable back-end infrastructure is necessary to enable the connection of decentralized stored content and to create a transparent view on a global information database. These are rising costs that have to be allocated by the involved parties.

On this account, it is comprehensive that investments related to hardware and software development and maintenance of Car2X systems can only be intended with a valid business case that has to be carried out by each partner.

15.4   Generic Deployment Paradigms

To meet the goal of a successful market introduction of Car2X technology, several ideas for a feasible marketing strategy have already been developed. This section will give an overview about the existing concepts and present their advantages and disadvantages. Generally, it is possible to force the market introduction from three different aspects:

- Content
- Technology
- Government
- Polities

15.4.1   Content-Driven Deployment

One possibility to persuade customers into buying a Car2X system is the offering of value-added services. The services within the content-driven market introduction approach can be divided into three basic content classes:

- Safety-related applications
- Traffic efficiency-related applications
- Infotainment applications and noncooperative services

Safety-related applications Safety applications enhance a driver's horizon and reduce or even avoid accidents, for example, by bad weather condition warnings, end of traffic jam warnings, or vehicle breaking in front warnings. Those examples can be further classified in event-driven applications, like danger notifications and cooperative awareness applications.

Event-driven applications notify the driver in case of upcoming dangers. This kind of information has usually a long lifetime and will be disseminated within a defined geographical area by using so-called Georouting algorithm. For this kind of routing scenarios a store and forward approach is needed. Within the lifetime the warning message will be forwarded to all neighbors that are in communication range. Hence, the vehicles are not just acting as message consumers but also as a provider and mobile storage for valid information. Simulations have shown that based on the store and forward idea it is possible to generate a user benefit by a system penetration rate of approximately 10 percent [16]. However, it is necessary to run optimized hazard detection algorithms within the vehicles, which sometimes could cause a problem regarding the information quality. Hence, it is not always easy to detect a specific hazard like an obstacle on the road or a friction situation by just using the vehicle's on-board sensors. In some situations, addition sensors like radar, lidar, or video cameras are needed, which of course raise the overall system costs.

Cooperative-awareness applications are probably the most relevant application type to increase safety. Thereby, the key focus lies on gaining knowledge about the position of nearby vehicles to avoid critical proximity. Therefore, usually a common set of data is exchanged among nearby vehicles, like the vehicle's speed, position, and direction, and has a lifetime of just a few seconds or even less. In general, those kind of messages will not be routed by an intelligent algorithm but use an ordinary one-hop broadcast approach. The messages have to be processed immediately after they have been received otherwise the information expires. Here, applications like forward collision warning, lane change warning, and extended break light warning can be listed. Because of a low information lifetime, those applications have high requirements on the system penetration rate. Thus, considering that the penetration rate will grow slowly in the initial stage of market introduction this type of application is not reasonable for a first deployment scenario. Obviously, the more vehicle participation in such cooperative systems, the higher is the reliability. Thus, in general, the penetration rate should be significantly higher than 10 percent.

Traffic efficiency-related applications: Different surveys have shown that customers are especially interested in accurate and up-to-date traffic information and route guidance (see, e.g., [19,20]). The advantage of this kind of information is the existence of the basic content at the road operators' side. However, the dissemination of traffic information takes about 15 minutes today by using regular broadcast technology like TMC or TMCPro. This can be improved by the Car2X technology and a propagation delay less than a minute can be achieved. Nevertheless, in the initial phase traffic and route services are more based on Car2I than Car2X communication. Hence, roadside units have to be mounted and back-end integration into the existing systems has to be done. However, the advantage of traffic information compared to active safety application is a lower requirement on the penetration rate, which is set to 5 percent [16].

Infotainment applications and noncooperative services: Third and last, infotainment applications will bring more comfort and entertainment into the driver's everyday life, providing, for example, car-to-home connectivity, map downloads and update, POI notifications, drive through payment, or any kind of personalized mobile services. The big advantage of these applications can be derived from the low requirements on cooperativeness. This means that the customer has a noticeable value from the system right after buying, which can be a big inducement for the customer to invest in the Car2X technology [11]. Although the system penetration rate for the vehicle's side is almost zero, of course, the service endpoints for the above mentioned applications are needed. This means that services have to be provided by pursuant parties and should be configurable for the customers.

Figure 15.3 exemplarily classifies several applications and services that can be offered to customers with respect to the necessary penetration rate of C2X-equipped vehicles and infrastructure. Based on this assessment it can be decided what kind of service can be provided to the customer in what stage of system availability. Thus, it is recommended to start with a service like car-to-home, which does not require any penetration rate. Further services like traffic and speed limit information can be rolled out in the next steps, successively, until the penetration reaches a higher value. However, highly cooperative services can only be offered in one of the final roll-out phases.

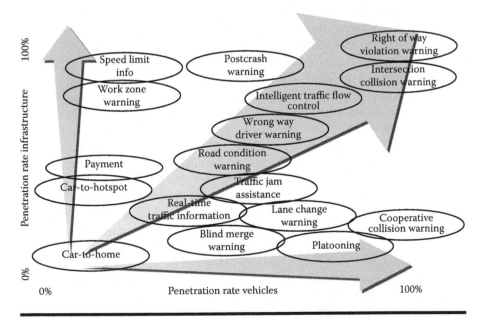

Figure 15.3 Classification of services with respect to the required penetration rates.

15.4.2 Technology-Driven Deployment

For a successful introduction of Car2X communication, it is necessary to consider the possibilities of existing and desired technologies. As mentioned, Car2X communication is widely considered as solely based on ad hoc communication, in particular, on the evolving IEEE 802.11p standard. However, introducing a Car2X system only by using ad hoc connectivity would imply a high financial investment without the assurance of customers' interest in the opportunities. For that reason, it is reasonable to consider the usage of cellular networks like GPRS or UMTS in a hybrid approach, too.

Figure 15.4 shows the possibilities and advantages of cellular and ad hoc network communication in parallel. Basically, in the first step of information dissemination where low real-time characteristics are required and the latency is not a core criterion cellular networks are even more recommended than short-range communication for the deployed applications. In this context, applications like traffic and weather information are intended. However, the more the probability for vehicle collisions is increasing and critical safety application like danger warning and emergency break applications are utilized, which comes with high real-time requirements, the more the short-range communication is reasonable. Although flexible pricing models like flat rates are already offered to minimize the communication costs, usually cellular communication is not free of charge. By looking at already existing telematics services and products like BMW's Connected

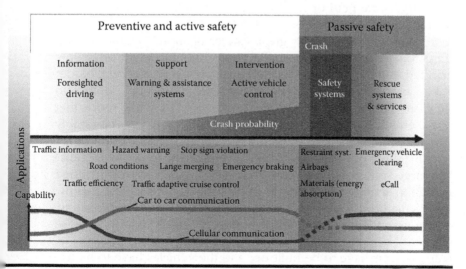

Figure 15.4 Comparison of technological potentials of different communication paradigms [6].

Drive and TomTom's HD traffic, a market introduction based on cellular communication seems to be a feasible way. It is even recommended to use these two communication channels in parallel and complementary ways, which can be applied for applications like eCall.

15.4.3 Political-Driven Deployment

Last, it seems to be reasonable that cooperative telematics systems are advanced by public authorities. It is very likely that in particular services with respect to increased traffic safety or traffic efficiency will effectuate sufficient public benefit that rationalizes preceding investments. Both minimizing road congestion and in turn environmental pollution and minimizing the number of fatalities and casualties is one political key concern. Hence, public authorities have already invested and will still invest in the future into respective measures. Car2X systems can thereby provide similar effects with lower costs and therefore rationalize a wide-spread system deployment that is subsidized by public authorities. However, the effective benefit of Car2X systems once introduced can hardly be quantified, because there are too many different influencing factors. A significant statistical analysis would require an isolated evaluation over years, apart from other developments like further improved driver assistance systems, the development of the overall number of vehicles and the total mileage, or improved traffic routing. In addition, obviously the expected benefits must be reliably derived prior to the deployment to rationalize the necessary investments. For this purpose, there are basically two main strategies: simulations and large-scale field operational tests. Therefore, the significance of simulations strongly depends on the exploited driver models. Realistic driver models take into account a great variety of input parameters and are therefore complex to compute. In the past, driver models have mainly been developed to simulate traffic performance. However, it is even harder to derive reasonable models with respect to safety decisions, because in critical situations people tend to act irrationally. Hence, while estimating the respective effects of Car2X systems on traffic efficiency can reasonably be done by simulations, the effects on traffic safety are hard to simulate.

To overcome these problems and also to optimize the respective simulative models, large-scale field operational tests (FOTs) are currently planned or in preparation worldwide. Since accidents happen very infrequently, again it is particularly hard to statistically quantify the effect on traffic safety. Hence, setting up and conducting significant FOTs are expensive because a large number of vehicles have to be equipped with respective hardware, infrastructure systems have to be built up, and the vehicles have to be frequently moved by typical drivers under usual conditions for a sufficient amount of time.

Last, the decision process within public authorities is most often complex and long-winded, because a variety of different stakeholders are involved in the decision process. In addition, the responsibilities are typically shared among different authorities. As a consequence, the investments of individual authorities may not pay off even in case there is a certain significant total benefit and hence a total return on investment. Taking these considerations into account, it seems necessary that national governments raise central funds to consolidate the necessary investments with the overall social advantages. Assuming that the social impact can be reliable, quantified, and rationalizes public investments, a politically driven deployment can significantly push Car2X technology and cooperative services into market. Hence, the FOTs currently in preparation will make a substantial contribution with respect to market introduction.

15.5 Preconditions and Dependencies

In the current state of the Car2X technology, both market and industry are located in a complex situation. After years of research and standardization it is still uncertain what the right strategy for market introduction is. However, besides the question of the right strategy there is still the question of the right timing and the coordination of the involved partners. As briefly mentioned before, it is unlikely that customers will buy a communication system without having the opportunity to consume services. However, on the other hand, no one will provide services without being sure that the services will be consumed.

Figure 15.5 shows the mutual dependencies from a rather abstract point of view. As mentioned, initially high investments in research and development of new technologies like communication hardware, embedded systems, and standardization activities are necessary. These investments have to be done at least to be able to estimate if a technology is worthwhile to be developed further. The conclusions of this phase provide input for new applications and services that are dependent on the technical and functional results, for example, what kind of services can be realized with the available processor power and communication bandwidth. Due to intelligent definition and composition of services and applications, customer acceptance can be established. In case customers are convinced about the added value of the overall Car2X system, they are also willing to pay for the technology, and in turn the system penetration rate will continuously increase. Thus, the industry can come up with new ideas for business cases. Thereupon the Car2X technology can be optimized and readiness to invest in further development will rise. This section provides an overview of the involved parties, their tasks, and the existing dependencies within the deployment process.

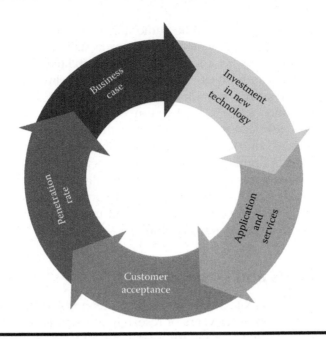

Figure 15.5   Mutual dependencies of key issues with respect to market deployment.

15.5.1   Car Manufacturers and Suppliers

In the very beginning of the Car2X research phase, the development has been driven almost only by car manufacturers and research institutes. There have been a lot of open issues regarding the standardization and feasibility of Car2X architecture and protocols. However, meanwhile the standardization has reached a high stage of maturity and a first consensus looms to be achieved. The next stage should be a more production-oriented vehicle integration of the single system components and the assurance of the operability of the overall system. It is important to start with such considerations as early as possible due to the fact that hardware development and integration cycles can last up to several years. Furthermore, car manufacturers can even take advantage of external devices by offering a well-integrated system. Another important aspect is a trade-off between car manufacturers regarding the type of basic services they want to offer to their customers. This is relevant for the introduction of cooperative services like collision and lane change warning that are strongly related to a high penetration rate.

15.5.2   Road Operators

As previously described, road operators are one of the most important traffic content providers within the realm of Car2I communication. Therefore it is necessary for the road operators to centralize their knowledge of road activities, so that they can deliver a homogeneous view of road conditions. Further the development and integration of a bidirectional Car2I communication interface and the installation of roadside units along their highways and rural roads should be another goal. Although the installation of roadside units seems to be a disproportionately high investment just to deliver content to a couple of drivers, road operators can take advantage of the bidirectional communication to enhance the quality of their traffic information by using the received data from the passing vehicles. Moreover, high-quality real-time traffic information can also be sold to the service users for a reasonable price.

15.5.3   Mobile Device Manufacturers

Beside regular vehicle integration, it is also important to run the Car2X applications on mobile devices like cell phones or mobile navigation systems. This would have a financial advantage for the customer and could present a low-budget solution with maybe minimized functionality. For these reasons, manufacturers have to enable wireless communication on their mobile solutions and guarantee the required hardware resources to integrate the Car2X functionality and protocols.

In addition, several mobile device manufacturers could assemble strategic partnerships with car manufacturers to enhance their application quality and functionality by getting access to vehicles' data and offer a cobranded after-market solution.

15.5.4   Research Institutes

After the first roll-out of the Car2X technology it is still necessary to keep up research activities. After the first deployment phase the penetration rate will grow continuously and different issues will be determined that have to be solved in a fast way. For these reasons, new optimized algorithms (e.g., proposed by Adler et al. [15]) will be needed. Hence, continuous research activities are necessary.

15.5.5   Governments and Public Authorities

The government is a powerful party within the Car2X community. On the one hand, it can help to finance the road-side unit infrastructure that would help to realize pursuant services. On the other hand, it could enact a law for the utilization of Car2X safety solutions, like one in discussion for the eCall system now.

15.5.6 Service and Content Providers

In the initial stage of market introduction, different system providers will offer self-developed services that most probably will not be compatible with each other. However, this is a feasible way for a quick system roll-out. Nevertheless for future reasons, it is significant to have the opportunity to consume applications that are offered by different companies and are independent from particular car manufacturers or mobile device suppliers. For that reason, a general common open-service and content provisioning architecture has to be designed; for example, as has been developed for the NGTP project (see Figure 15.6) [21].

The architecture should be flexible and extensible to guarantee a transparent service and content offering, as well as an easy to use service composition concept where customers can choose their preferred telematics service providers and the type of service implementation they would like to use. The architecture has to allow an easy service provisioning and should switch between the different providers in a transparent manner, for example, to solve issues like crossing of country borders where a roaming functionality is needed. Furthermore, it opens opportunities to design one's own services and introduce one's own content to enlarge the service diversity and thus to address a wider range of customers.

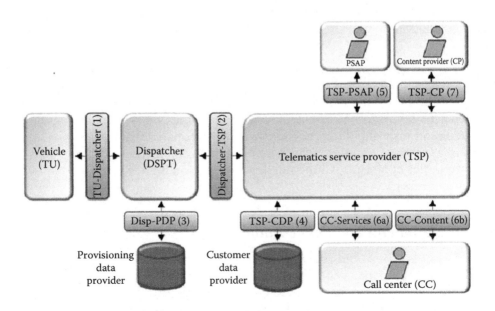

Figure 15.6 NGTP service architecture [21].

15.6 Dynamic Deployment Strategy

This paragraph concludes the chapter regarding the current state of market development, the involved parties, and several market introduction concepts. Based on this knowledge, a multistage and dynamic deployment approach for the introduction of Car2X technology will be presented.

As mentioned in Section 15.2, the cooperative applications based on intervehicle communication are altruistic. This means, in particular, each system node basically sends information to their neighbors without having any benefit until there is no communication partner in range where other messages could be received from. The benefit for the overall community grows with the penetration rate of the technology.

The basic issue of an altruistic approach is the opposite character especially toward the customer attitude. Customers typically follow an egoistic approach that expects added value from the very beginning after investing in new technologies—certainly they are entitled to do so. However, to solve these issues a win-win strategy for system providers and customers is necessary.

15.6.1 Vision

Concerning the variety of stakeholders that are involved in the development and integration of the Car2X system and the high complexity of the overall system itself, it can be assumed that a successful market introduction cannot be performed by just one stakeholder on his own. Therefore, a general overall commitment of all participating stakeholders as it has been listed before is necessary. This is also indicated by a statement of the C2C-CC: "only a joint initiative of all European vehicle manufacturers, suppliers, scientific organizations and standardization bodies will lead to an economically promising and successful market introduction" [16]. However, the stakeholders have to commit to one common terminology, which has to be endorsed by all involved parties to inspire customers about the opportunities of the new Car2X technology. Further, individual stakeholders should even start some kind of marketing cooperative to promote the capabilities of the technology with the advantage of minimizing advertisement costs. With this strategy, the market would be enriched by different sectors, and several customer groups would be addressed with the effect of market sensitization and awareness of the new technology.

15.6.2 Content and Technology

Regarding a valuable strategy for market introduction, a multistage and technology-mashed approach has to be defined. As was shown in Section 15.4, there are basically three types of services, namely, active safety, traffic

Table 15.1 Communication Technologies

	Cellular Networks	DSRC Networks
Penetration rate	Almost 100 %	Not deployed yet
Customer acceptance	High	Low
Latency	Middle–High	Low
Bandwidth	Low	High
Data rate	Low	High
Customer costs	Middle	Low

efficiency, and infotainment, and two kinds of technology: the short range and the cellular communication. The challenge is the specification of a reasonable and flexible composition of content and technology.

Table 15.1 shows a comparison between cellular and short-range communication networks. It can be seen that cellular networks are already deployed areawide and highly accepted by customers, for example, regarding connection availability. With respect to network availability, cellular communication currently has advantage over ad hoc networks, although the communication costs are higher. However, to enable the offering of services that need high data rates, high communication bandwidth, and fast response time as is required by Car2I services, cellular networks are not recommended. Here the DSRC technology is applicable. Nevertheless, the two technologies are not implicitly a discrepancy. Rather it can be assumed that cellular networks are one kind of infrastructure and hence appropriate for the usage of Car2I services. Furthermore, by introducing Car2I services via cellular technology it can easily be switched to DSRC communication in a next step because all necessary back-end systems will already exist at that point in time and can be used with minimal adaptations.

Table 15.2 shows the three different application classes. For an initial deployment stage it is recommended to start the market introduction with services that require low system penetration. Therefore active safety applications are not appropriate. Moreover, infotainment and traffic efficiency services have to be offered. However, it has to be kept in mind that some of those services require high bandwidth and hence need a fast connection, where also communication costs should be considered.

Table 15.2 Application Classes

	Active Safety	Traffic Efficiency	Infotainment
Operational mode	Cooperative	Cooperative/Infrastructure	Infrastructure
Penetration rate	$\geq 10\%$	$\geq 5\%$	$\leq 1\%$ (Car2Home)
Bandwidth	Low/Middle	Low/Middle	High
Content provider	Neighbor	Neighbor/Back-end	Back-end

15.6.3 Basic Service Set

It seems to be practical to introduce a Car2X system by an initial "basic service set," comprising a bundle of both communication paradigms and different services. It is therefore reasonable to combine the technologies and services in the following manner:

- Traffic efficiency services based on cellular communication
- Stand-alone infotainment services based on short-range communication

Hence, customers would have high profit from the very beginning because the traffic-related information already exists and customers are also familiar with them because of the TMC and TMCPro technology. Furthermore, there would be additional new services that can be used by customers from the very beginning like map updates, play list synchronization, or other Car2Home applications. For that reason, penetration rate would inherently increase in parallel. After reaching a critical penetration rate, cooperative service could be offered and existing services that were based on cellular networks could be switched to ad hoc technology. It is important to react on the development of the overall system topology and arrange the service offering in a dynamic manner depending on existing market demands. Nevertheless, a coexistence of both technologies is mandatory as it was depicted in Figure 15.4. For safety related services, it is always reasonable to have a redundant communication channel. Further, most probably there will always be an area where no short-range cellular connectivity is available and vice-versa.

Concerning the technological preconditions to realize a dynamic service offering concept, basically, the network and the application protocol have to be standardized. In addition, concerning the further development and optimization of Car2X systems it is important to start the first stage of introduction with a highly flexible communication framework architecture. These aspects would enable easy software updates and guarantee compatibility of services with different versions. Based on this assumption an initial set of basic applications could be offered and subsequently continuously enhanced.

15.6.4 Back-End Architecture

Basically, it is not necessary to define exactly the right two or three services before starting the introduction phase. Rather, a dynamic composition of services for a fast reaction on the market demand is necessary. Therefore, it has to be committed on an overall open service and content-provider architecture. This would allow personalized service configurations and would meet the individual expectations of each customer. Furthermore, an extensible, flexible, and standardized service architecture would gain customer acceptance by enabling the offering of individual and personalized customer-desired content and the idea of a real Car2X community could arise. Whether that stage is obtained, the growth of the penetration rate and a high service variety will result automatically.

15.6.5 After-Market Solutions

To reach a desired penetration of 10 percent, it is not practical to offer just fully integrated Car2X systems. As shown in Section 15.2, market deployment will take too long in the case when Car2X technology is only offered for new admitted upper and upper-middle-class vehicles. The goal should be the achievement of the required 10 percent penetration rate as soon as possible. The only feasible way to attain this goal seems to be an after-market solution. Hence, car manufacturers should cooperate more closely with suppliers and probably create cobranding strategies to enrich the market from different directions. An after-market solution would have the additional advantage regarding the technology life cycle of the system. More precisely, the technological potentials of the Car2X system would be decoupled from the life cycle of the vehicle. Thus, the Car2X system could even be offered in a low-budget version and be exchanged every couple of years.

15.7 Conclusions

This chapter presented a high-level view of the Car2X landscape, providing insight into the existing technologies, existing and potential services, and their required market penetration. In addition, we discussed the market development and the complexity and the existing dependencies within the Car2X community.

Based on the specific characteristics and challenges of intervehicle communication, a feasible deployment strategy for an initial market introduction was shown.

The main challenge, thereby, is to achieve an immediate noticeable added value for potential customers on one hand, and prepare the overall system design for future cooperative services on the other hand. To accomplish this task and gap the resulting conflicting goals, a hybrid approach mixing up content and technologies is recommended that starts with a set of basic services comprising cooperative and noncooperative services, while exploiting ad hoc and cellular networks in parallel. In addition, customers have to be convinced of the advantages of the vision of connected vehicles.

Concerning the further development and optimization of Car2X systems, it is important to start the first stage of introduction with a highly flexible and extensible framework architecture and protocol design. This would enable easy software updates and guarantee compatibility of services with different versions.

Last, once the overall social benefit of safety and traffic-related applications is reliably proven, public authorities have a great leverage to speed up market introduction of cooperative telematics systems by subsidizing vehicle equipment and road infrastructures accordingly. Keeping this in mind, bringing communication-based cooperative systems into market is a highly complex task that is subject to a variety of different influencing factors. In this sense, technology seems to play a minor part, the cooperation of stakeholder an important part, public authorities an enabling part, and the customers the decisive part.

References

[1] BMW. http://www.bmw.com/com/en/insights/technology/connecteddrive/assist.html (accessed November 14, 2008).

[2] TomTom. http://www.tomtom.com/hdtraffic (accessed November 11, 2008).

[3] ASFINAG Maut Service GmbH. http://www.go-maut.at/go/default.asp (accessed November 25, 2008).

[4] McKinsey & Company, "A Road Map For Telematics—The McKinsey Quarterly", http://www.mckinseyquarterly.com/A_road_map_for_telematics_1180 (accessed December 03, 2008).

[5] General Motors Corporation. http://www.onstar.com. (accessed December 1, 2008).

[6] Car 2 Car Communication Consortium,.http://car-2-car.org/ (accessed December 2, 2008).

[7] K. Taga and S. Maric, "The Arthur D. Little Global Telematics Report 2006" Arthur D. Little, 2006.

[8] Federal Office for Spatial Development."Aggregierte Verkehrsprognosen Schweiz und EU." http://daten.clearingstelle-verkehr.de/218/ (accessed December 4, 2008).

[9] AKTIV project. http://www.aktiv-online.org/ (accessed July 15, 2008).

[10] CALM. http://www.calm.hu/ (accessed July 15, 2008).

[11] Network-on-Wheels Project. http://www.network-on-wheels.de/ (accessed July 15, 2008).

[12] IEEE, "IEEE 802.11 "Wireless local Area Networks," The Working Group for WLAN Standards. http://grouper.ieee.org/groups/802/11/ (accessed November 6, 2008).

[13] COMeSafety project. http://www.comesafety.org/ (accessed July 15, 2008).

[14] SAE International, "SAE J2735 dedicated short range ommunications (DSRC) message set dictionary." http://www.sae.org/ (accessed November 1, 2008).

[15] C. Adler, S. Eichler, T. Kosch, C. Schroth, and M. Strassberger, "Self-organized and context-adaptive information diffusion in vehicular ad hoc networks," (ISWCS 2006), Spain, 2006.

[16] Car 2 Car Communication Consortium, "CAR 2 CAR Communication Consortium manifesto—version 1.1." http://car-2-car.org/. (accessed October 22, 2008).

[17] DEUFRAKO, "IVHW system concept and issues relevant for standardization," DEUFRAKO Project Consortium, 2002.

[18] K. Matheus, R. Morich, I. Paulus, C. Menig, A. Luebke, B. Rech, and W. Specks, "Car-to-car communication—market introduction and success factors," In *Proc. of ITS 2005: 5th European Congress and Exhibition on Intelligent Transport Systems and Services*, 2005.

[19] M. Killi and H. Samstad, "Travelers' valuation of traffic information with respect to trips to work," Institute of Transportation Economics Technical Report, Norway, 2002.

[20] R. Schwarz, W. Schaufelberger, L. Raymann, H. Merz, F. Zaugg, T. Kloth, and P. Farago, "Wirksamkeit und Nutzen von Verkehrsinformation," Forschungsauftrag. SVI 2000/386 auf Antrag der Vereinigung Schweizerischer Verkehrsingenieure (SVI), Switzerland, 2004.

[21] Next Generation Telematics Protocol (NGTP), http://www.ngtp.org (accessed July 15, 2008).

Index